To Tom and Sherry
with love
from
Helen.   28 October 1989.

Cartographical Innovations

Cartographical Innovations

An International Handbook

of

Mapping Terms to 1900

Edited by Helen M. Wallis and Arthur H. Robinson

Helen Wallis

Published by Map Collector Publications (1982) Ltd. in association with the
International Cartographic Association, 1987

ISBN 0-906-430-04-6

Frontispiece – Decorative frontispiece by Jan van Vianen in Le Neptune français, *Paris, Hubert Jaillot,
1703. By permission of the British Library. (Copper engraving, 7.0920a)*

Typesetting by PJD Photoset, High Wycombe, Bucks
Printing by Campfield Press, St Albans, Herts

Contents

List of Illustrations

President's Preface

The ICA is particularly fortunate to have provided the support for this volume prepared by two of the pre-eminent scholars on the history of cartography. Dr Wallis and Professor Robinson are unequalled in the English speaking regions of the world in their knowledge of and rigorous scholarship in this area of the discipline of cartography. This work, instituted under the present ICA Standing Commission on the History of Cartography, has taken a number of years to research and produce and has utilized the input from many experts in the world's cartographic community. *Cartographical Innovations* brings together much material that previously has been located in widely scattered sources and was in danger of being lost to the world community of scholars. Through their tireless efforts, cartography now has an indispensable addition to its long and glorious history. We in the cartographic profession owe a debt of gratitude to Dr Wallis and Professor Robinson for their efforts.

JOEL MORRISON
President
International Cartographic Association

International Cartographic Association
List of Commission Members

Members of the Commission F on the History of Cartography of the International Cartographic Association, from 1976 to 1984, and of the Standing Commission on the History of Cartography, 1984 to date

FULL MEMBERS

Rudolf Habel (German Democratic Republic) 1976–
Mireille Pastoureau (France) 1976–1984
Monique Pelletier (France) 1984–
Alexei V. Postnikov (USSR) 1976–
Arthur H. Robinson (USA) 1976–

Richard I. Ruggles (Canada) 1976–
Francisco Vazquez Maure (Spain) 1976–1982*
Wolfgang Scharfe (German Federal Republic) 1976–
Helen Wallis (United Kingdom) Chairman, 1976–

CORRESPONDING MEMBERS

Charles Beattie (United Kingdom) 1984–
Eugenia Bevilacqua (Italy) 1984–
M.L. Chopra (India) 1976–1980
László Csendes (Hungary) 1976–
Istvan Czigany (Hungary) 1980–
Arthur Dürst (Switzerland) 1980–1984
Akio Funakoshi (Japan) 1980–
K.L. Khosla (India) 1980–1984
Cornelis Koeman (Netherlands) 1976–1984
Olga Kudrnovska (Czechoslovakia) 1985–
C.G.C. Marlin (South Africa) 1980–

Leena Miekkavaara (Finland) 1986–
Ludovik Mucha (Czechoslovakia) 1982–
Dorothy F. Prescott (Australia) 1980–
Taisto Saarentaus (Finland) 1980–1986
Edward Schnayder (Poland) 1976–
Günter Schilder (Netherlands) 1984–
Antoine de Smet (Belgium) 1976–1980
T. Ukita (Japan) 1976–1980
Franz Wawrik (Austria) 1980–
John A. Wolter (USA) 1976–
David Woodward (USA) 1976–

Acknowledgements

Two people stand pre-eminent for acknowledgement. First is my co-editor Arthur H Robinson, who as President of the International Cartographic Association from 1972 to 1976 gave great encouragement to the Working Group on the History of Cartography. He devised the alpha-numeric system for the text and compiled the general index. His knowledge and judgement have been invaluable. Secondly, I am grateful to our copy editor and research assistant in London, Mary Alice Lowenthal, who has been indefatigable and provided the index to authors cited in Sections C.

Next I owe thanks to the many contributors and correspondents (p.327) who have given us the benefit of knowledge and experience in their various specialities. Working groups were organized – in Poland by Edward Schnayder, in Japan by Muroga Nobuo, and their members are not individually named. Arthur Robinson in the United States and I in the United Kingdom also called upon a wide circle of colleagues, among whom I would mention the ever helpful Richard A Gardiner, Keeper of Maps at the Royal Geographical Society, who died in 1978.

To turn now to the International Cartographic Association, sponsor of the project, I wish to thank the presidents who followed Arthur Robinson, namely, Professor F J Ormeling, who has done so much to advance the Association's interest in the history of cartography during his two periods of office as president (1976 to 1980, and 1980 to 1984), and Joel Morrison, who from 1984 has carried on the tradition with distinction. To members of the Working Group on the History of Cartography, 1972 to 1976, and of the Commission, 1976 to 1980, and 1980 to 1984, and of the Standing Commission from 1984 (p. viii), are also due our thanks. One member who has been sadly missed following his untimely death in 1982 is Francisco Vazquez Maure (Spain).

The chairmen of the ICA Publications Committee, Harold Fullard (1976 to 1984) and Roger Anson (1984 to date), have been unstinting in the time and care which they have bestowed on our arrangements for publication. Valerie Scott, director of Map Collector Publications (1982) Ltd., our publishers, deserves much credit for so expertly seeing this work through the press.

The scene of much activity was the Map Library of the British Library in London. My thanks therefore are due to my secretaries, and to Professor Robinson's secretary at the University of Wisconsin, for their generous help and to my colleagues in the Students Room where so many works were consulted. In so far as possible we checked every original work and reference cited. The usefulness of *Imago Mundi*, the learned journal on the history of cartography, should be mentioned, and also *The Map Collector*. The British Library's Photographic Service kindly provided the illustrations.

Finally, we commend the results of our labours to the attention not only of historians but also of modern cartographers. In this present period of rapid innovation, they are producing the maps which will be the source material for future historians of cartography.

HELEN WALLIS
Chairman
Standing Commission on the History of Cartography
International Cartographic Association

Foreword

In recognition of the importance of the history of mapmaking for the science of cartography the Fourth General Assembly of the International Cartographic Association (Ottawa, 1972) established a Working Group on the History of Cartography. Under its chairman Dr Helen Wallis (UK) the working group took up as its first task the compilation of an historical glossary of cartographic techniques (later changed to "innovations") and their diffusion up to 1900. From the outset the glossary was conceived as an international project in accordance with the aims and objectives of the association, with contributions from international specialists on cartographic innovations in their respective countries and cultures.

At the Fifth General Assembly (Moscow, 1976) the Working Group presented a preliminary study entitled *Map Making to 1900: An Historical Glossary*, a kind of prototype of what was intended for a later date. Encouraged by the favourable reception of this study, and by the promotion of the Working Group by the same General Assembly to the status of a full Commission, the chairman Dr Wallis selected a steering group out of her commission members and with their assistance set to work on the realization of her worldwide and long-term project. She recruited as co-editor Professor Arthur H Robinson (USA), distinguished Past President of the Association.

With great satisfaction the International Cartographic Association welcomes the completion of the work resulting in the publication of this valuable glossary. Having witnessed personally, as a member of the ICA Executive and of the Publication Committee, the ups and downs of the project for a number of years, I gladly congratulate Dr Wallis and Professor Robinson, and I compliment the whole team of commission members for the expertise and perseverance with which this ambitious project has been brought to a successful end. At the same time the strong encouragement and assistance given by Harold Fullard (UK), past chairman of the ICA Publications Committee, and his successor Roger W Anson (UK) should be recognized.

Cartographical Innovations documents a wealth of innovative ideas and concepts which led to the advance of the art, craft, and science of mapmaking in the past. The book covers all aspects of the field ranging from types of maps to concepts and from techniques of production to those of symbolization. As Arthur Robinson puts it, *Cartographical Innovations* is "a kind of museum of mental developments in mapmaking."

In conformity with its original design and scope, the book is truly international. It contains contributions on a wide variety of innovations submitted by scholars, craftsmen, and scientists from many countries. The International Cartographic Association is pleased to add this glossary to its growing list of co-operative publications!

F J ORMELING
Past President
International Cartographic Association

Introduction

Cartography, as defined by the International Cartographic Association's *Multilingual Dictionary of Technical Terms in Cartography*, includes the study of maps as scientific documents and as works of art. An important facet of such a study is the history of map making. It was appropriate therefore that the Commission on the History of Cartography established by the International Cartographic Association in 1976 should take up as its first major project the preparation of a handbook on cartographic innovations. Our purpose has been to identify on a worldwide scale the main points of advance and change in the science and art of cartography.

The project was itself innovative in the attempt to show how ideas developed, how processes and techniques began, when materials were first used, and how knowledge of innovations was diffused and transmitted. The work has been international in scope, covering the main cultures in which cartography has been a significant activity. The results show how universal are many of the images and styles adopted as the language of maps. Similar elements to those in current use appear in the artefacts of ancient civilizations and also in the maps of more recent cultures, literate and nonliterate, which have developed independently of the traditional centres of map making. Modern cartographers may also be surprised to learn that many of the techniques employed today originated a century or more earlier than is generally supposed. Our handbook thus reveals something of the essential nature of cartography.

The organization of the work has been international, with contributions invited from experts all over the world, and from working groups set up in various countries. We have arranged the 191 entries (mapping terms) in eight groups. Each entry is divided into sections, comprising definition (Section A), details of innovation and diffusion (Section B), and bibliography (Section C). A numerical system for the terms provides a logical order and an easy system for indexing (see "Guide to Arrangement and Use," below, p.xvii).

The first three groups, "Types of Maps," "Maps of Human Occupation and Activities," and "Maps of Natural Phenomena," present major themes. Surveying, one of the oldest professions, was an activity of the ancient literate cultures of Mesopotamia, Egypt, Greece and China. Professional map making thus began some 4000 years ago. One of the earliest surveyors comes down to us as Gudea, ruler of the Sumerian city-state of Lagash, *c.* 2200 BC, represented in two statues now in the Louvre, Paris. The plan of the temple of Ningirsu which Gudea holds on his lap is perhaps the earliest known map drawn to scale (Plan, 1.1630; Figs. 3 and 4).

Another prototype mapmaker is the mythical topographer and engineer, the Emperor Yü of China. The Yü Kung (Tribute of Yü), believed to date from the 6th century BC, is the earliest geographical account of China and describes how Yü crossed the empty territories of China, laying down mountains and determining the courses of the rivers. The map of the tracks of the Great Yü, the "Yü Chi Thu," was carved on stone in 1137 and is now in the Pei Lin Museum, Sian. Joseph Needham, the leading Western authority on Chinese science and civilization, describes it as "the most remarkable cartographic work of its age in any culture."

XI

To the Greeks God was a great geometer. The Pythagoreans of the 6th century BC were by repute the first to express nature in quantitative mathematical terms. Anaximander of Miletus (early 6th century BC), a contemporary therefore of Yü, is believed to have made the earliest Greek world map (World Map, ancient, 1.2320a). Eudoxus of Cnidus (fl. 365 BC) made an early celestial globe and gave the first account of the movements of the heavenly bodies; his description of the stars formed the basis of the *Phaenomena*, the astronomical poem of Aratus of Soli (fl. 270 BC), which became a major source for medieval Europe (Astronomical map, 3.011).

In medieval thinking all aspects of life were symbolical of God, and Greek classical science was assimilated to this concept. Through the powerful testimony of Aristotle and the Aristotelians, God in medieval Christendom features as a surveyor measuring the world which he has made, as in the "Bible moralisée" of the 13th century (preserved in the Nationalbibliothek, Vienna). In similar vein, Christ embraces the earth, as represented in various mappae mundi (1.2320b; Fig.7).

Another theme is the continuity of techniques and traditions. The square grid (4.191) originated in China in the first century when the famous astronomer Chang Heng (79-139) "cast a network about heaven and earth, and reckoned on the base of it." The rectangular grid was in continuous use in China from the time of Phei Hsiu (224-271) to the late 16th century when the European missionaries arrived in China. The "Yü Chi Thu," 1137, is one of the notable examples of a grid map. In the Western world the square grid was used in Greek and Roman surveying and then seems to have disappeared, emerging in the maps of the Holy Land made by Pietro Vesconte in Venice, *c.* 1327 (Fig. 17). It has been suggested that travellers to the east may have brought back information on the Chinese grid. The Franciscan William of Rubruck, for example, on his travels to Mongolia, 1251-1254, learned details of Chinese or Korean maps from the Mongols, which he passed on to Vincent de Beauvais and Roger Bacon. Marco Polo returned from China to Venice in 1295 with maps whose details were incorporated into the Catalan map of 1375. Square grids are also found in 12th-century medieval copies of Roman surveying manuals from about AD 50 (Non-co-ordinate reference systems, 4.141).

In celestial mapping China likewise was foremost for continuity in tradition (Astronomical map, 3.011). The oldest extant Chinese map of the stars dates from 940. In the west, standard works of the classical world, the *Phaenomena* of Aratus of Soli (fl. 270 BC) and the Latin *Poeticon astronomicon* of Caios Julius Hyginus (1st century BC), were the source of a manuscript depiction, the "Harley Aratus" (Harley MS 647), drawn in the scriptorium of Lupus of Ferrières (805-862), in France. This manuscript passed to Canterbury in 1000, subsequently to be copied more than once. Thus the medieval scriptoria preserved and disseminated the scientific knowledge of classical Greece.

Some of the greatest medieval world maps were derived from Agrippa's survey of the Roman empire, 12 BC, with a Christian iconography superimposed (Mappae mundi, 1.2320b). These Christian elements followed the archetype of the Spanish monk Beatus Liebanensis (8th century), who inspired a main tradition of Christian cartography in the following centuries. T-O maps in the Beatus style were incorporated in the manuscripts illustrating the Apocalypse of St John. Illustrations of the island of Patmos where he wrote the Apocalypse may also have contributed to the development of the *isolario* or island atlas, first known from that of Cristoforo Buondelmonte, 1420 (Island atlas, 8.0110c).

Breaks in transmission are also significant. Of the broken traditions the best known concerns the scientific works of Claudius Ptolemy (*c.* 90-*c.* 168), written at Alexandria in Egypt, which recorded the essence of Greek geography and astronomy. Ptolemy's *Mathematical Syntaxis*, written between 141 and 147 and comprising almost the only complete work on Greek astronomy to be passed down, came into the hands of the Arabs, who called it "the Greatest," the *Almagest* (Greek *megiste*, Arabic prefix *al*). When the Arab conquests reached India, Hindu astronomers revitalized the work through the introduction of Hindu mathematics with its numeral system. Thence the Arabs brought the *Almagest* to Europe, where for almost a millennium knowledge of Hellenistic science had been lost. The *Almagest* with its translations provided astronomical concepts of apparent consistency and inner perfection. Despite faults that were revealed in the 16th century, Ptolemy's system established mathematical methods which in the hands of Nicolaus Copernicus, Tycho Brahe, and Johann Kepler were to revolutionize astronomical theory.

The Arabs also preserved and in due course transmitted to Europe Ptolemy's *Geography*, *c.* 150. Maximus Planudes (*c.* 1260-1310), monk of the Chora monastery near Constantinople, was the agent who sought out copies for the emperor Andronicus II Palaeologus (1282-1328). The developing contact between the Byzantine world and the west in the 14th century led to the circulation of Greek manuscripts of the *Geography* in Italy. The first translation into Latin was undertaken by Jacobus Angelus in 1406. Once printed with maps from 1477 onwards, the work became an inspiration for renaissance mapmakers. Ptolemy's *Geography* was erroneous and out of date; nevertheless its mathematical system provided the basis for modern cartography. The identification of locations by means of co-ordinates and the establishment of the graticule of latitude and longitude lines were the essential elements of the scientific geography of the later 15th and 16th centuries. Printed Ptolemy atlases of the 15th century onwards were forerunners of the modern atlas as interpreted by Abraham Ortelius in his *Theatrum Orbis Terrarum* (1570) and Gerard Mercator with his *Atlas* (the first so named) of 1595 (Atlas, 8.0110).

Another break in tradition concerns the heritage of Rome. The plan drawn to scale was a well-known type of Roman mapping, as illustrated by the plan of Rome carved in marble, "Forma Urbis Romae," late 1st century AD (Plan, 1.1630). Thereafter the making of this kind of plan disappeared in Europe and did not begin again until the early 15th century. The earliest plan drawn explicitly to scale after Roman times is one of Vienna of 1422, now known in a mid-15th century copy. Land surveys normally gave area in terms of customary reckoning and only rarely were "by measure" (*per mensuram*). Some of the earliest medieval local plans seem especially to be related to the practical needs of monasteries and religious houses, as illustrated by the oldest known in England, the plans of the water works at Canterbury, 1153-61, and of the water conduits at Wormley, Hertfordshire, 1220-30.

The renaissance of the 15th and 16th centuries was the most remarkable period of innovation in map making, as in other practical, artistic, and intellectual pursuits. Europeans believed themselves at the opening of a new epoch of human endeavour, although conscious of their debt to the ancient world. The editors of Ptolemy's *Geography*, for example, combined respect for his authority with an increasing willingness to make original additions in the form of new maps, indexes (6.0920), new conventional signs (5.033), new types of point symbol (5.1610), and so on.

The invention of printing, moreover, brought about a revolution in the making and use of maps. From about 1460 techniques of reproduction from copperplates and woodcuts (Relief technique, 7.1810; Metal engraving, 7.1810b, Woodcut, 7.1810c) made possible what William M Ivins, Jr., has described as "the exactly repeatable pictorial statement," constituting a breakthrough for the diffusion of geographical and scientific knowledge.

A mapmaker's craft developed and an educated public supplied its market. Whereas the atlas (8.0110) and the globe (1.0710) were for three centuries the most popular cartographic forms, new types of map also appeared. Erhard Etzlaub, mathematical instrument-maker of Nuremberg, made not only the first printed road map, *c*. 1500 (1.1810d), but also maps which rank as ethnographical (2.052) and linguistic (2.122). The bird's eye view, by the 16th century an established style for town plans, achieved the highest degree of artistic excellence in such a work as Jacopo de' Barbari's great six sheet woodcut plan-view of Venice, 1500. At the other end of the scale are the tidal maps (3.201) devised by Guillaume Brouscon of Le Conquet, Brittany, *c*.1546, for the benefit of unlettered seamen. His printed pocket books conveyed simply and effectively the essential information sailors needed, and Sir Francis Drake and other men of rank secured copies for their own use (Fig.12).

As mapmakers in the 16th and 17th centuries sought greater accuracy of measurement and topographical detail, the language of maps became more elaborate. Groups 4 to 6, "Reference Systems and Geodetic Concepts," "Symbolism," and "Techniques and Media," reflect this process. The conventional signs (5.033) of military (2.1310) and battle maps (2.1310a) illustrate the increasing complexity of symbols (Fig. 9). Special maps of mining and mineral resources were also made, as illustrated by the plans of the Wieliczka salt mines of Poland by Marcin German, engraved by Willem Hondius, Danzig, 1645 (Mineral map, 3.131; Figs. 10 and 11).

The beginnings of scientific cartography in the form of the thematic map date from the later years of the 17th century. Thematic maps (1.202) differ from descriptive and special maps in depicting not reality but ideas about reality. Edmond Halley's innovative thematic maps arose from his search for a general rule derived from a series of observations about physical phenomena, notably winds, tides and magnetic variation. His earliest thematic map, a world map of trade winds published in 1688, achieved immediate fame (Wind map, 3.232). Wind systems were henceforward to be included on many English and Dutch charts. Even the gentleman's pocket globe (1.0710c) displayed the trade winds.

Halley's second thematic map was a chart of the Atlantic Ocean, published in 1700, showing magnetic variation by means of lines of equal variation, i.e. isogonic lines (Magnetic variation, 4.133; Fig. 16). It is generally accepted as the first isarithmic map, thus marking a great advance in cartography (Isarithm, 5.0910). Instead of representing a set of isolated observations, the chart depicted a corpus of information systematically recorded and organized. Halley employed a technique by which he co-ordinated the observations and expressed them by means of an abstraction, the line of equal value. His map was in a true sense a distribution map, and as Sydney Chapman has said, it had a remarkable progeny.

In this respect Halley was far in advance of his time. Despite the widespread copying of his maps of magnetic variation, the logical implications of the line of equal value were not appreciated. It was not until 1817 that the concept was carried a step further, when Alexander von Humboldt employed the Halleyan technique for isotherms (5.0910g). Like Halley Humboldt devised the technique to demonstrate a theory. He was concerned to show and explain the distribution of heat throughout the world, for which "the general circumstances have long been known, but not rigorously determined nor submitted to exact calculation."

The principle of Humboldt's isotherm line was rapidly taken up and extended to other phenomena. The technique was an ideal tool for the developing geographical sciences. Heinrich Berghaus's *Physikalischer Atlas*, Gotha, 1845, and the English version *The Physical Atlas* of A K Johnston, Edinburgh, 1848, illustrate the range of distributions of temperature, pressure, and other phenomena, now expressed in cartographic and statistical form. Since then innumerable applications of the isarithm have been made in a wide variety of fields.

Another type of thematic mapping emerged in the 1820s under the title of "Moral statistics" (2.132) but failed to gain until recent years its true place as a cartographic genre. It dealt with the mapping of what would be called today social phenomena. Charles Dupin in his *Forces productives et commerciales de la France*, Paris, 1827, included a map of the distribution and intensity of illiteracy in France, designed to advance his argument for education as one of the most necessary conditions of a prosperous country. Joseph Fletcher adopted the same techniques to deal with the moral statistics of England and Wales. In 1847 he was content to rely on tables of statistics and one index map. At the behest of Prince Albert he produced twelve elaborate maps to illustrate his further study of 1849 (Fig. 15).

If the mapping of moral statistics remained outside the mainstream of geographical progress, thematic mapping as a whole became widely accepted. Even Alexander von Humboldt did not anticipate the range of its development when he wrote in his *Political Essay on the Kingdom of New Spain* (London, 1811, p. cxxxii), "It would be ridiculous to express by curves, moral ideas, the prosperity of nations, or decay of their literature; but whatever relates to extent and quantity may be represented by geometrical figures; and statistical projections which speak to the senses without fatiguing the mind, possess the advantage of fixing the attention on a great number of important facts."

In the early 1890s the officers of the Geographical Society of Finland were presumably inspired by such sentiments, enlivened by patriotic fervour (for Finland was then a Grand Duchy of the Russian Empire), when they set out to express their country's national identity in maps. They prepared the maps for the Sixth International Geographical Congress, held in London in 1895. The display, which was designed to give "une représentation du pays et des conditions générales de sa civilisation," aroused such a lively interest that the Society went on to publish at Helsinki in 1899 a selection with additions in the form of an Atlas of Finland. Thus the "national atlas" as later defined unobtrusively made its debut.

These are but a few of the topics and themes which our study of innovations reveal. Some general conclusions can be drawn. The importance of the use of numbers and of mathematical concepts is illustrated by the dominance of Greek cartography

List of Terms

Group 1
Types of Maps

1.031	Cartogram
1.0320	Chart
1.0320a	_____, Coastal
1.0320b	_____, Harbour
1.0320c	_____, Nautical
1.0320d	_____, Portolan
1.033	Chorographic map
1.034	Choropleth map
1.041	Dasymetric map
1.042	Dissected map
1.043	Distance map
1.044	Dot map
1.061	Fan map
1.062	Flow map
1.0710	Globe
1.0710a	_____, Celestial
1.0710b	_____, Lunar
1.0710c	_____, Pocket
1.0710d	_____, Terrestrial
1.091	Imaginary map
1.131	Map games
1.161	Perspective/Panoramic map
1.162	Pictorial map
1.1630	Plan
1.1630a	_____, Plat
1.1630b	_____, Town Plan
1.164	Planisphere
1.1810	Route map
1.1810a	_____, Canal/River chart
1.1810b	_____, Cycling
1.1810c	_____, Railway
1.1810d	_____, Road
1.191	Spherical chart
1.192	Symbolic map
1.201	Tactile map
1.202	Thematic map
1.203	Topographical map
1.204	Typographic map
1.231	Wall map
1.2320	World map
1.2320a	_____, Ancient
1.2320b	_____, Mappa mundi

Group 2
Maps of Human Occupation and Activities

2.011	Archaeological map
2.0210	Boundary map
2.0210a	_____, Administrative
2.0210b	_____, Cadastral
2.0210c	_____, Estate
2.0210d	_____, Jurisdictional
2.0210e	_____, Property
2.041	Disease map
2.051	Enclosure map
2.052	Ethnographic map
2.081	Historical map
2.091	Insurance map
2.121	Land-use map
2.122	Linguistic map
2.1310	Military map
2.1310a	_____, Battle
2.1310b	_____, Reconnaissance
2.132	Moral statistics map
2.161	Population map
2.162	Poverty map
2.163	Product map
2.164	Promotional map
2.181	Religious map
2.191	Sanitary map
2.192	Satirical map
2.201	Tithe map
2.202	Tourist map
2.203	Traffic map

Group 3
Maps of Natural Phenomena

3.011	Astronomical map
3.021	Block diagram
3.022	Botanical map
3.031	Compass declination map
3.051	Eclipse map
3.071	Geological map
3.081	Hypsometric map
3.131	Mineral map
3.132	Morphological map
3.151	Ocean current map
3.161	Precipitation map
3.201	Tidal map
3.231	Weather map
3.232	Wind map
3.261	Zone map
3.262	Zoological map

Group 4
Reference Systems and Geodetic Concepts

4.031	Cardinal direction
4.032	Compass direction
4.041	Date line
4.051	Ellipsoid
4.071	Geoid
4.072	Graticule
4.073	Grid
4.121	Latitude
4.122	Longitude

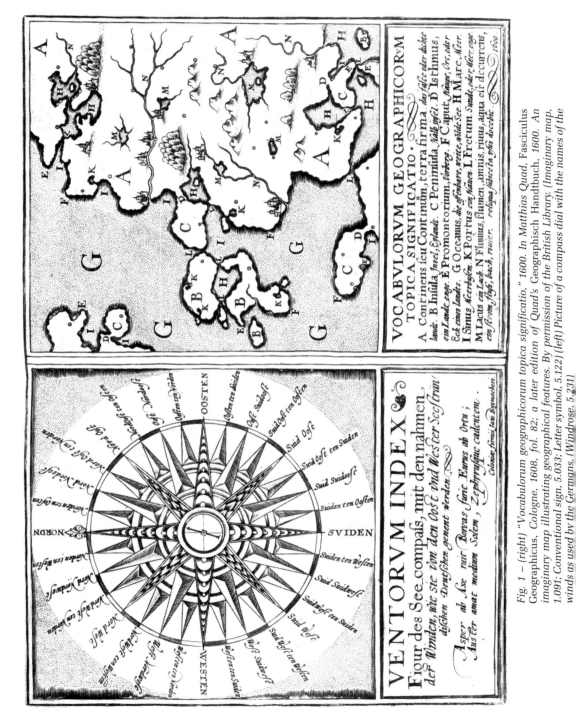

Fig. 1 – (right) "Vocabulorum geographicorum topica significatio," 1600. In Matthias Quad, Fasciculus Geographicus, Cologne, 1608, fol. 82: a later edition of Quad's Geographisch Handtbuch, 1600. An imaginary map illustrating geographical features. By permission of the British Library. (Imaginary map, 1.091; Conventional sign, 5.033; Letter symbol, 5.122) (left) Picture of a compass dial with the names of the winds as used by the Germans. (Windrose, 5.231)

Cartographical Innovations

GROUP 1

Types of Maps

1.031 Cartogram

A (MDTT 823.17; 431.5) An imprecise term used for a variety of kinds of maps, the only common denominator being the cartographic representation in some way of numerical data. (a) In continental Europe the term is often used to refer to any representation of statistical data on a geographical base, especially when the latter is highly generalized. (b) In the United States the term is also used to refer to a map whose geographical base is distorted to accord to the geographical arrangement of some phenomenon, either merely to simplify complex geographical relationships or especially to substitute some other quantity for geographical area to portray statistical relationships. Also known as *anamorphoses* (French) and *vari-valent* maps (USSR).

B Cartograms had their beginnings in the diagrammatic representation of statistics in the later 18th and early 19th centuries by William Playfair (1786), August Friedrich Wilhelm Crome (1785), and others who used proportional circles and squares to portray the relationships of geographical statistics. The *cartes figuratives* by Charles Dupin, Adrien Balbi and André Michel Guerry, and A M Guerry *et al.*, in the 1820s and 1830s (see Density symbol, 5.041), and by Charles Joseph Minard *et al.*, from 1845 to 1869 (see Flow line, 5.061), are the earliest cartograms in sense (a). In sense (b) cartograms began to appear in the latter part of the 19th century perhaps inspired by Émile Levasseur who in 1868 ". . . used for . . . illustrations . . . squares proportional to the extent of surfaces, population, budget, commerce, merchant marine of the countries of Europe, the squares being grouped about each other in such a manner as to correspond to their geographical position" (Funkhouser, p. 355). The next flurry of activity in the development of this type of cartogram seems to have taken place in the statistical service in Berlin, which produced a number of such maps around the turn of the century (Mayet, 1905).

The early history of graphic statistical representation is treated by Funkhouser and by Beniger and Robyn. The early history of statistical mapping is summarized in Robinson (1982). The relation of cartograms (b) to map projections was pointed out by Tobler.

C BENIGER, J R, and ROBYN, D L (1978): "Quantitative Graphics in Statistics: A Brief History," *American Statistician, 32,* 1-11.

FUNKHOUSER, H Gray (1938): "Historical Development of the Graphical Representation of Statistical Data," *Osiris, 3,* 269-404.

MAYET, P (1905): "Die schematisch-statistischen Karten des Kaiserlichen Statistischen Amtes zu Berlin," *Bulletin de l'Institut International de Statistique, 14,* 3ᵉ livraison, 214-22.

NORDBECK, Stig (1964): "Statistiska Kartor från 1800-talets mitt.," *Svensk Geografisk Arsbok, 40,* 7-18.

ROBINSON, Arthur H (1967): "The Thematic Maps of Charles Joseph Minard," *Imago Mundi, 21,* 95-108.

ROBINSON, Arthur H (1982): *Early Thematic Mapping in the History of Cartography.* Chicago, University of Chicago Press.

TOBLER, Waldo R (1963): "Geographic Area and Map Projections," *Geographical Review, 53,* 59-78.

1.0320 Chart

A A map designed primarily for navigation.

B The *periplus* of the ancient classical world was a coasting pilot, giving descriptions of a voyage and sailing directions (Coastal chart, 1.0320a). No charts survive from this period, and it is arguable that the reference to "drawn peripli" did not refer to charts. Extant Arabic charts date from the 10th century. In the late 13th century the so-called portolan chart (1.0320d) came into use in southern Europe, marking a major advance in medieval and hydrographic cartography. Its influence persisted for some 400 years. One of the innovations incorporated was the use of "arabic" (Hindu) numerals.

An innovation of the later medieval period in Mediterranean cartography was the *isolario*, or island atlas (8.0110c), of which the earliest known is that of Cristoforo Buondelmonte, 1420. This created a prototype for a popular genre still in vogue at the end of the 17th century. In northern Europe the sea-book, or rutter, evolved from the 14th to the 16th centuries as the northern counterpart to the Mediterranean portolano. The first printed sea atlas, *Spieghel der Zeevaerdt* of Lucas Janszoon Waghenaer, published at Leyden, 1584-85, comprised a pilot book illustrated by charts (see Atlas, 8.0110).

The coastal chart gained new importance in the 16th century with the exploration of long navigation and trade routes overseas. The *roteiros* of the Portuguese governor of India João de Castro, 1538 to 1542, rank as masterpieces of their kind. Harbour charts (1.0320b), as a special type of coastal chart, also developed as necessary aids to navigation – the "haven-finding art." Tidal maps (3.201) were the invention of Breton sailors in the 1540s, and progressed from simple cartograms to the scientific maps of tides in the English Channel by the British astronomer Edmond Halley, 1702.

The nautical chart (1.0320c) improved in various ways with the rediscovery of Ptolemy's Geography in the 15th century and advances in navigational science. The projectionless plane (or plain) chart was marked with lines of latitude in the early 16th century, and was adapted with some ingenuity to take account of the variation of the compass. With Gerard Mercator's invention of the Mercator projection, 1569, charts became available on which sailors could draw a "nautical triangle," showing latitude and longitude, direction and course correctly (see Map projection, 4.133). The plane chart, however, remained in production until the early 18th century, and atlases such as Johannes van Keulen's *Zee-Fakkel* (1695 edition) would advertise their charts as "both plane and increasing," i.e. on Mercator's projection.

From the time of the Italian and Catalan schools in the 14th century, charts became much more than aids to navigation. They were "the key to the empire, the way to wealth," providing a great range of information on inland areas. Their vivid vignettes were often of eye-witness coastal scenes recorded on voyages overseas. Such works were sometimes made on commission as presentation pieces for royal and noble persons or as intelligence reports for merchant houses. The great renaissance world maps thus developed from the tradition of the portolan chart as well as being, for regions beyond the known world, an extension of the scholastic medieval world map (see World map, 1.2320).

In China the importance of waterways in the organization of local and regional activities explains the long tradition in chart making. River mapping (1.1810a) developed from the sixth century AD in the Chin dynasty, while coastal charting came later, in the 15th century. The Japanese excelled in the production of decorative coastal charts of the "Inland Sea" for screens and picture scroll maps of sea and land routes (Coastal chart, 1.0320a).

Charts of other cultures include Micronesian sailing charts of the 18th century and Inuit charts of the coast of Greenland (Nautical chart, 1.0320c; Map surface, 6.1310).

1.0320A Coastal Chart

A A hydrographic chart used for navigation when approaching coastal waters, or when sailing between ports on the same stretch of coast.

B The earliest known Greek *periplus*, or coastal pilot, is a pilot of the Mediterranean Sea, attributed to Skulax (Scylax) of Karuanda, *c.* 500 BC but the existing text was probably written much later, *c.* 350 BC. No charts survive from this or any other Greek *periplus*. The portolano, or pilot-book, of southern Europe and the associated portolan charts (1.0320d) dating from the 13th century onwards were concerned partly, but not exclusively, with coastal navigation.

In northern Europe the 14th century German *Seebuch* was illustrated with coastal views (see Profile, 5.162). The first printed sea-book with charts as well as coastal views was the "Caerte van die Oostersche See," by Cornelis Anthoniszoon of Amsterdam, 1544. Lucas Janszoon Waghenaer's *Spieghel der Zeevaerdt,* Leyden, 1584-85, was the first printed pilot book illustrated by charts (paskaarten) to form a maritime atlas. It thus combined two hydrographic traditions, the sea atlas and the pilot-guide. The charts cover the coasts of northern and western Europe and are of a standard scale large enough for pilotage. They are marked with all the features necessary for navigation, such as soundings, shoals and sea-marks. Waghenaer's *Spieghel* has been described as the greatest single advance in the history of hydrographic publication. The English edition, published in London in 1588, introduced into the language the term "Waggoner" for a sea atlas.

The Portuguese manuscript *roteiros* of the early 16th century were also notable for innovations in coastal charting, though of limited influence because of the restrictions imposed on the dissemination of nautical information. The manuscript "Book of Francisco Rodrigues," 1512 (incorporated in the "Suma oriental" of Tomé Pires), includes coastal charts of the route to India and panoramic views as well as sketches of the coasts of the East Indies (Bibliothèque de la Chambre des Députés, Paris, now in the Bibliothèque Nationale; Cortesão and Teixeira da Mota, Vol. 1, pp. 79-84). João de Castro's three roteiros of the route from Lisbon to Goa, from Goa to Diu, and of the Red Sea, 1538 to 1542, were masterpieces of hydrographic observation and also included panoramic views (see Profile, 5.162). What is now reckoned as the finest, that of the Red Sea, was purchased by Sir Walter Raleigh for £60 and is now in the British Library (Cotton MS. Tiberius D.IX). The roteiro was ideally suited for the depiction of an oceanic empire with long lines of communication by sea, and remained in vogue well into the 17th century. The books and atlases of the "State of Oriental India" ("Estado da India Oriental") by Antonio Bocarro, 1635, Pedro Barreto de Resende, 1636(?), and others were made to the order of Philip III of Portugal and include many charts as well as panoramic views. The one which represents the genre in its most complete form is the manuscript codex of 1646, with drawings by Pedro Barreto de Resende and charts mainly by Pedro Berthelot (1635), now in the British Library (Sloane MS. 197).

In England one of the earliest coastal charts, probably of 1514, depicts the north Kent coast and the river Swale from Faversham to Margate (British Library, Cotton MS. Augustus I.ii.75). It is drawn on a narrow paper strip 25 feet long. Coastal features and defences are shown in pictorial detail, indicating a military purpose — defence against France (Tyacke and Huddy, pp. 11-12). Information about depths (written in words) qualifies it to be counted as an early example of a chart with soundings. The earliest English chart to show soundings by number is a chart of the Yorkshire coast from Hull to Scarborough, dated about 1569 (British Library, Royal MS. 18.D.III.f.69v-70). A notable chart for its depiction of channels and sandbanks of the Thames estuary is that by Richard Cavendish, *c.* 1533 (British Library, Cotton MS. Augustus I.i.53). These charts, preserved in the collection of the Elizabethan antiquary Sir Robert Cotton, and other charts at Hatfield House, Hertfordshire, the home of Lord Burghley, Lord High Treasurer to Elizabeth I, illustrate the extensive charting of the English coasts for official purposes during the reigns of Henry VIII, Edward VI and Elizabeth I.

In the 17th century the Dutch publishing houses of Blaeu, Colom, Doncker, Goos, Jacobsz/Lootsman, and Van Keulen (to name the most outstanding) dominated the chart trade. Their preferred type of publication, the atlas of charts combined with a "rutter," or set of sailing directions, followed the style of Waghenaer's *Thresoor der Zeevaert,* printed at Leyden in 1592, and regarded as more suitable to seamen's needs than the expensive *Spieghel* (1584-85). The coastal chart and pilot in volume form thus remained the most popular type of publication (see Nautical chart, 1.0320c). *The English Pilot* of John Seller, London, 1671, which was based primarily on Dutch charts, was the first of a series which established English publishers in competition with the Dutch and served English seamen on navigations throughout the world. *Great Britains Coasting Pilot* by Greenvile Collins, London, 1693, constituted the first systematic survey of British coasts, made on the orders of the Masters of Trinity House.

In the 18th century Murdoch Mackenzie in his charting of the Orkneys, 1744-49, achieved a new standard of accuracy by establishing a framework of shore stations fixed by triangulation from a measured baseline. His survey was published in *Orcades, or a Geographic and Hydrographic Survey of the Orkney and Lewis Islands,* London, 1750. For North America, the surveys of Joseph Frederick Wallet Des Barres, a Swiss who became a British subject, published in *The Atlantic Neptune* (1777 onwards), were remarkable for the range of coastal depiction. Differences in soils and types of beaches were distinguished, indications of landscape near the shore provided, together with views and profiles. Some 290 plates were engraved for the work. The reviewer in *L'Esprit des Journaux,* Paris, 1784, commented that it was "one of the most remarkable products of human industry that has ever been given to the world through the arts of printing and engraving . . . the most splendid collection of charts, plans and views ever published" (quoted in Cumming, p. 56).

In the procedures for charting from running coastal surveys on exploring expeditions in the 18th century Captain James Cook's achievements on his three Pacific voyages, 1768 to 1771, 1774 to 1777 and 1778 to 1780, were preeminent. The level of precision achieved set the standard for the British Hydrographic Department of the Admiralty, established in 1795.

These advances in the 18th century were made for governments and seamen. A very different project, undertaken by the Italian Luigi de Marsigli, was intended for the learned world. Establishing himself at Montpellier in France, he undertook researches into the coasts of Languedoc and Provence, "pour se rendre compte de l'anatomie des mers." His charts and profiles, published in his *Historie physique de la mer,* Amsterdam, 1726, rank as the first scientific description of the Gulf of Lyons (Dainville, pp.99-100).

In China coastal charts developed rather later than maps of inland waterways (see Charts, 1.0320). One of the most celebrated is the Mao K'un map which records the route and observations of Chêng Ho, the Chinese admiral and ambassador who made his first voyage (to Hormuz) in 1415 and his last in 1433. The map, which was printed in the second or third quarter of the 16th century, extends about 18 feet over 40 pages and depicts in a long narrow strip, coastlines (often much compressed) and sea routes from Nanking via Java to Hormuz, a total of 7460 miles (Mills, p. 251;

Needham, p. 560). The associated text by Ma Huan, geographer to the admiral, was entitled *Ying-yai Sheng-lan* ("The overall survey of the Ocean's shores").

Silk scroll maps of the Chinese coasts date from the 16th to the 19th centuries, and were designed to give a comprehensive view of the coast for military and administrative purposes. A series of maps, for example, is included in Chhen Lun-Chhiung's *Hai Kuo Wên Chien Lu* ("Record of things seen and heard about coastal regions") (see Needham, p. 517). A scroll of the coasts and islands of China, 1840, is a late example derived from 16th century prototypes (British Library, Maps 162.o.3.).

Japanese chart-making was notable both for its indigenous skills and for its assimilation of European techniques. The importance of maritime activity on the "Inland Sea" (Seto-naikai) led to the depiction of the sea in screen and scroll paintings. Examples are the manuscript "Map of the Inland Sea of Seto" on an eight-fold screen of the early 17th century and Dōchū Emakimono's picture scroll of the sea and land routes from Edo to Nagasaki, mid 17th century (both are in Nanba's collection, Nishinomiya).

C CORTESÃO, Armando, and TEIXEIRA DA MOTA, Avelino (1960): *Portugaliae Monumenta Cartographica*. Lisbon, Comemoracões do V Centenário do Morte do Infante D. Henrique.

CUMMING, W P (1972): *British Maps of Colonial America*. Chicago and London, University of Chicago Press.

DAINVILLE, François de (1961): *Cartes anciennes du Languedoc*. Montpellier, Société Languedocienne de Géographie.

MILLS, J V G, ed. (1970): *Ma Huan: Ying-Yai Sheng-Lan* ("The overall survey of the Ocean's shores") [1433]. Cambridge.

NANBA Matsutaro, MUROGA Nobuo, and UNNO Kazutaka (1973): *Old Maps in Japan*. Osaka, Sogensha.

NEEDHAM, Joseph (1959): *Science and Civilisation in China, Vol. 3*. Cambridge, At the University Press.

ROBINSON, A H W (1962): *Marine Cartography in Britain*. Oxford, Leicester University Press.

SKELTON, R A (1954): "Captain James Cook as a Hydrographer," *The Mariner's Mirror, 40,* 92-119.

STEVENS, Henry (1973): *Notes Biographical and Bibliographical on The Atlantic Neptune*. London, Henry Stevens, Son & Stiles.

TYACKE, Sarah, and HUDDY, John (1980): *Christopher Saxton and Tudor Map-Making*. London, The British Library.

1.0320B Harbour Chart

A A hydrographic chart used for navigation and anchorage in harbours and inland waterways. A port plan is a special type intended to show details of port installations rather than for navigation in the ordinary sense.

B The harbour chart seems to have developed in Europe comparatively late. In the Middle Ages, and even as late as the 16th century, the practice of working in and out of a port was normally learned from observation and was passed on orally. As trading voyages became more ranging, however, literate seamen began to keep notes which became codified into sailing directions and rutters, a development which took place in the 13th century in Italy and in the 14th century in northern Europe. Descriptions of harbours were thus included in *portolani*, as the name (*porto*, harbour) implies. Harbours also featured in some detail in *isolarios* (see Island atlas, 8.0110c). Local pilots, however, tended to guard the secrets of navigation to themselves. In England, for example, the pilots of the larger ports formed guilds to protect their interests. It may be significant that early English charts of harbours found in the collection of Lord Burghley, Treasurer to Elizabeth I, cover areas where guilds existed, such as Sandwich Haven and the Downs, 1548, the Humber, *c.* 1569, the Medway, *c.* 1580, and the Thames, 1580. The earliest dated chart of any harbour in the British Isles is that of Orwell Haven by Richard Lee, 1533-34 (British Library, Cotton MS. Augustus I.i.56). The chart of the Humber mentioned above, *c.* 1569, is believed to be the earliest English map to show soundings by number (British Library, Royal MS. 18.D.III.f. 62v-63).

Other English harbour plans were made by engineers engaged on harbour works, as at Dover, where various projects for port improvements are recorded on plans of 1531-32, 1545, etc. in the Cotton collection (British Library). Other plans were made for constructing coastal defences, as at Poole in Dorset (Skelton and Summerson, p. 33).

Harbour charts also featured as a major element on coastal charts. Thus the manuscript *roteiros* of João de Castro, 1538 to 1542, were comprised primarily of harbour plans. De Castro stated his purpose as being "to make *távoas* (surveys, plans or charts) of each place and river, which comprise the view of the land, shoals, reefs, routes, and how they must be entered . . ." (Cortesão and Teixeira da Mota, vol. I, p. 135; see Coastal chart, 1.0320a). A fine example is the "Tavoa da Aguada do Xeque," the Portuguese fort and anchorage of Suk at the Bender Dibni on the north coast of Socotra, depicted in the "Roteiro of the Red Sea," 1542 (British Library, Cotton MS. Tiberius D.IX.ff 11ᵛ-12). Even the charts in Waghenaer's *Spieghel der Zeevaerdt* (1584-85), the first printed sea atlas, were basically a series of harbour plans linked by imperfectly surveyed tracts of coast (Skelton and Summerson, p. 33). The Spanish rutters of the South Sea, which were copied by English chartmakers such as William Hack in the 1680s, comprised mainly harbour charts, with the shores largely in profile, linked by generalized indications of the intervening coasts. Thus William Hack's manuscript "Waggoner of the Great South Sea," 1682 (British Library, K. Mar. VIII.15) carries the subtitle "A Description of all the Ports Bays Rivers Harbours Islands Sands Rocks & Dangers from the mouth of Calafornia [*sic*] to the Straights of Le Maire."

John Thornton, the only member of the Drapers' Company School to publish printed charts, included harbour plans in his *Atlas Maritimus*, London, *c.* 1685. They were standard items likewise in Johannes van Keulen's *Zee-Atlas*, Amsterdam, from 1680, and *Zee-Fakkel*, from 1681. Jacques Nicolas Bellin, Ingénieur Hydrographe de la Marine in Paris, 1721 to 1772, produced in *Le Petit Atlas Maritime, recueil de cartes et plans de quatre parties du monde,* Paris, 1764, 580 charts of the whole world, many of them harbour charts. Alexander Dalrymple, Hydrographer of the East India Company from 1779, was indefatigable in seeking out and publishing all the harbour and coastal charts available. His *Collection of Plans of Ports in the East Indies . . . ,* London, 1774-75, was an experiment in marine chart publication, in which the harbour plans were accompanied wherever possible with views of land, often from different sources, engraved on the same plate. A contemporary work applauded for its magnificent harbour charts and views was J F W Des Barres's *The Atlantic Neptune,* London, from 1777. In this, as in many other maritime atlases and charts, the harbour plans were often shown as insets.

C COOK, Andrew (1981): "Alexander Dalrymple's 'A Collection of Plans of Ports in the East Indies (1774-1775),'" *Imago Mundi, 33,* 46-64.

CORTESÃO, Armando, and TEIXEIRA DA MOTA, Avelino (1960): *Portugaliae Monumenta Cartographica.* Lisbon, Comemoraçoes do V Centenário da Morte do Infante D. Henrique.

DALRYMPLE, Alexander (1774-75): *The Collection of Plans of Ports in the East Indies.* London, A. Dalrymple.

HODSON, Donald (1978): *Maps of Portsmouth before 1801. A Catalogue . . .* Portsmouth, City of Portsmouth.

HOWSE, Derek, and SANDERSON, Michael (1973): *The Sea Chart.* Newton Abbot, David & Charles.

SKELTON, R A, and SUMMERSON, John (1971): *A Description of Maps and Architectural Drawings in the Collection Made by William Cecil, First Baron Burghley, Now at Hatfield House.* Oxford, The Roxburghe Club.

1.0320C Nautical Chart

A A chart designed to assist navigation at sea or on other waterways.

B No charts survive from the ancient classical world (see Coastal chart, 1.0320a). With the development of Arabic sciences and the expansion of Islam from the 8th century onwards, Arabic charts were made for use at sea or for inclusion in geographical treatises. The so-called "Atlas of Islam," a set of 21 maps dating from the 10th century, which was copied and revised over a long period, normally included three sea-charts (the Mediterranean, the Persian Gulf and the Caspian Sea). Examples are the three manuscripts by Abu Zaid al Balhi, born at Samistigan, Balh, died 934 (Bagrow, p. 55; Miller, I). The charts have a schematic geometrical style, which is also found in the Turkish world map by al-Kashgari, 1074. Marco Polo recorded the use of charts by the seamen of Ceylon and in the Indian Ocean in general in the second half of the 13th century.

At a later period, in the early 16th century, the geographical discoveries of Europeans influenced other cultures, notably the Ottoman Empire, which had close ties with Greece and Italy. Piri Re'is, the Turkish admiral and cartographer, produced the *Kitab-i-Bahriye*, or guide to the navigation of the Mediterranean, in 1521 and 1525 versions, which included many plans and charts. The *Bahriye*, of which 29 examples survive, has been described as "the greatest geographical compendium of the time" (A. Adivar, *Osmanli Turklerinde Ilim,* Istanbul, Maarif Matbaasi, 1943).

A major innovation in the development of the plane chart was the introduction of a latitude scale. The earliest charts to show such a scale appear to be the anonymous King Hamy map (Huntington Library, San Marino, California), and the planisphere of Nicolaus de Caverio, both *c.* 1505. A concept of longitude, as yet normally omitted from the markings, was implicit. Surviving manuscript world charts of the Portuguese Diego Ribeiro, who became "cosmographer and master maker of charts" for Spain, which are dated 1525 (anonymous), 1527 and 1529 (two in number), were based on the "padron real," the official Spanish chart, and rank today as among the earliest to display the circuit of the globe between the Polar circles. They were thus true planispheres. The loxodromic or rhumb lines still found on most renaissance charts were now properly described as compass lines and intended for use as such. Jean Rotz, the Dieppe hydrographer, explained how to plot a course from them in his "Boke of Idrography," 1542 (British Library, Royal MS. 20.E.IX). He also tried to solve the problem of magnetic variation by using differential scales of latitude on the western and eastern margins of the charts. A different attempt at solution was the oblique meridian on Pedro Reinel's chart of the North Atlantic, 1505, a device followed by some later chartmakers even as late as *c.* 1592 (e.g. "Nova Francia," by Peter Plancius, Amsterdam).

The application of printing to maritime cartography was a further innovation. The first printed portolan atlas, although somewhat crude in style, was the *Portolano* of the Venetian Pietro Coppo, published at Venice in 1528. The first printed portolan-style sea chart, of a more elaborate kind, was the woodcut chart of the Mediterranean of Giovanni Andrea di Vavassore, Venice, 1539 (see also Island atlas, 8.0110c).

The plane chart was not drawn to a mathematical projection and thus took no account of the convergance of the meridians towards the poles, as the Portuguese marine scientist Pedro Nuñes commented in the 1530s. Gerard Mercator's invention of the projection named after him and its demonstration on his celebrated world chart of 1569 marked a major advance, although the plane chart continued in use over the years, to the despair of scientific experts such as the astronomer Edmond Halley, who censured seamen for their ignorant use of plane charts (Letter to Samuel Pepys, 17 February 1696, in "Papers of Mr. Halley's & Mr. Greave's touching on imperfect knowledge in Navigation &c.," Magdalene College, Cambridge, Pepys Library, MS.2185. A copy is in the British Library, Add. MS.30221, fol. 183 et seq). The force of tradition was particularly strong in England, where the Drapers'

Company School of chartmakers, working in manuscript on vellum, dominated private practice in the chart trade, and continued to produce work in the typical portolan style. The innovator who was the exception to the rule was William Hack (active 1671 to 1708), who replaced the rhumb lines and compass roses with the rectangular grid of the projectionless plane chart, combined with a quarter compass rose (Campbell, p. 94), as illustrated by his general map in "A Description of all Ports . . . in the South Sea of America 1698" (British Library, K.Mar. VIII. 16; 7 Tab 122).

Dutch publishing houses dominated the market in the chart trade from the late 16th century until the mid 18th century. Waghenaer's *Spieghel der Zeevaerdt* (1584-85), despite its useful innovations, proved too costly for seamen (see Coastal chart, 1.0320a). His more handy pilot book, *Thresoor der Zeevaert,* first issued in oblong folio at Leyden in 1592, followed the tradition of the early rutters. It provided a more acceptable model which was followed in the early 17th century by the celebrated cartographer Willem Janszoon Blaeu of Amsterdam and his successors, also by Blaeu's rival Johannes Janssonius and by Jacob Aertsz Colom (from 1632), and later in the 17th century by the famous publishing house founded in 1678 by Johannes van Keulen I, which was still in business as late as 1885. The *Zee-Fakkel* ("Sea Torch") first issued in 1681 and the *Zee-Atlas* (from 1680) of Johannes van Keulen were masterpieces of hydrographic publication. Van Keulen advertised what he regarded as his finest edition, that of 1695, as "a Large New Augmented Sea-atlas or Water-world containing all the coasts of the world in excellent charts, both plane and increasing . . . In five languages." The charts were made by Claas Jansz Vooght, "Geometra." The house of Van Keulen has been described as the "largest unofficial hydrographic office in the world" (Koeman, 1972, p. 2) and was without rival until the hydrographic departments of the French Marine and the British Admiralty were established in 1720 and 1795 respectively.

English contributions to hydrographic publication lagged behind the Dutch until the late 17th century. A notable exception is the work of the mathematician Edward Wright, who applied a mathematical method to the construction of the Mercator projection in his chart of the Azores, 1599, and his world chart 1599. Sir Robert Dudley in his *Dell' Arcano del Mare* ("Secrets of the Sea," Florence, 1646-7), produced the earliest sea-atlas with all the charts drawn on the Mercator projection. The inclusion of soundings, figures for magnetic variation, and notes on currents made it an outstanding work for its day, but it never became widely known. John Seller's *The English Pilot,* first part issued in London in 1671, was the earliest of the commercial sea atlases regularly used by the Royal Navy and mercantile marine until the Admiralty was established in 1795. The scientific innovations of Edmond Halley in the form of lines of magnetic variation (from 1701) and winds (1688) were incorporated in these works and the resulting charts were widely copied by chartmakers on the continent of Europe (see Thematic map, 1.202).

In the 18th century a further innovation was introduced by William Gerard de Brahm in his charting of the Gulf Stream, one of the major features in his "Hydrographical Map of the Atlantic Ocean," in *The Atlantic Pilot* (London, 1772). The American statesman and scientist Benjamin Franklin also made a map of the Gulf Stream, reputedly in 1768, but the extant maps date from 1785-86 . One of the most remarkable hydrographic surveys of the period was that of the coasts of North America accomplished by Joseph Frederick Wallet Des Barres, published in *The Atlantic Neptune* from 1777 (see Coastal chart, 1.0320a).

Peoples of non-literate cultures have devised ingenious forms of navigational charts. The Micronesian sailing charts used for navigation in the Marshall Islands before the advent of Europeans in the Pacific in the 18th century reflect the remarkable skills of Pacific islanders in oceanic navigation. The charts were constructed as a lattice-work from midribs of leaves of the coconut palm, with shells denoting islands. They depict the direction of the principal seasonal wave swells in relation to the islands. An example of the "Meddo" class, representing a part of the Marshall Group, is preserved in the British Museum, London (Department of Ethnography, No. 2289). The Greenland eskimoes (Inuits) made charts of the coasts in carved wood creating the effect of a relief model.

C BAGROW, L (1964): *History of Cartography.* Revised and enlarged by R A Skelton. London, C A Watts; Cambridge, Mass., Harvard University Press.

BROWN, Lloyd A (1949): *The Story of Maps.* Boston, Little, Brown and Co.

CAMPBELL, Tony (1973): "The Drapers' Company and Its School of Seventeenth Century Chart-Makers," in Wallis, Helen, and Tyacke, Sarah, *My Head is a Map.* London, pp. 81-99.

CRONE, Gerald Roe (1978): *Maps and Their Makers.* 5th ed. Folkestone, Kent, Dawson; Hamden, Conn., Archon Books.

DAVENPORT, William (1960): "Marshall Islands Navigational Charts," *Imago Mundi,* 15, 19-26.

DE VORSEY, L (1976): "Pioneer Charting of the Gulf Stream. The Contributions of Benjamin Franklin and William Gerard De Brahm," *Imago Mundi, 28,* 105-20.

KOEMAN, Cornelis (1972): *The Sea on Paper. The Story of the Van Keulens and Their "Sea-torch."* Amsterdam.

KOEMAN, Cornelis (1980): "The Chart Trade in Europe from Its Origin to Modern Times," *Terrae Incognitae, 12,* 49-64.

MILLER, Konrad (1926): *Mappae Arabicae, Arabische Welt- und Länderkarten des 9.-13. Jahrhunderts.* Stuttgart, Selbstverlag des Herausgebers.

ROBINSON, A H W (1962): *Marine Cartography in Britain.* Oxford, Leicester University Press.

SKELTON, R A, ed. (1964): *Lucas Jansz Waghenaer. Spieghel der Zeevaerdt.* Amsterdam, Meridian Publishing Co.

SMITH, Thomas R (1978): "Manuscript and Printed Sea Charts in Seventeenth-Century London. The Case of the Thames School," in Thrower, Norman J W: *The Compleat Plattmaker. Essays on Chart, Map, and Globe Making in England in the Seventeenth and Eighteenth Centuries.* Berkeley, University of California Press, 45-100.

WALLIS, Helen (1981): *The Maps and Text of the Boke of Idrography Presented by Jean Rotz to Henry VIII.* Oxford, The Roxburghe Club.

YULE, Col. H, ed. (1903): *The Book of Ser Marco Polo, the Venetian, concerning the Kingdoms and Marvels of the East,* 3rd edition, revised by Henri Cordier. London.

1.0320D Portolan Chart

A The term "portolan" (from Italian *portolano*) refers to written sailing directions for sailors. Since the middle of the 19th century, however, it has been extended incorrectly to the atlases and charts which some commentators considered owed their origin to those written directions. The American historian Justin Winsor in *Christopher Columbus* (Boston and New York, 1891, p.530), referred to "those atlases of sea-charts, which have come down to us under the name of 'portolanos.'" The term portolan map seems to have been first used by A E Nordenskiöld in 1895 (*Report of the Sixth IGU Congress,* p. 694), though it was Franz R Wieser who established the phrase *portolan karten* (*Petermanns Geographische Mitteilungen,* vol. 45, 1899, pp. 188-94). Since that time "portolan chart" has been the term used mainly by those writing in English to describe the superficial characteristics of a class of early coastal charts capable of navigational use. Among alternative expressions employed at various times are "rhumb line charts," "loxodromic charts" and "compass charts." The last two terms may be regarded as questionable in attributing to the compass an essential role in both the origin and use of the charts that is not universally accepted. The fact that a compass bearing (or loxodrome) can be represented as a straight line only on Mercator's projection would seem to imply (anachronistically) that the apparently projectionless portolan charts had anticipated that development. The terms used by other European authorities, such as *cartes nautiques, carte nautiche, Seekarten, cartas de marear, cartas nauticas* &c., as the equivalent of "sea" or "nautical" charts (1.0320c), do not refer exclusively to those with portolan features.

The features generally though not inevitably present, and regarded as characteristic, are (1) a network of interconnecting lines (usually termed "rhumb lines") springing from sixteen equidistant points around the circumference of one, or sometimes two, "hidden circles"; this network will often be extended beyond the circles, and one or more of the intersection points might be elaborated into a compass rose; (2) place names written inland at right angles to the coast, following one another in an unbroken sequence that invariably sets some of them, e.g. those on Italy's Adriatic coast, "upside down"; (3) the charts are drawn in ink on one or more skins of prepared vellum, following certain colour conventions: the more important places are named in red, the rest in black; the lines for the eight main winds are in black or brown, for the eight half-winds in green, and for the sixteen quarter winds in red; (4) the coastlines are shown in a generalized fashion with bays and headlands exaggerated and navigational hazards, such as rocks and shoals, marked with crosses and dots.

B The earliest surviving portolan chart is the "Carte Pisane," named from the fact that it belonged to a family of Pisa. It is thought to have been made in Genoa and probably dates from the end of the 13th century (Paris, Bibliothèque Nationale, Cartes et Plans, Rés. Ge. B. 1118). The earliest dated chart is that of Pietro Vesconte of Genoa, 1311 (Florence, Archivio di Stato); and the earliest atlas – essentially an arrangement of the standard chart on several sheets, sometimes prefaced by a lunar calendar for tidal calculations – is Pietro Vesconte's of 1313 (Paris, Bibliothèque Nationale, Cartes et Plans, Rés. Ge. DD. 687) (Fig.2). The standard area embraced by those charts and atlases that followed during the next three centuries was the

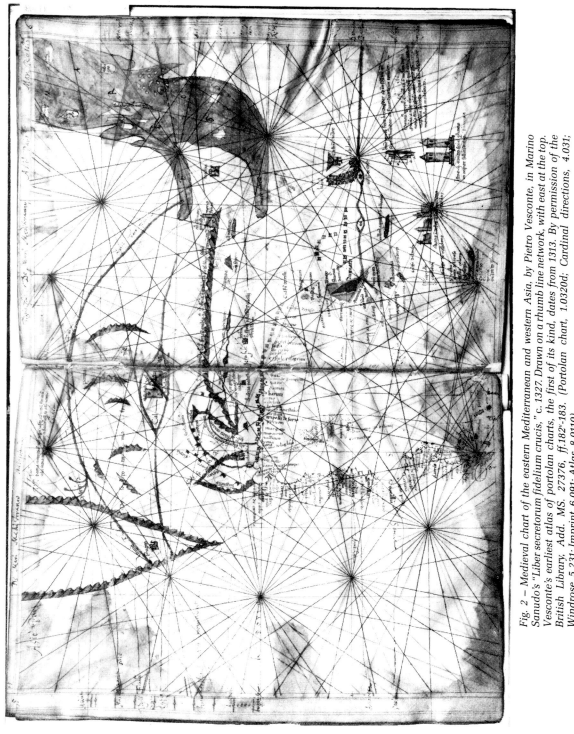

Fig. 2 – Medieval chart of the eastern Mediterranean and western Asia, by Pietro Vesconte, in Marino Sanudo's "Liber secretorum fidelium crucis," c. 1327. Drawn on a rhumb line network, with east at the top. Vesconte's earliest atlas of portolan charts, the first of its kind, dates from 1313. By permission of the British Library. Add. MS. 27376, ff.182"-183. (Portolan chart, 1.0320d; Cardinal directions, 4.031; Windrose, 5.231; Imprint, 6.091; Atlas, 8.0110)

Mediterranean and Black seas, with the Atlantic coasts from Denmark to Morocco and including the British Isles. This coverage might be extended in any of six ways: (a) to take in the Baltic; (b) to include interior and decorative details, as exemplified by Catalan-style charts (some of which were produced by Italians); (c) to take in real and legendary Atlantic islands (some of which have been interpreted as representing a pre-Columbian discovery of America); (d) to document the Portuguese discoveries along the west coast of Africa after 1415; (e) to extend eastwards to the Caspian Sea and beyond (as on the Catalan Atlas of *c*.1375), though these exemplars are as much world maps as portolan charts; and (f) to show discovered parts of America and Asia after 1500 (these 16th century examples, some of which incorporate latitude scales, could be regarded instead as plane charts).

Despite Nordenskiöld's belief in their essentially unchanging nature, the portolan charts exhibit extensive hydrographic development in the earliest stages, as well as a significantly evolving toponymy up to the middle of the 15th century. The main centres of production during this period were Venice and Majorca, and to a lesser extent Genoa. Though the lavish nature of some productions would obviously have precluded such use, portolan charts were intended as navigational aids. An Aragonese ordinance of 1354 commanded that two charts be used on each war galley.

The question of the portolan chart's origin has exercised scholars for a century and a half without any sustainable hypothesis having so far emerged. The earliest reference to a chart dates only from 1270 (shortly before the presumed date of the "Carte Pisane") and was made in connection with the journey of Louis IX of France from Aigues Mortes to Tunis. Forced ashore at Cagliari, he was shown a *mappamundi* (as recorded in the *Gesta Ludovici IX* by Guillaume de Nangis), and this is generally considered to denote a portolan chart. Assertions that the charts stem from the classical world or from Arab models, such as those of al-Idrisi (1099-1164), lack firm supporting evidence, and the toponymic differences between the "Carte Pisane" and the earliest surviving *portolano* (the mid-13th century "Compasso da Navigare") cast doubt on Motzo's claim that the first derives from the second. Nor can any reliance be placed on Nordenskiöld's argument that the charts represent a synthesis of earlier local charts, since the marginal illustrations to the manuscript cosmography *La Sfera*, *c*. 1400, of Leonardo Dati or his brother Gregorio (or Goro), which Nordenskiöld saw as reflections of these original "coasting charts," are no more than crude copies of the charts they are supposed to predate. To Charles de la Roncière (1:37-42) the prototype chart was the work of the Genoese admiral Benedetto Zaccaria (fl.1298), but this theory has found little favour. Much of the earlier writing on the subject forms part of an extended and inconclusive controversy (involving Caraci and Winter in particular) over the conflicting claims that the Catalan (Majorcan) or Genoese practitioners were the originators of the portolan charts, or the main spurs to their improvement.

Though each chart carries its own scale, the lack of any statement of value has led to much discussion of the unit involved. It is clear that up to about the year 1400 different units were employed in constructing the Mediterranean and Atlantic portions of the chart, with the result that Atlantic distances were understated by about 16 per cent.

Various nautical, geographical and decorative innovations in the evolution of the portolan chart may be distinguished. The compass rose is first found in the Catalan

Atlas *c.* 1375 (Paris, Bibliothèque Nationale, MS. espagnol 30), but it is not placed at an intersection point of the interconnecting lines (see Windrose, 5.231). The first instances of intersection points being elaborated into compass roses occur in the first half of the 15th century. An example is the 1439 Gabriel de Valseca chart made in Majorca (Barcelona, Biblioteca de Catalunya). The first fleur de lys with a north pointer appears on the chart of Jorge de Aguiar, 1492, which is notable as the earliest signed and dated Portuguese chart known (Yale University Library). Compass roses grew in number and complexity on charts produced in the 16th and 17th centuries. The chart of Angelino Dalorto, 1325 or 1330 (Prince Corsini, Florence) and that by Angelino Dulcert (probably the same man), 1339 (Paris, Bibliothèque Nationale, Rés. Ge. B. 696), are among the first to provide extensive inland detail, thus combining the form of the portolan chart with the concept of a world map. The Catalan Atlas, which embraces the known world of Europe, Africa and Asia, is the most famous work of this type.

The invention of printing introduced further technical innovations. The world map by Francesco Rosselli, *c.* 1508, with its rhumb line network, may be considered the earliest printed portolan chart (Greenwich, National Maritime Museum, and Florence, Biblioteca Nazionale Centrale). The first printed portolan chart of the Mediterranean is that by Giovanni Andrea di Vavassore, Venice, 1539. In the early 16th century latitude scales with equidistant degrees were added to portolan charts but these were not apparently part of the construction. In the 17th century, charts with a rhumb line network were provided with a set scale of increasing degrees of latitudes (i.e. were drawn on the Mercator projection). These charts in their external features nevertheless qualify as portolan charts. The chartmakers of the Drapers' Company (or Thames School) in London, established *c.* 1590, continued the portolan tradition into the second decade of the 18th century (see Nautical chart, 1.0320c).

Besides charts produced by northern Mediterranean Christians, or in the case of Majorca by a school of Jewish cartographers, a few early Arabic examples have survived. The oldest known, the so called "Maghreb Chart" (Biblioteca Ambrosiana, Milan), covers only the western Mediterranean and the Atlantic coasts as far as the British Isles and the Netherlands. It is probably an early 14th century extract from an Italian or Catalan work. Charts lettered in Greek are known from the 16th century. The Japanese with their skills of assimilation produced in the 17th century portolan charts of the Indian Ocean and south-east Asia based on Portuguese models; an example is preserved in the Tokyo National Museum, Division 1, no. 784, one of about ten such charts now known in Japan. Another type of portolan chart of the Japanese islands is Western in style but based on indigenous Japanese cartography as illustrated by a chart of *c.* 1625 in the Tokyo National Museum.

C ANDREWS, Michael C (1925-26): "The Boundary between Scotland and England in the Portolan Charts," *Proceedings of the Society of Antiquaries of Scotland, 12,* 36-66.

BRICE, William C (1977): "Early Muslim Sea Charts," *Journal of the Royal Asiatic Society of Great Britain and Ireland, 1,* 53-61.

CAMPBELL, Tony (1986): "Census of Pre-Sixteenth-Century Portolan Charts," *Imago Mundi, 38,* 67-94.

CAMPBELL, Tony (1987) "Portolan Charts from the Late Thirteenth Century to

1500," in Harley, J B, and Woodward, David, eds., *The History of Cartography, vol. 1*. Chicago, Chapter 19.

CARACI, Giuseppe (1959): *Italiani e Catalani nella Primitiva Cartografia Nautica Medievale*. Rome.

CORTESÃO, Armando (1969-71): *History of Portuguese Cartography*. Coimbra. 2 vols.

DESTOMBES, Marcel (1952): "Cartes Catalanes du XIVᵉ Siècle," *Union Géographique Internationale. Rapport de la Commission pour la Bibliographie des Cartes Anciennes, Fascicule 1*.

FISCHER, Theobald (1886): *Sammlung Mittelalterlicher Welt- und Seekarten italienischen Ursprungs und aus italienischen Bibliotheken und Archiven*. Venice.

FONCIN, Myriem, DESTOMBES, Marcel, and LA RONCIÈRE, Monique de (1963): *Catalogue des cartes nautiques sur vélin conservées au Département des Cartes et Plans*. Paris, Bibliothèque Nationale.

FRABETTI, Pietro (1978): *Carte nautiche italiane dal XIV al XVII secolo conservate in Emilia-Romagna*. Florence, Olschki.

GROSJEAN, Georges (1978): *Mapamundi, the Catalan Atlas of the Year 1375*. Dietikon-Zurich, Urs Graf.

KAMAL, Youssouf (1936-53): *Monumenta Cartographica Africae et Aegypti*. Cairo, vols. IV & V.

KELLEY, James E, Jr (1979): "Non-Mediterranean Influences That Shaped the Atlantic in the Early Portolan Charts," *Imago Mundi, 31*, 18-35.

KRETSCHMER, Konrad (1909): *Die italienischen Portolane des Mittelalters, ein Beitrag zur Geschichte der Kartographie und Nautik*. Berlin.

LA RONCIÈRE, Charles de (1924-25): *Découverte de l'Afrique au Moyen-Âge. Cartographes et explorateurs*. Cairo, 2 vols.

MOLLAT DU JOURDIN, Michel, and LA RONCIÈRE, Monique de (1984): *Les Portulans: Cartes marines du XIIIᵉ au XVIIᵉ siècle*. Fribourg, Office du Livre. English language edition (1984): *Sea Charts of the Early Explorers 13th to 17th Century*, translated by L le R Dethan, New York, Thames and Hudson.

MOTZO, Bacchisio R (1947): "Il Compasso da Navigare, opera italiana della meta del secolo XIII," *Annali della Facoltà di lettere e filosofia della Università di Cagliari, 8*.

NANBA Matsutaro, MUROGA Nobuo, and UNNO Kazutaka (1973): *Old Maps in Japan*. Translated by Patricia Murray. Osaka, Sogensha.

NORDENSKIÖLD, Adolf E (1897): *Periplus: An Essay on the Early History of Charts and Sailing Directions*. Stockholm.

PAGANI, Lelio (1977): *Pietro Vesconte. Carte Nautiche*. Bergamo, Grafica Gutenberg.

PELHAM, Peter T (1980): "The Portolan Charts, Their Construction and Use in the Light of Contemporary Techniques of Marine Survey and Navigation," MA thesis in Geography at the Victoria University of Manchester.

REVELLI, Paolo (1937): *Cristoforo Colombo e la Scuola Cartografica Genovese.* Genoa.

REY PASTOR, Julio, and GARCÍA CAMARERO, Ernesto (1960): *La Cartografía Mallorquina.* Madrid.

STEVENSON, Edward Luther (1911): *Portolan Charts, Their Origin and Characteristics with a Descriptive List of Those Belonging to the Hispanic Society of New York.* New York.

TAYLOR, Eva G R (1956): *The Haven-Finding Art.* London.

UZIELLI, G, and AMAT DI S. FILIPPO, P (1882): *Mappamondi, Carte nautiche, portolani ed altri monumenti cartografici specialmente italiani, del secoli XIII-XVIII.* Rome.

VERNET-GINÉS, J (1962): "The Maghreb Chart in the Biblioteca Ambrosiana," *Imago Mundi, 16,* 1-16.

WINTER, Heinrich (1940): "Das katalanische Problem in der aelteren Kartographie," *Ibero-Amerikanisches Archiv, 14,* 89-126.

1.033 Chorographic Map

A (MDTT 813.4) Any map representing large regions, countries, or continents on a (relatively) small scale.

B In Strabo's *Geography,* completed AD 17-23, chorography is described as a branch of geography, and a chorographic map represents a part of the oecumene or inhabited world. Ptolemy's Geography, *c.* 150, gives the same definition. The term was generally adopted in Europe when the Geography was rediscovered through Jacobus Angelus's Latin translation (1406).

In the 16th century geographical writers followed Ptolemy's definition: "as if a painter should set forth the eye or eare of a man, and not the whole body, so . . . Chorographie consisteth rather in describying the qualitie and figure, then the bignes, and quantitie of any thinge" (William Cunningham, *The Cosmographical Glasse,* 1559, p.7). The use of the term chorography became less precise in the following centuries. In the *Encyclopédie* Diderot and d'Alembert defined it as follows: "Chorographie: l'art de faire la carte, ou la description de quelque pays ou province." Dainville (p. 37) writes: "les cartes chorographiques, celles qui représentent avec un plus grand détail une contrée, un diocèse, une province, une élection en France, un shire en Angleterre." At the end of the 18th century the French Dépôt de la Guerre distinguished the scales of the Ancien Système as Topographie de détail, Topographie générale, Chorographie — "2 l. = 1000 t.,1:432 000, réduction de la topographie complète en chorographie par le dessin et la gravure" — and, finally, Géographie (1:1 728 888) (Dainville, p. 54).

C DAINVILLE, François de (1964): *Le langage des géographes.* Paris, Éditions A et J Picard, 3, 37, 54.

DIDEROT, Denis, and D'ALEMBERT, Jean LeRond (1753): *Encyclopédie ou diction-naire raisonné des sciences, des arts et des métiers,* III. Paris, 374.

ECKERT, Max (1921): *Die Kartenwissenschaft.* Berlin and Leipzig, Walter de Gruyter, Vol. 1, 51.

STRABO: *Géographie.* Texte etabli et traduit par Germaine Aujac (1969), I, 2ᵉ partie. Paris, Les Belles Lettres, 154, 180-1.

WESTERMANN (1968): *Lexikon der Geographie* (ed. by Wolf Tietze), I. Braunschweig, 677.

1.034 Choropleth Map

A (MDTT 443.6) A map symbolizing statistical values, usually by classes, with tones or patterns applied to enumeration areas or aggregates of them.

B The first choropleth map appears to have been made by Baron Charles Dupin in 1826 to illustrate an address to the Conservatoire des Arts et Métiers in Paris. The map shows the number of persons per male child in school for each département and is the first moral statistics map (2.132). It was included in Dupin's *Forces productives . . . ,* published in 1827. Funkhouser, and several others referring to Funkhouser, are probably mistaken in stating that the map had been produced in 1819 (Robinson, p. 232, note 276). In 1829 Adrien Balbi and André Michel Guerry published choropleth maps of several classes of data. These early choropleth maps did not separate the unit data into classes, instead they attempted (unsuccessfully) to use graded tones to fit each value. The first choropleth map to show numerical classes is a Prussian manuscript population map dated 1828. After 1830 almost all choropleth maps showed classed data (Fig. 15).

After the mid 1830s large numbers of choropleth maps on a great variety of topics were made, since by that time much census data had become available.

C BALBI, Adrien, and GUERRY, André Michel (1829): "Statistique comparée de l'état de l'instruction et du nombre des crimes dans les divers Arrondissements . . . de France" (Map). Paris, Jules Renouard.

DUPIN, Charles (1827): *Forces productives et commerciales de la France,* 2 vols. Paris, Bachelier.

FUNKHOUSER, H Gray (1938): "Historical Development of the Graphical Representation of Statistical Data," *Osiris, 3,* 300.

PRUSSIA (1828): *Administrativ-Statistischer Atlas vom Preussischen Staate.* Berlin, map 13.

ROBINSON, Arthur H (1982): *Early Thematic Mapping in the History of Cartography.* Chicago, University of Chicago Press.

1.041 Dasymetric Map

A (MDTT 443.4) A refinement of the simple choropleth technique (1.034) in which the density value (usually, but not necessarily, population) of an enumeration area is distributed unevenly within the area on the basis of supplemental geographical information. The lines separating the new density regions are neither isopleths (5.0910f) nor necessarily the boundaries of the enumeration areas; instead, the lines occur in zones where the average density changes more or less abruptly to the different density of an adjacent area.

B Although the present term "dasymetric" was coined and first used by the Russian Benjamin Semenov-Tian-Shansky in the title of a population map of European Russia in 1922, the technique was first used about a century earlier.

The first dasymetric map, albeit robustly general and crude, is apparently also the first attempt at a world map of population density. It occurs as a frontispiece in a book on political economy by George Poulett Scrope published in 1833. It divides the earth into three broad density classes. Scrope's map, about which there is not a word in the book with which it appeared, is dasymetric in fact but no doubt by accident. Even if Scrope had had any idea of the technical aspects of the innovation, he could not have worked from detailed enough data to have allowed the usual computations needed to make the portrayal more nearly like reality.

The next dasymetric map to be made is as precise and sophisticated as Scrope's is crude. It was prepared by Henry Drury Harness for an atlas published in 1838 to accompany a report by commissioners to recommend a railway system for Ireland. The procedure involved in refining the choropleth method was thoroughly appreciated and is carefully described in an appendix to the commissioner's report. The map itself is engraved, detailed, and employed the aquatint etching method (7.0920b) to obtain the shading.

Small, highly generalized maps of the dasymetric class were not uncommon after 1840. Heinrich Berghaus included one in his *Physikalischer Atlas* (7th Abteilung, Anthropographie No. 1, 1848), and August Petermann prepared one for C F W Dieterici, 1859. The dasymetric modification of the simple choropleth technique was strongly promoted during the remainder of the 19th century by Georg Mayr, director of the Bavarian Statistical Bureau, and was greatly influenced by Friedrich Ratzel and the growth of interest in anthropogeography.

C ANDREWS, John H (1966): "An Early World Population Map," *Geographical Review,* 56, 447-48.

DIETERICI, Carl F W (1859): "Die Bevölkerung der Erde, nach ihren Totalsummen, Racen-Verschiedenheiten und Glaubensbekentnissen," *Petermanns Mitteilungen, 5,* 1-19.

HARNESS, Henry Drury (1838): "Report from Lt. H D Harness, R.E., Explanatory to the Principles on which the Population, Traffic and Conveyance Maps Have Been Constructed." In Railway Commissioners (1838): *Second Report of the Commissioners Appointed to Consider and Recommend a General System of Railways for Ireland,* Dublin, HMSO, Appendix 3.

MAYR, Georg (1871): "Zur Verständigung über die Anwendung der 'geographischen

Methode' in der Statistik," *Zeitschrift* (koeniglichen Bayerischen statistischen Bureaus), *3*, 179-82.

MAYR, Georg (1874): "Gutachten über die Anwendung der graphischen und geographischen Methode in der Statistik," *Zeitschrift* (koeniglichen Bayerischen statistischen Bureaus), *6*, 36-44.

McCLEARY, George F (1969): "The Dasymetric Method in Thematic Cartography," Ph.D. dissertation, University of Wisconsin-Madison.

ROBINSON, Arthur H (1982): *Early Thematic Mapping in the History of Cartography.* Chicago, University of Chicago Press.

SCROPE, George Poulett (1833): *Principles of Political Economy, Deduced from the Natural Laws of Social Welfare, and Applied to the Present State of Britain.* London, Longmans.

1.042 Dissected Map

A A map mounted on board and cut into a number of irregular pieces to be put together again as a geographical exercise; a jigsaw puzzle map. It should be noted that the term "dissected" is sometimes also used to describe a map cut into sections for convenience of handling or storage, but this does not refer to a type of map and therefore is not relevant here.

B The first dissected maps were made in the early 1760s by John Spilsbury, an engraver and mapmaker of Russell Court in London. He took advantage of the popular interest in geography and its consequent importance in the schoolroom by devising an early form of visual aid. Using ordinary maps from his stock, he mounted them on thin board and then cut them along appropriate administrative or political boundaries. These were then advertised "for teaching geography," so that children might learn both the shapes of the individual countries or divisions and their relative positions by putting the pieces together again. The earliest of his maps extant is Europe, *c.* 1766. By the time Spilsbury died in 1769, dissected maps had not yet become widely known, and his immediate successors had no particular interest in this side of the business.

The man chiefly responsible for popularizing dissected maps was John Wallis, a leading London publisher of puzzles and prints for children, who from the 1780s produced dissections not only of maps and historical tables but also of more frivolous subjects. Despite the introduction of less educational puzzles, maps retained an important share of the market, and even late into the 19th century 20 percent of the puzzles made in England were cartographical.

Dissected puzzles were introduced on the continent of Europe towards the end of the 18th century, but seem never to have become as popular as they were in England. The German-speaking nations rapidly developed elaborate interlocking systems for their puzzles, but rarely used maps; the French also used interlocking pieces, but less elaborate ones. In America, dissected maps and puzzles were imported from Europe by the early 19th century, but they were not manufactured there until about 1830.

By the end of the 19th century, interlocking pieces were becoming common in England and America, thus reducing the educational aspect of map dissections. The invention of the cutting die at about this time made possible the mass-production of puzzles, which, after various trade-names had been used, became simply and universally known as "jigsaws."

C HANNAS, Linda (1972): *The English Jigsaw Puzzle 1760-1890.* London, Wayland Publishers.

HANNAS, Linda (1980): "When Maps Were . . . Cut into Pieces," *The Map Collector, 12,* 18-21.

HANNAS, Linda (1981): *The Jigsaw Book.* London, Hutchinson.

1.043 Distance Map

A A cartographic representation on which the distances between points on the surface of the Earth are expressed in figures in a variety of forms. The distances may be noted on the map itself or in tabular form on the sheet. A distance map need not necessarily show actual routes, e.g., road, river, rail, etc.

B Maps showing distances between places developed as a form of road map (1.1810d). The earliest distance map is the Peutinger map of the Roman Empire, 4th or 5th century, on which figures are written along the roads. The earliest example for Great Britain is the Gough map, *c.* 1350. Erhard Etzlaub's map "Das ist der Romweg . . . ," Nuremberg, *c.* 1500, which shows routes by dotted lines with each dot representing one German mile, ranks as the first printed distance map. The English cartographer John Norden (1548-1625?) invented (as he rightly claimed) the triangular distance table, which he included in his book *England. An intended Guyde for English Travailers. Shewing in generall, how far one Citie, & many Shire-Townes in England, are distant from other,* London, 1625. This useful device is still in use today. Norden's tables were copied in *A Direction for the English Traviller,* London, 1635, the earliest English road book with maps. The 37 plates, engraved probably by the Dutchman Jacob van Langren who signs the title-page, display the distance tables together with a small thumb-nail map of the county.

John Ogilby in the first English road atlas, *Britannia,* London, 1675, gave on his strip maps the "direct horizontal" distance, as the crow flies, between two points, the "vulgar computed distance" along the road, in local customary miles, and the "dimensurated" road distance from his survey with a wheel dimensurator. These mileages provided John Adams, barrister and topographer of London, with the information necessary for the construction of his innovative "distance map" of England. He advertised in June and July 1677 "Distances without Scale and Compass: A New Large Map of England full six Foot Square wherein Computed and Measured Miles are entered in Figures" (Tyacke, pp.19-20). Adams produced in 1679 a new version on two sheets with additional roads shown as straight parallel lines, with computed distances in circles, computed and measured distances in ellipses. The map was revised and still in print *c.* 1770, published by Carington Bowles in London as "Bowles's Map of England." Adams included a reduced version of his distance map in his *Index Villaris: or, an Alphabetical Table of all the Cities, Market-Townes . . . in England and Wales,* London, 1680. This type of map became accepted as a standard form. The Quarter Master Generals Office at the Horse Guards, London, issued in 1854, presumably for military use, a "Distance Map of England and Wales. Compiled and drawn by T K King and W J Kelly," made at a scale of 10 miles to the inch (1:633,600). The distances were presumably established not from road measurements but from the trigonometrical survey.

Another style of distance map, derived from the post map, was the "Influenzkarte der Eilpost-Diligenze und Packwagens-Course in dem Kaiserthume Österreich," by Franz Raffelsperger, Vienna, 1834. This shows postal stations in circles which are linked by strips bearing postal information including time and distance (Elias, p.29).

With the introduction of the Hansom cab, patented in 1834, a generation of Londoners benefited from the publication of a variety of distance maps designed to protect them from the wiles of cabmen. F G Harding had provided the first, "The Arbitrator or Metropolitan Distance Map," London, 1833, which gives distances in each street and square, and M A Leigh improved on it in 1834. Joachim Friederichs produced c. 1847 a more complicated map which he claimed to have invented, "The Circuiteer," with the addition to the title, "Distance Map of London." A note on a later edition described it as one of a series of distance maps of the principal towns of the United Kingdom. Simpkin, Marshall and Co. issued in 1851 "A New Distance Map of London for ascertaining Cab Fares," its surface criss-crossed with diagonal lines. The popular London publisher Edward Mogg, a specialist in transport, produced some of the most handy map-guides, including "Mogg's Cab Fare Distance Map and London Guide," by William Mogg, 1858, which was issued in various forms to 1876 (Hyde, pp.14-17).

Another type of distance map shows distances from a chosen centre, usually a town, thus pandering to civic and tourist interest and to local worthies. One of the first English examples is "An actual survey of the City of Bath, in the County of Somerset, and of Five Miles Round. Surveyed by Tho: Thorpe in the year 1742," dedicated to the nobility and gentry, published at Bath 1743-44. Made on a scale of one mile to 1000 mm, the map is circular, concentric circles mark each mile's distance from the centre, and the circumference is graduated with a scale of miles. Smaller maps of the area five miles round, derived from Thorpe's survey, were published in 1773, 1787, and 1800. Benjamin Donn published at Bristol a similar map of the area eleven miles round Bath in 1790; he had already produced a fine map of the country eleven miles round Bristol in 1769. This type of map remained in vogue well into the 19th century.

C ELIAS, Werner (1982): "Maps and Road Books of Europe's Mail Coach Era 1780-1850," *The Map Collector, 20,* 24-29.

FORDHAM, Sir Herbert George (1912): *Notes on British and Irish Itineraries and Road-books.* Hertford, Stephen Austin and Sons.

HYDE, Ralph (1979): "Maps That Made Cabmen Honest," *The Map Collector, 9,* 14-17.

TAYLOR, E G R (1941): "Notes on John Adams and Contemporary Map Makers," *Geographical Journal, 97,* 182-4.

THROWER, Norman J W, ed. (1978): *The Compleat Plattmaker: Essays on Chart, Map, and Globe Making in England in the Seventeenth and Eighteenth Centuries.* Berkeley and Los Angeles, University of California Press.

TYACKE, Sarah (1978): *London Map-Sellers 1660–1720.* Tring, Map Collector Publications.

TYSON, Blake (1985): "John Adams's Cartographic Correspondence to Sir Daniel Fleming of Rydal Hall, Cumbria, 1676-1687," *Geographical Journal, 151,* 21-39.

1.044 Dot Map

A (MDTT 823.14) A map showing the distribution of some phenomenon by individual marks, usually dots, each of which represents one or a given number of items. As generally used, the term refers to the portrayal of discrete statistics, such as population or hectares in wheat, gathered by enumeration districts. Ideally, each dot is placed at the location of the item, if representing one, or at the mean position (centre of gravity) of the group it represents within the enumeration district to which the statistics refer. If the dots are evenly distributed within the enumeration districts, they comprise regular patterns and thus constitute a choropleth map (1.034).

B The locations of individual items, such as towns, mines, fortified places, etc., have been shown on maps from the beginning. The dot map proper, prepared to portray the character of a distribution, probably was first employed to show the occurrence of cases of yellow fever and cholera in their epidemic outbreaks from the latter part of the 18th century to well after the mid-19th. Numerous examples are listed in Stevenson and Jarcho. The most famous such map is one by Dr. John Snow in 1855 who demonstrated by the map that water from a particular well in London was a source of cholera infection (see Disease map, 2.041).

The first printed dot map to show a general distribution (population) was one of France by A Frère de Montizon in 1830. A dot map combining graduated circles was published by August Petermann in 1857 ("Physikalisch-geographisch-statistische Skizze von Sieben-bürgen," *Petermanns Mitteilungen, 3,* 508-513, Map: Plate 25). This was followed by a map of Scandinavia in 1859 by T A von Mentzer. The dot map, however, did not become popular until the early 20th century.

C JARCHO, Saul (1970): "Yellow Fever, Cholera, and the Beginnings of Medical Cartography," *Journal of the History of Medicine and Allied Sciences, 25,* 131-41.

KANT, Edgar (1970): "Über die ersten absoluten Punktkarten der Bevölkerungsverteilung," *Lund Studies in Geography, Ser. B. Human Geography No. 36.*

ROBINSON, Arthur H (1982): *Early Thematic Mapping in the History of Cartography.* Chicago, University of Chicago Press.

SNOW, John (1855): *On the Mode of Communication of Cholera,* 2d ed. London.

STEVENSON, Lloyd G (1965): "Putting Disease on the Map: The Early Use of Spot Maps in the Study of Yellow Fever," *Journal of the History of Medicine and Allied Sciences, 20,* 226-61.

1.061 Fan Map

A Map drawn or printed on folding fans.

B (Japanese) The first folding fan, in so far as is known, was devised in Japan in the early 9th century and was used as an important accessory to the ceremonial attire worn by courtiers and aristocracy. The early folding fans were made by joining together thin slats of *hinoki* (Japanese cypress), but toward the close of the 10th century, a new type of folding fan made by fixing paper onto wooden or bamboo ribs began to be produced. These folding fans spread gradually among all classes of society, and a good many of them were exported to China. Eventually in the 16th century they went as far as Europe, carried by Portuguese traders.

Most folding fans had a brightly painted picture or design of some sort to make them decorative, and some bore a map design, although the number of such fans was small. The oldest extant fan with a map drawn on it belonged to Toyotomi Hideyoshi, the Emperor's chief minister in the late 16th century, who had ambitions for Japan to dominate East Asia. On this fan is a map of East Asia, rendered in brilliant colours.

Not until much later, well into the 19th century, did printed maps appear on fans. Coinciding with the wave of popular curiosity about the world that was aroused by the successive visits of Western ships to Japan, a folding fan with a hemispheric map of the world printed from a copperplate appeared on the market. Copper-engraved landscapes and maps had been produced in Japan since the end of the 18th century by the techniques learned from the Dutch. People at that time had a fondness for the precision and intricacy that those techniques allowed in contrast with woodblock printing, and exquisite little maps on small areas of folding fans were especially appealing to popular tastes. Later, maps of Japan and road maps were also printed on fans. By the end of the 19th century, when new Western methods replaced copperplate printing, folding fans with maps virtually disappeared.

Fan maps also must have been made in China, but information about them is lacking. In Europe fan maps became popular in the 18th century, and some were travelling accessories; for example, "The Ladies Travelling Fann of England and Wales," London, 1768 (Hannas Collection), and Léon Pouillot's "Évantail cycliste," carrying a street map of Paris, *c.* 1898.

C AKOIKA Takejiro (1955): *Nihon Chizu-shi* (History of Japanese Cartography). Tokyo, Kawade Shobo.

NANBA Matsutaro, MUROGA Nobuo, and UNNO Kazutaka (1973): *Old Maps in Japan.* Osaka, Sogensha.

1.062 Flow Map

A (MDTT 823.18.) A map in which directions and/or routes of movement are represented by lines or bands of varying widths, the widths being proportional to the statistical values of the movements along them. The statistical values may be of any kind, such as numbers of items, volume of material, or monetary amounts. The lines may follow routes or be generalized to show only origin and destination. Many traffic maps (2.203) employ the symbolism of flow lines. By this definition flow maps are shown to differ from maps showing ocean or wind currents and rivers on the land.

B The first published flow maps were two by H D Harness dated 1837, showing for Ireland (a) the number of passengers carried by public conveyances and (b) the relative quantities of traffic. They appeared in an atlas accompanying a report of the Railway Commissioners. Harness, in his "Report from Lt. H. D. Harness, R.E., Explanatory to the Principles on which the . . . Maps Have Been Constructed" (Appendix III to the *Second Report of the Commissioners . . .*), describes in detail the manner in which the data were derived and the maps made. However, the Harness maps seem not to have become generally known at the time.

In 1845 both C J Minard in France and Alphonse Belpaire in Belgium made flow maps showing the movements of goods and passengers. It was the flow maps of Minard that made known and popularized the technique. Between 1845 and 1869 he made some 50 flow maps, ranging from the export of British coal, to the movements of freight in France, to the passenger traffic on European railways.

C *Atlas to Accompany the Second Report of the Commissioners appointed to consider and recommend a General System of Railways for Ireland* (1838): Dublin, HMSO.

ROBINSON, Arthur H (1955): "The 1837 Maps of Henry Drury Harness," *Geographical Journal, 121,* 440-50.

ROBINSON, Arthur H (1967): "The Thematic Maps of Charles Joseph Minard," *Imago Mundi, 21,* 95-108.

1.0710 Globe

A A spherical structure on whose surface is depicted the geographical configurations of the earth (terrestrial globe), or the arrangement of the constellations (celestial globe). O.E.D.: The triangular or lune shaped pieces that form the surface of a terrestrial or celestial globe are known as globe-gores.

B Once the heavens and earth were conceived to be spherical in shape, celestial and terrestrial globes were regarded as ideal instruments for their depiction. The earliest recorded celestial globe was made by the Greek astronomer Eudoxus of Cnidus (fl.365 BC). Although it does not survive, an idea of it may be gained from the earliest extant globe, the celestial globe borne on the shoulders of Atlas and forming part of the marble statue, Atlante Farnesiano, in Naples (see Celestial globe, 1.0710a). The stoic philosopher Crates of Mallos, royal librarian at Pergamon *c.* 150 BC, is the first geographer known to have constructed a globe of the earth. In ancient China terrestrial globes were made as early as AD 260.

The Arabs as heirs to Greek scientific learning carried on the tradition of Greek geography and astronomy. The earliest extant Arabic celestial globe from the western part of the Islamic world dates from 1085. The earliest from the eastern part is a Persian globe 1140-41. No terrestrial globe of Arabic origin survives.

Although no medieval European globes survive, scholars such as Bishop Gerbert (*c.* 940-1003), later Pope Sylvester II, used celestial globes and armillary spheres in teaching astronomical science. The manuscript treatise, "Le Livre du Ciel et du Monde" by Nicolas Oresme, 1377, shows the earth as a sphere in various illustrations (Bibliothèque Nationale, MS français 565). In a French Bible Moralisée of the 13th century God is depicted measuring with a pair of dividers the universe which he has created, comprising a spherical earth surrounded by the ocean and firmament (National Bibliothek, Vienna, Codex 2554).

Specific references to celestial globes date from *c.* 1400, and the earliest extant European one is that made by Nicolas of Cusa at Nuremberg in 1444. In the same city nearly forty years later Martin Behaim made his "Erdapfel," the earliest surviving terrestrial globe, 1492 (see Terrestrial globe, 1.0710d). Following the invention of printing and engraving techniques in the later years of the 15th century, cartographers as well as instrument-makers had the opportunity to take up globe-making. Francesco Rosselli (1445-1520), the printer at Florence, is believed to have made globe-gores (Fiorini, pp.93-98). If so these would pre-date Martin Waldseemüller's printed terrestrial globe-gores of 1507, now the earliest known. The artist Albrecht Dürer (1471-1528) proposed in 1525 the globe-biangle as a means of transferring the sphere on to a flat surface. The Swiss Henricus Loritus Glareanus, professor of mathematics at Basle, was in 1527 the first to describe how to construct globe-gores. In his *Poetae laureati di Geographia Liber unus,* Basle, 1527, with later editions to 1542 (Paris), chapter XIX, "De inducendo papyro in globum," describes the method, illustrated by a diagram. It is not known if Glareanus himself invented the construction, which is simple but not completely correct mathematically (Nordenskiöld, pp.74-75).

Geographers in the early and middle years of the 16th century were concerned how best to express the relationship between the terrestrial and the celestial spheres. The generally accepted solution was a matching pair of terrestrial and celestial globes accompanied by a book of instruction, and these remained for some 300 years the main instruments and method of geographical teaching. Gemma Frisius, professor of mathematics at the University of Louvain, made the earliest pair of globes now extant, 36.5 cm in diameter, the terrestrial dated 1536, the celestial 1537. These were the best known globes in Europe before Gerard Mercator's pair, the terrestrial 1541 and the celestial 1551, 41 cm in diameter, secured the market.

Metal globes of fine workmanship were also constructed in the late 15th and 16th centuries, as for example, the gilt terrestrial globe, *c.* 1528 (Bibliothèque Nationale), entitled "Nova et integra universi orbis descriptio." A famous later terrestrial globe was that made by Isaac and Josias Harbrecht under the direction of Conrad Dasypodius for the Strasbourg Cathedral clock, completed in 1573.

Craftsmen, astronomers, and geographers also turned their hand to a very different type of production, the ephemeral globes which featured as cosmographical devices in renaissance pageants. Spheres of the moon and sun decorated the streets for the entry of the Spanish Infanta Catherine of Aragon into the City of London on 12 November 1501. Henry VIII had a great hall constructed at Calais for a banquet to entertain the Emperor Charles V on 12 July 1520 after the Field of the Cloth of Gold. It was 800 feet round and contained "the whole sphere" on its roof. In May 1526 Henry's astronomer Nicholas Kratzer designed for the King's banquet at Greenwich a roof to show "the whole earth ... like a very Mappe or Carte" (Wallis, 1968, pp.204-5)

In the 17th century the greatest globemakers were the Blaeus of Amsterdam, whose pairs of large 68 cm globes, made between 1617 and 1645-48, were celebrated throughout the world. Towards the end of the century the Blaeus were surpassed by Father Vincenzo Coronelli of Venice (1650-1718), who claimed, with justification, to be the greatest globemaker of all time. His pair of manuscript globes, 15 feet in diameter, made for presentation in 1683 to Louis XIV of France, are the finest of any age (now preserved in the Orangerie at Versailles). The derived pairs of printed globes, 3½ feet (110 cm) in diameter, were published in Venice between 1688 and 1705, with Parisian versions engraved by Jean Baptiste Nolin, 1688 to 1693. Coronelli in his concern to document his globes, to make them available to a wide clientele, and to challenge the authority of those who disputed his pre-eminence, invented the "book of globes," the *Libro dei Globi,* first edition published at Venice in 1697, second edition 1701. He was not the first to publish globe gores as maps. His originality lay in putting the gores of his globes together in the form of an atlas. All these achievements were acclaimed as breaking new records in the art and craft of globe-making. Coronelli also used a system of advanced subscriptions for selling his globes and atlases, circulating the information through the publishing society he had founded, the "Accademia Cosmografica degli Argonauti," the first geographical society in the world.

The mathematical practitioner Joseph Moxon of London (1627-1700) introduced two innovations. First was the pocket globe with its case bearing representations of the celestial hemispheres (1.0710c), a device which remained popular until the early years of the 19th century. The second, called the "English Globe," 1679, comprised a terrestrial globe with a celestial planisphere at its base. In the 18th century other adaptations were devised, such as George Adams's "Planetary" globe, *c.* 1790, which displayed various planetary motions. A much more elaborate "planetarium" to show these motions was invented by John Rowley in 1719, following an earlier idea of George Graham, and named the "orrery" after Charles Boyle, Fourth Earl of Orrery. From 1731 onwards the orreries of Thomas Wright (1711-1786), optician and mathematical instrument maker of London, were regarded as the finest of their day.

The first globe of the moon (1.0710b) was made by Sir Christopher Wren at Oxford in 1661 but does not survive. John Russell's globe, London 1797, ranks as the earliest published and the earliest extant lunar globe. A globe of Mars, "Globe Géographique de la Planète Mars d'apres Camille Flammarion par E. Antoniadi," was published at Paris by Bertaux about 1896 (Wawrik, p.11).

C BAYNES-COPE, A D (1985): *The Study and Conservation of Globes.* Vienna, Internationale Coronelli-Gesellschaft.

FAUSER, Alois (1967): *Die Welt in Händen.* Stuttgart, Schuler Verlagsgesellschaft.

FIORINI, Matteo (1899): *Stere terrestri e celesti di autore italiano oppure fatte o conservate in Italia.* Rome.

KROGT, Peter van der (1984): *Old Globes in the Netherlands.* Utrecht, HES Uitgevers.

MURIS, Oswald, and SAARMAN, Gert (1961): *Der Globus im Wandel der Zeiten.* Berlin, Columbus Verlag.

NORDENSKIÖLD, A E (1889): *Facsimile-Atlas to the Early History of Cartography.* Stockholm (reprinted 1961, New York, Kraus Reprint).

PAULY, August Friedrich von, and WISSOWA, Georg (1919): *Paulys Real-Encyclopädie der classischen Altertumswissenschaft, 20* (Karten). Col.2145-8. Stuttgart, J B Metzler.

SCHRAMM, Percy Ernst (1958): *Sphaira, Globus, Reichsapfel.* Stuttgart, Anton Hiersemann.

STEVENSON, Edward Luther (1921): *Terrestrial and Celestial Globes,* 2 vols. New Haven, Yale University Press for The Hispanic Society of America.

WALLIS, Helen (1962): "A Chinese Terrestrial Globe, AD 1623," *British Museum Quarterly, 25,* 83-91.

WALLIS, Helen (1962): "Globes in England up to 1660," *Geographical Magazine, 35,* 267-79.

WALLIS, Helen (1967): "Thematische Elemente auf frühen Englischen Globen (1592-1900)," *Veröffentlichungen des Staatlichen Mathematisch-Physikalischen Salons* (Berlin), *5,* 75-85 *(Der Globusfreund, 15-16* (1966-67), 75-85).

WALLIS, Helen (1968): "The Use of Terrestrial and Celestial Globes in England," *Actes du XIe Congrès International d'Histoire des Sciences.* Wroclaw, Ossolineum Maison d'Édition de l'Académie Polonaise des Sciences, *4,* 204-12.

WALLIS, Helen, ed. (1969): *Vincenzo Coronelli: Libro dei Globi.* Amsterdam, Theatrum Orbis Terrarum.

WALLIS, Helen, and PELLETIER, Monique (1980): "Les Globes du roi soleil," in *Cartes et figures de la terre.* Paris, Centre Georges Pompidou.

WAWRIK, Franz (1981): "Lunar- and Mars-Globes," *Internationale Coronelli-Gesellschaft für Globen- und Instrumentenkunde,* Nr. 4.

1.0710A Celestial Globe

A A sphere representing the heavens, i.e., the visible sky, and usually showing the positions of stars and the forms of the constellations. Compare with Astronomical map, 3.011.

B Celestial globes date from Greek antiquity and may have originated in Babylon. There is a tradition, recorded by Cicero (*De Republica,* 1.14) that Thales of Miletus (*c.* 580 BC) made a celestial globe in the form of a simple solid sphere. Anaximander of Miletus (early 6th century BC) who was Thales's pupil is said to have constructed a sphere which may, however, have been a map (see Astronomical map, 3.011; Kahn, p.89). According to Cicero, the globe first invented by Thales was later elaborated by the astronomer Eudoxus of Cnidos (fl.365 BC), and this is regarded as the best authenticated of the early Greek celestial globes. Eudoxus's catalogue of stars was used by the Hellenistic poet Aratus of Soli (fl.270 BC) in his *Phaenomena,* a poetic description of the heavens which seems to have been written with a celestial globe in mind (Needham, 382). The astronomer Hipparchus (fl.146-127 BC) and the geometrician Archimedes of Syracuse (*c.* 287-212 BC) are both reported to have constructed celestial globes.

The oldest extant celestial globe is that borne on the shoulders of Atlas, the statue known as "Atlante Farnesiano," previously in the Farnese Palace and now in the Museo Nazionale at Naples. This is a Roman copy of a Greek statue and globe. It shows the figures of the constellations but no stars. Thiele believed the original of the globe to have been made between *c.* 130 and *c.* 100 BC and the figure of Atlas to have been designed not much later, alleging that they derive from the scholars of the great Greek centre of learning at Pergamum in Asia Minor (Thiele, pp.19-26, taf.II-VI). The constellations, however, can be shown to be based on those of Claudius Ptolemy of Alexandria. His "Syntaxis," the "Almagest" (as it was called by the Arabs), *c.* AD 150, provided a star catalogue and described how to construct a celestial globe.

The Arabs as heirs to Greek scientific learning carried on the tradition of Greek geography and astronomy. The earliest extant Arabic celestial globe from the western part of the Islamic world was made at Valencia in Spain by Ibrahim Ibn Said-as-Sahli in 1085 and is preserved in the Istituto di Studi Superiori, Florence. The earliest from the eastern world is a Persian globe made of brass, 1140-41, now in the Iran Bastan Museum, Tehran. This marks stars but not constellations. The earliest extant Persian globe depicting constellations dates from 1144-45.

Interest in globes revived in Europe in the 15th century. An anonymous manuscript in the Jagiellonian Library, Cracow, dated *c.* 1400, describes the construction of a celestial globe (Rosińska). The earliest celestial globe preserved in Germany (at Nuremberg) is that of the mathematician and humanist Cardinal Nicolas of Cusa (1401-1464). It is made of wood and glued cloth, with constellations drawn in ink. The largest and most decorative of metal globes constructed in Europe before 1500 is the celestial globe, dated 1480, by the astronomer and mathematician Martin Bylica of Olkusz (1433-1493), now in the Jagiellonian University Museum.

After a pair of engraved terrestrial and celestial globes was made in 1536-37 by Gemma Frisius, professor of mathematics at the University of Louvain, it became common practice to issue globes in matching pairs (see Terrestrial globe, 1.0710d; Pocket globe, 1.0710c).

It has been an accepted convention that celestial globes show the heavens as they would appear to an observer in space looking inwards. The constellations are therefore shown in the reverse direction of their appearance when observed in the sky above. The great Venetian globemaker Vincenzo Coronelli (1650-1718) made his celestial globes in two versions, concave and convex, as reported by Carlo Malavista in a lecture on the development of astronomy at the Accademia Fisico-Mathematica in Rome, 7 December 1692. In the concave version the globe was divided in two and the constellations were reversed (Wallis, 1969, ix). The celestial hemispheres of pocket globes are based on the same principle. Coronelli published not only globes but books of globes, comprising the unmounted paper gores of his globes, celestial and terrestrial. The first edition of his *Libro dei Globi* was published in 1697 at Venice as part of his great atlas series, the *Atlante Veneto*.

Celestial globes (*hun hsiang*) were probably made in China in the time of Chhien

Lo-Chih (AD 435), after Chhen Cho had constructed (310) a standard series of star maps. Before this time the astronomical instruments referred to in literary references of the Chhin or early Han periods were probably demonstrational armillary spheres with earth models (Needham, 382-3, 389). None of the early Chinese celestial globes has survived.

CAMEISENOWA, Zofia (1959): *Globus Marcina Bylicy z Olkusza i mapy nieba na wschodzie i zachodzie* [*The Globe of Martin Bylica of Olkusz and Celestial Maps in the East and in the West*]. Wroclaw, Polska Akademia Nauk.

DILKE, O A W (1985): *Greek and Roman Maps.* London, Thames and Hudson.

FAUSER, Alois (1967): *Die Welt in Händen.* Stuttgart, Schuler Verlagsgesellschaft.

KAHN, Charles H (1960): *Anaximander and the Origins of Greek Cosmology.* New York, Columbia University Press.

MURIS, Oswald, and SAARMAN, Gert (1961): *Der Globus im Wandel der Zeiten.* Berlin, Columbus Verlag.

NEEDHAM, Joseph (1959): *Science and Civilisation in China,* vol 3. Cambridge, At the University Press.

PAULY, August Friedrich von, and WISSOWA, Georg (1919): *Paulys Real-Encyclopädie der classischen Altertumswissenschaft, 20* (Karten). Col.2145-8. Stuttgart, J B Metzler.

ROSIŃSKA, Grazyna (1973): "Les descriptions de la construction du globe céleste . . . à Cracovie au XVᵉ siècle," *Der Globusfreund* (Wien), *21-23,* 120-27.

SCHRAMM, Percy Ernst (1958): *Sphaira, Globus, Reichsapfel.* Stuttgart, Anton Hiersemann.

STEVENSON, Edward Luther (1921): *Terrestrial and Celestial Globes,* 2 vols. New Haven, Yale University Press for The Hispanic Society of America.

THIELE, Georg (1898): *Antike Himmelsbilder: Mit Forschungen zu Hipparchos, Aratos und seinen Fortsetzern und Beiträgen zur Kunstgeschichte des Sternhimmels.* Berlin, Weidmannsche Buchhandlung.

WALLIS, Helen (1962): "A Chinese Terrestrial Globe, AD 1623," *British Museum Quarterly, 25,* 83-91.

WALLIS, Helen (1967): "Thematische Elemente auf frühen Englischen Globen (1592-1900)," *Veröffentlichungen des Staatlichen Mathematisch-Physikalischen Salons* (Berlin), *5,* 75-85 *(Der Globusfreund, 15-16* (1966-67), 75-85).

WALLIS, Helen (1968): "The Use of Terrestrial and Celestial Globes in England," *Actes du XIᵉ Congrès International d'Histoire des Sciences.* Wroclaw, Ossolineum Maison d'Édition de l'Académie Polonaise des Sciences, *4,* 204-12.

WALLIS, Helen, ed. (1969): *Vincenzo Coronelli: Libro dei Globi.* Amsterdam, Theatrum Orbis Terrarum.

1.0710B Lunar Globe

A A spherical structure on whose surface is depicted the configurations of the moon.

B Once the astronomer Johannes Hevelius of Danzig (1611-87) had published the first detailed maps of the moon in his *Selenographia* (Danzig, 1647), the construction of a lunar globe was a logical next step (see Astronomical map, 3.011). Hevelius himself included an illustration of such a globe in his *Selenographia*, but there is no evidence that he ever made one. In England Sir Christopher Wren, as Savilian Professor of Astronomy at Oxford in 1661, was the first to make a lunar globe, constructed for King Charles II at the request of the Royal Society. It was a relief globe, showing "not only the spots and various degrees of whiteness upon the surface of the moon, but the hills, eminences, and cavities of it moulded in solid work" (Birch; Gunther). No trace of the globe survives since its sale by the Bodleian Library, Oxford, in 1749. In France the astronomer Philippe de la Hire was reputed to have begun (and then abandoned) the construction of a lunar globe (*c.* 1680). Tobias Mayer, working for the firm of J B Homann in Nuremberg, planned in 1750 the publication of a lunar globe, but only the prospectus was published, illustrated with two plates depicting the globe and including a subscription sheet (Nuremberg, 1750). A drawing of Mayer's lunar hemisphere and a description were published by Georg Christoph Lichtenberg (the elder) at Göttingen, 1775.

The first lunar globe to be made in the 18th century was the work of John Russell, RA (1745-1806), the fashionable English portrait artist who was also an amateur scientist. After eighteen years' work in drawing maps of the moon and consultation of the works of Hevelius, Mayer, and Jacques Cassini (Paris, 1740), he published in London in 1797 a 12-inch globe of the moon in a specially designed frame, calling the complete device the "Selenographia." He also offered relief globes and relief segments made to order; and although none of these appears to have survived, the map publisher William Faden, John Russell's brother-in-law, stated that one such globe was "fitted up" for Sir Henry Englefield, the antiquary and scientific writer. Ignorant of Wren's globe of 1661, Russell claimed his moon globe to be "the only work of this nature ever submitted to the Public." It ranks nevertheless as the earliest published and the earliest extant moon globe. Six examples survive in British collections, together with copies of the accompanying pamphlet, published by W Faden.

The existence of Russell's globe was overlooked by the next generation. Wilhelmine Witte (née Bottcher, 1777-1854), a friend of Caroline Herschel, sister of the astronomer Sir William Herschel, made a relief globe of the moon in 1830, and a second at the request of Alexander von Humboldt for King Wilhelm III of Prussia. According to the *Report of the British Association for 1845*, John Herschel, son of Sir William, demonstrated Madame Witte's relief globe under the impression that it was the first to have been made.

The moon globe made at Bonn in 1849 on a scale of 1:600,000 achieved a record for size.

C BIRCH, Thomas (1756): *The History of the Royal Society of London, I*. London, 21.

31

CASSINI, Jacques (1740): Tables astronomiques du soleil, de la lune, des planètes . . . Paris.

FISCHER, K (1967): "Beiträge zur Geschichte der Mondgloben," *Veröffentlichungen des Staatlichen Mathematische-Physikalischen Salons* (Berlin), *5,* 103-22 *(Der Globusfreund 15-16* (1966-67), 103-22).

GUNTHER, R W T (1923): *Early Science in Oxford, II.* Oxford, 263.

MAFFEI, Paoli, (1962): "Carte lunari di ieri e di oggi," *L'Universo, 5,* 912-52; *6,* 1043-66.

MEINE, K-H (1982): "Globen des Erdmondes und der Planeten unseres Sonnensystems: Ihre kartographischen Grundlagen und Stufen ihre Entwicklung." *Internationale Coronelli-Gesellschaft für Globen- und Instrumentenkunde,* Nr. 6, 17-48.

RUSSELL, John, RA (1797): *A Description of the Selenographia.* London.

RUSSELL, William (1809): *The Lunar Planispheres, Engraved by the Late John Russell.* London.

RYAN, W F (1966): "John Russell RA and Early Lunar Mapping," *The Smithsonian Journal of History* (Washington, DC), *1,* pt.1, 27-48.

WAWRIK, Franz (1981): "Lunar- and Mars-Globes," *Internationale Coronelli-Gesellschaft für Globen- und Instrumentenkunde,* Nr. 4.

1.0710C Pocket Globe

A A small terrestrial globe, about 3 inches (7 cm) in diameter, fitted with a case normally bearing concave representations of the celestial hemispheres.

B Joseph Moxon (1627-1691), after his return from the Netherlands, was the first globemaker in England to make pocket globes and probably was their inventor. He describes how he showed Roger Palmer, earl of Castlemaine, at the beginning of 1672 "(knowing that anything new and ingenious would be acceptable to him) one of my 3 inch Terrestrial Globes, with the Stars described in the inside of its Case, which when his Lordship had considered, and bin inform'd by me, that its only Use was to keep in memory the situation of Countries, and Order of the Constellations and particular Stars, He intimated, that certainly much more might be done by it" (Moxon 1679, sig.A.2.r-v; Wallis, pp.6-8).

Despite these strictures, the pocket globe remained in vogue well into the 19th century. The London mapmaker Herman Moll made a pocket globe showing the trade winds, a popular feature through the publication of Edmond Halley's wind map of 1688 and its derivatives (see Wind map, 3.232). The trade and monsoon winds became standard elements of English pocket globes, such as those of Charles Price, London, *c.* 1705, of Price in partnership with John Senex and John Maxwell, London, 1710-14, of Price with Senex, 1714 to 1718, and of John Senex alone *c.* 1730 and (published by his widow), *c.* 1750. In the later years of the 18th century it was common for English pocket globes to display the track of the circumnavigator George (later Admiral) Anson, 1740-44, as on N Hill's globe, 1760?, and the tracks of the three voyages of Captain Cook, which are marked, for example, on the anonymous "A Correct Globe with the new Discoveries," London, *c.* 1780. The case displays the "New Constellations of Dr Halley &c." Another example of a pocket globe carefully marked with all the latest discoveries is that of J & W Cary, 1791. The case bears not the celestial hemispheres but "The WORLD as known in CAESAR'S Time agreeable to D'Anville," showing the northern hemisphere. The other hemisphere contains "A TABLE OF Latitudes and Longitudes of Places not given on this GLOBE" (Krogt, pp.77-8). It is an indication of the popularity of this type of globe that N Lane's of 1776 was still being revised and published in 1818, 1819 and as late as 1840 (published by Burbidge).

In the early 18th century Johannes Deur (1667 or 68–1734) of Amsterdam engraved pocket globes which were sold by the bookseller Hendrik de Leth. The only publisher of globes in the Netherlands in the century, the firm of Gerard Valk (1652-1726) and his son Leonard (1675-1746), included in their range of productions pocket globes made between 1707 and 1728, examples of which are known. The globe factory later came into the hands of Cornelis Covens (1764-1825), who revised and re-issued the pocket globes *c.* 1790 (Krogt, pp.220-4).

C KROGT, Peter van der (1984): *Old Globes in the Netherlands.* Utrecht, HES Uitgevers.

WALLIS, Helen (1978): "Geographie Is Better than Divinitie. Maps, Globes and Geography in the Days of Samuel Pepys," in Thrower, Norman J W, ed.: *The Compleat Plattmaker.* Berkeley, etc., University of California Press, pp.1-43.

1.0710D Terrestrial Globe

A A sphere representing the earth.

B The Greek philosopher Crates of Mallos made the earliest terrestrial globe known in the western world, *c.* 150 BC. The geographer Strabo (*c.* 50 BC to AD 25) commented, "whoever would represent the real earth as near as possible by artificial means should draw a sphere like that of Crates, and upon this draw the quadrilateral within which his chart of geography is to be placed." He said that to show the oecumene clearly a globe not less than ten feet in diameter was required (Strabo, Bk.II. cap.V. §10). In ancient China terrestrial globes were made *c.* AD 260. These were small models of the earth constructed as part of armillary spheres and used for purposes of demonstration.

No terrestrial globe of medieval origin survives from either the Arabic or the Western world. The earliest terrestrial globe extant is Martin Behaim's "Erdapfel" (earth apple), made in Nuremberg in 1492. The globe is *c.* 50 cm in diameter and drawn in manuscript on vellum. The earliest extant printed globe-gores are the woodcut set, 1507, by Martin Waldseemüller of St Dié, Lorraine. Johann Schöner, the Nuremberg mathematician and geographer (1477-1547), was the first to undertake globe production on a considerable scale. His woodcut globe made in 1515 is the earliest printed terrestrial globe known.

Gemma Frisius of the University of Louvain issued in 1529 his first terrestrial globe, which he described in his book *Gemma Phrysius de principiis astronomiæ et cosmographiæ Deque usu Globi ad eodem editi* . . . (1530), but no example is known. His second terrestrial globe, 37 cm in diameter, was published in 1536, with its celestial partner following in 1537. These became the best known globes in Europe until superseded by the larger pair of his pupil Gerard Mercator. Mercator's terrestrial globe, 1541, 41 cm in diameter, incorporated two innovations. Loxodromes or rhumb lines were marked from the numerous compass roses as a guide for seamen, a feature adopted by later Dutch globemakers, such as Jacob Florensz van Langren, William Jansz Blaeu and his son Joan. The second device was the depiction of stars at various places on land and sea, to help the traveller orient himself at night. This did not inspire others to follow suit.

The first terrestrial globe published in England, Emery Molyneux's of 1592, was notable for its record of England's discoveries overseas and her claims to empire. The routes of the circumnavigations of Sir Francis Drake (1577-80) and Thomas Cavendish (1586-88) are marked, and a long dedication to Elizabeth I surmounted by her arms is set across the continent of North America. The globe was made as a statement of imperial achievements and designs. Jodocus Hondius, engraver of the Molyneux globe, on his move from England to the Netherlands, c. 1595, was similarly inspired in making terrestrial globes promoting the Dutch cause, and later Dutch globemakers kept up the tradition. The large terrestrial globes of Willem Jansz and Joan Blaeu, 68 cm in diameter, issued from 1617 to 1645-8, were the most famous of their day and "the most up-to-date and detailed picture of the world then available in a single document" (Campbell, p.26). Joan Blaeu used regularly an ingenious method of revising the globes and adapting the dedications, by pasting on slips as overlays with new geographical information and new inscriptions. Jodocus Hondius may have been the first to apply this procedure, as a pasted revision slip is found on his terrestrial globe of 1613 (National Maritime Museum, G.167).

The Blaeu globes inspired the Venetian cosmographer Vincenzo Coronelli (1650-1718) to still greater feats of globemaking. He made on request for Louis XIV of France in the years 1681-83 the manuscript 15 foot terrestrial globe with its celestial partner. The printed version, 3½ feet in diameter (110 cm), was published in Venice in 1688, and a new version in Paris in 1693. The gores of all his globes were published in the *Libro dei Globi,* Venice, 1697 (see Globe, 1.0710).

Instrument makers of the 17th century were aware of the limitations of matching pairs of globes in representing the place of the earth in the heavens. The London geographer and mathematician Joseph Moxon constructed to the specification of Roger Palmer, earl of Castlemaine, and published in 1679 the "English Globe." The globe was fixed to a pedestal carrying a dial marked with a stereographic projection of the stars, like an astrolabe (Wallis, 1978a, p.8). It was described by Moxon in an accompanying handbook entitled *The English Globe. Being a Stabil and Immobil one, performing what the Ordinary Globes do, and much more. Invented and described by the Earl of Castlemaine,* London, 1679. Although Coronelli singled out the globe for special mention in his lectures at Venice, published in his *Epitome Cosmografica,* Venice, 1693, the idea was not taken up by other globemakers. Moxon's advertisements of his globes, significantly, included spheres Ptolemaic and Copernican, an indication of the continuing strength of the Ptolemaic tradition.

The 18th century was the greatest period of English globemaking. Pairs of globes were acquired for every well-appointed library. Globemakers such as George Adams the Younger, instrument maker to George III, invented various pieces of mechanical equipment to improve the terrestrial globe's performance. Adams's "planetary globe," *c.* 1790, was designed, he claimed, to remedy the defects of terrestrial globes "mounted in the common way." It showed the apparent movement of the earth in relation to a planet and thus solved "in an easy natural manner, the several phenomena of the sun, moon and earth" (Adams, *Lectures on Natural and Experimental Philosophy,* London, 1794, vol. IV). Another globemaker (probably William Bardin) demonstrated the phenomena of terrestrial magnetism by suitable equipment added to an example of "Ferguson's Terrestrial Globe, Improv'd by G. Wright . . . Made and sold by Wm Bardin." The sphere, *c.* 1791 in date, is designed to be taken apart, and has been provided with iron bars within it so that the magnetic pole can be moved and bearings taken from any point by fixing a compass on to a movable arm. (This globe is in private ownership in England.)

In the 19th century in response to popular interest globes became objects of spectacular display, comparable with their role in 16th century pageants (Globe, 1.0710). In association with the Great Exhibition in London, 1851, Wyld's Monster Globe was built at Leicester Square by James Wyld Junior (1812-1887), Geographer to Queen Victoria. It comprised a model of the earth, 60 feet in diameter, with 10,000 square feet of surface, which the public viewed from its interior, and was intended to initiate "the commencement of a new era in geographical instruction." Alexander Keith Johnston of Edinburgh, also Geographer to the Queen, celebrated the occasion with an exposition of thematic cartography in global form: "Johnston's Geological & Physical Globe, showing the structure of the earth, currents of the ocean and lines of equal temperature . . . Edinburgh, 1851." Following up his English version of Heinrich Berghaus's *Physicalischer Atlas,* 1848, he was thus the first in Great Britain to add isotherms and geology to the terrestrial globe.

Philippe Vander Maelen of Brussels (1795-1869) undertook a novel project of a different kind in designing a globe from his *Atlas Universel,* printed by lithography and published at Brussels, 1825-27, in six volumes. All the sheets were made on the same scale (1:1,641,836) and on the same projection, so it was possible to mount them to form a globe *c.* 7.775 m in diameter, made from about 1000 gores. Vander Maelen included in each volume of the atlas a "Carte d'Assemblage," and invited subscriptions either for the atlas or for the sheets to cover the globe. In the event only the Bibliothèque de Bourbon at Naples subscribed for a globe in preference to the atlas (Vander Maelen, vol.1, p.3). Vander Maelen himself constructed the only globe known to have been made and installed it in a special hall in the "Établissement géographique de Bruxelles," which he had founded in 1830. Thus, whereas Coronelli in 1693 had converted his globes into an atlas (Globes, 1.0710), Vander Maelen reversed the process and converted his atlas into a globe.

Map publishers working at the more popular end of the market also profited from the new methods of production which made possible the publication of cheaper globes for school room use. Edward Mogg published in London, 1812, "Moggs Dissected Globe," comprising four shaped pieces of pasteboard, printed with maps. F J Waygand, a publisher at Amsterdam and The Hague, made *c.* 1825 a dissected terrestrial globe comprising six gores, connected by strings, to be stored in a cover inscribed "Globe artificiel et mecanique à l'usage du Petit Géographe," Amsterdam (Krogt, p.258). John Betts, the publisher of dissected puzzles, issued in 1852 his "Portable Globe," made up of gores. William Stokes, Teacher of Memory, ventured into a more curious style of globe making with "Stokes's Capital Mnemonical Globe," London, 1868, "by which the relative portions of the principal geographical places in the world may be learned as an amusement in a few hours." The Bristol schoolmaster Ebenezer Pocock, exploited the new processes of lithography to patent in 1830 and publish an inflatable globe made of 12 thin paper gores designed to be inflated by hot air or an air pump to a circumference of 12 feet, yet when rolled "may be placed in a gentleman's side pocket." "John Betts's Portable Terrestrial Globe," London, *c.* 1860, was printed on linen and could be opened by means of an umbrella-like mechanism. The Japanese equivalent was entitled "Chikyu-gi, shintei" [A terrestrial globe revised]. It was published at Tokyo, *c.* 1880, and was compiled by an Englishman, "Hakudo," with the text translated into Japanese at Tokyo Middle School.

The earliest surviving Chinese terrestrial globe was made in 1623 by the Jesuits Nicolo Longobardi (1559-1654) from Sicily and Manuel Dias the Younger (1574-1659) from Portugal. The globe (British Library, Maps C.6.a.2.) is believed to have been made at the request of a Christian patron, perhaps for presentation to the Emperor, and may have belonged to the Imperial Palace, Peking. It is also the first globe made in China by European missionaries. The sphere is painted in lacquer on wood, 59 cm in diameter. The long explanatory legend records that "The earth is like the heavy turbid yolk of an egg concentrating in one place. Concentration in a sphere must have a limit, namely in a point which is the centre of the earth. Therefore the centre of the earth is the lowest point. All objects having mass by their nature tend towards it. A needle being attracted downwards to a loadstone is a rather inadequate explanation of this." The analogy of the yolk in an egg, already used by Father Matteo Ricci on his Chinese world map of 1602, was one of the oldest expressions in Chinese cosmological thought for the shape of the earth in the heavens. The reference to the effect of gravitational force, an early expression of the concept of terrestrial magnetism, which was in advance of European ideas, was entirely derived from Chinese work on the magnet. An idea of the original equipment of the globe can be gained from the picture of a Ch'ing terrestrial globe illustrated in *Huang-ch'ao li-ch'i t'u-shih* (1766), ch.2.f.17.

C ADAMS, George (1766); *A Treatise of Celestial and Terrestrial Globes.* London.

CAMPBELL, Tony (1976): "A Descriptive Census of Willem Blaeu's Sixty-eight Centimetre Globes, *Imago Mundi, 28,* 21-50.

GILBERT DE CAUWER, E (1970): "Philippe Vander Maelen (1795-1869), Belgian Map-Maker," *Imago Mundi, 24,* 11-20.

KROGT, Peter van der (1984): *Old Globes in the Netherlands. A Catalogue of Terrestrial and Celestial Globes.* Utrecht, HES Uitgevers.

MAELEN, Philippe Vander (1827): *Atlas Universel de Géographie.* Brussels, Imprimerie de Ode et Wodon.

STEVENSON, Edward Luther (1921): *Terrestrial and Celestial Globes.* New Haven, Yale University Press for The Hispanic Society of America.

WALLIS, Helen (1951): "The First English Globe: A Recent Discovery," *Geographical Journal, 117,* 275-90.

WALLIS, Helen (1955): "Further Light on the Molyneux Globes," *Geographical Journal, 121,* 304-11.

WALLIS, Helen (1962): "A Chinese Terrestrial Globe, AD 1623," *British Museum Quarterly, 25,* 83-91.

WALLIS, Helen (1965): "The Influence of Father Ricci on Far Eastern Cartography," *Imago Mundi, 19,* 38-45.

WALLIS, Helen (1967): "Thematische Elemente auf frühen Englischen Globen (1592-1900)," *Veröffentlichungen des Staatlichen Mathematisch-Physikalischen Salons* (Berlin), *5,* 75-85 (*Der Globusfreund, 15-16* (1966-67), 75-85).

WALLIS, Helen (1968): "The Use of Terrestrial and Celestial Globes in England," *Actes du XIe Congrès International d'Histoire des Sciences,* Wroclaw.

WALLIS, Helen (1978a): "Geographie Is Better than Divinitie. Maps, Globes, and Geography in the Days of Samuel Pepys," in Thrower, Norman J W, ed.: *The Compleat Plattmaker.* Berkeley, etc., University of California Press.

WALLIS, Helen (1978b): "The Place of Globes in English Education, 1600-1800," *Der Globusfreund, 25-27,* 103-10.

WELLENS-DE DONDER, Liliane (1970): "Le globe géant de Philippe Vandermaelen à l'établissement géographique de Bruxelles," *Der Globusfreund, 18-20,* 130-33.

1.091 Imaginary Map

A Although, literally speaking, an imaginary map is one that exists only in the imagination, the term is normally used to describe a map of an imaginary place; also known as a fantasy map; cf. Satirical map (2.192), Symbolic map (1.192).

B Precursors of the true imaginary map are found in the mapmaking of ancient Egypt and China. Landscapes drawn as guides for the departed soul were painted in diagrammatic form on wooden coffins dating from *c.* 2000 BC in Egypt and have been identified as the earliest "hieroglyphical" or fantastic maps, "Mappae imaginariae" (Bonacker, p.16; see also Route map, 1.1810). The Egyptian "Fields of the Dead," *c.* 1400 BC, drawn on papyri, represent idealized lands in the after-life, though their form evidently derived from real surveys of estates. In ancient China frescoes in the Chhien-Fo-Tung cave-temples near Tunhuang in Kansu depict fanciful topographical "story-pictures" drawn in the early Thang period, 7th century (Needham, 550-51).

Maps of deliberately invented places may be divided roughly into two groups –
those which illustrate the location of some legend or literary tradition, and those
which are independent conceits, usually having some internal allegorical
significance. The literary tradition of imaginary places goes back a long way, with the
distinction between the completely imaginary features and the picturesque
exaggeration of real features becoming increasingly confused, as for example in the
Odyssey, Homer's poem of Odysseus's wanderings after the fall of Troy, traditionally
dated *c.* 1850 BC. Although pictorial illustrations of imaginary or legendary places
may be traced back many centuries, especially in association with various religious
traditions, there do not seem to be any intentionally imaginary maps before the 16th
century.

The first and most famous was of the island of Utopia, printed in the satire by Sir
Thomas More, Lord High Chancellor of England, *Libellus vere aureus nec minus
quam salutaris festiuus de optimo reip. statu. deq. noua Insula Vtopia. . . ,* Louvain,
1516. The anonymous woodcut frontispiece entitled "Utopiae Insulae Figura" shows
a map-view of the island which the artist has based as closely as possible on the
author's detailed description, although the relative distances given in the text are in
fact geometrically impossible (Goodey). For the third edition published at Basel in
1518, Ambrosius Holbein, elder brother of the painter Hans Holbein, provided a map
more elaborate in form and less true to the text (for example, the island is no longer
crescent-shaped, as in the earlier version); and it incorporates additional symbolism
of the hermetic tradition. Later in the century, *c.* 1591, the otherwise conventional
cartographer Ortelius was persuaded to draw a rather more detailed map of Utopia;
his version also failed to tally with Sir Thomas's description on a number of points.

Matthias Seutter (1678-1757), map engraver and publisher of Augsburg, produced
a new version of Utopia based on the 16th century satire of Hans Sachs. Using
normal geographical conventions he depicts the land of cockaigne, a fool's paradise
of idleness, luxury, and vice: "Accurata Utopiæ Tabula. Das ist Der Neu-entdeckten
Schalck-Welt, oder des so offt benannten, und doch nie erkannten,
Schlaraffenlandes," Augsburg, *c.* 1730. Schlaraffenland comprises nineteen
countries such as Prodigalia, Mammonia, Magni Stomachi Imperium, each
representing a particular vice. A wealth of place-names is marked, nearly all of them
puns, many of them crude. This map was one of the most sought-after curiosities of
its day (Tooley, p.21).

In 17th century France, the vogue for *précieux* literature gave rise to a number of
highly allegorical imaginary maps. Madelaine de Scudéry's *Clélie* (1656-60)
introduced the most famous of these, the "Carte de Tendre," where distances are
measured in "Lieues d'amitié" and villages bear such names as "Tendresse" and
"Billet doux." Seutter also ventured his hand, producing a pseudo-military plan of
the attack of Love: "Representation Symbolique et ingenieuse projettée en Siege et en
Bombardement, comme il faut empecher prudemment les attaques de L'Amour,"
Augsburg, *c.* 1730. A fortress – the male stronghold – is under siege from love with all
her feminine wiles. Johann Gottlob Immanuel Breitkopf, a printer of Leipzig,
produced at Leipzig in 1777 the Kingdom of Love, "Das Reich der Liebe," in which
the pilgrim sets out from the Land of Youth to encounter the Stone of Warning. Six
other lands, such as the Land of Unhappy Love, with the Desert of Melancholy, await
him. Love has remained popular as a subject for imaginary maps, and the treatment

has ranged from the frivolous to the moralistic. A moralistic theme became widespread in the 19th century, when many allegorical maps were of a distinctly "improving" nature, with a high moral tone and often a religious bias; thus the "Illustrative map of human life, deduced from passages in sacred writ," published in New York in 1842.

A didactic form of imaginary map was the map of geographical terms. Matthias Quad, mapmaker and engraver of Cologne, was the author of "Vocabulorum geographicorum topica significatio," published in his *Geographisch Handtbuch*, Cologne, J. Bussemacher, 1600 (Fig.1). One of Seutter's inventive productions was an elaborate map of this kind: "Mappa Geographiae Naturalis sive Tabella Synoptica," Augsburg, *c.* 1730. A host of physical and political elements are displayed in an imaginary landscape with appropriate symbols and names in Latin and German, and some translations into French and Italian. The map is in effect a guide to conventional signs (see 5.033), some of which are explained in a key.

Another type of imaginary map is purely literary rather than allegorical or didactic, to illustrate a work of fiction. The rise in popularity of science fiction and fantasy literature dating from the 19th century brought forth maps of invented countries (Porter and Lukermann). In English literature the most famous is that in Robert Louis Stevenson's *Treasure Island,* first published in London in 1884. In this instance, the map was not made to illustrate the book; rather, the book was written to illustrate the map. Stevenson records how on holiday in the summer of 1881 "I made the map of an island; . . . the shape of it took my fancy beyond expression; . . . I ticketed my performance 'Treasure Island.'" Then "as I paused upon my map of 'Treasure Island,' the future character of the book began to appear there visibly among imaginary woods . . . The next thing I knew I had some papers before me and was writing out a list of chapters" (Hill, p.24).

Utopia (which means "nowhere") and Treasure Island had no real existence, although it has been said that literary Utopias lie just beyond the frontiers of man's knowledge (Porter and Lukermann, p.207). A different type of imaginary map is the pseudo-realistic. That in Joseph Hall's *Mundus alter et idem* (Frankfurt [i.e. London?], *c.* 1605), comprises a satirical representation of the hypothetical southern continent, Terra Australis Incognita, featured on many world maps of the day (see Satirical map, 2.192). The maps in Jonathan Swift's *Gulliver's Travels,* London, 1726, were based on contemporary maps, with Lilliput and Houyhnhnms Land, for example, located near the shores of New Holland (Australia).

C BONACKER, Wilhelm (1950): "The Egyptian 'Book of the Two Ways,'" *Imago Mundi, 7,* 5-17.

FILTEAU, Claude (1980): "Cartes sans territoire," *Cartes et figures de la Terre*. Paris, Centre Georges Pompidou.

GOODEY, Brian R (1970): "Mapping 'Utopia': A Comment on the Geography of Sir Thomas More," *Geographical Review, 60,* 15-30.

HILL, Gillian (1978): *Cartographical Curiosities.* London, British Library Board.

MANGUEL, Alberto, and GUADALUPI, Gianni (1981): *The Dictionary of Imaginary Places.* London, etc., Granada.

NEEDHAM, Joseph (1959): *Science and Civilisation in China*, vol.3. Cambridge, At the University Press.

PORTER, Philip W, and LUKERMANN, Fred E (1976): "The Geography of Utopia," in Lowenthal, David, and Bowden, Martyn J, eds.: *Geographies of the Mind*. New York, Oxford University Press.

POST, J B (1973; 1979): *An Atlas of Fantasy*. Baltimore, Md., Mirage Press; London, Souvenir Press.

TOOLEY, Ronald Vere (1963): *Geographical Oddities*. Map Collectors' Series. No. 1. London, Map Collectors' Circle.

1.131 Map Games

A Any game to which the use of a map is necessary; particularly, a board game played on the face of a map.

B The earliest known map games are related to a traditional gambling game, the "jeu de l'oie" or game of goose, which had been popular throughout Europe since the Middle Ages. Players would move their tokens along a numbered, spiral course or track, the length of the move being determined by the throw of dice or other similar means; various hazards would be met on the way, and the winner would be the first player to reach the end of the track. In the middle of the 17th century the French cartographer Pierre Du Val published a number of such games, often using small maps to decorate each position on the board; the first was "Le Jeu du Monde" in 1645, followed by "Le Jeu de France" and "Le Jeu des Princes de l'Europe." He also produced a draughts board with alternate squares mapped or blank.

Although much geographical information was incorporated in Du Val's maps, it was not until the 18th century that the idea was conceived of using maps as an essential part of the game. Instead of a simple spiral, the course was laid out as a real journey, however improbable, across the face of the map. The first known dated game, "A Journey through Europe," was published by John Jefferys in London in 1759; it was to set the pattern for map games over the next eighty years. Each game was accompanied by a descriptive list of the numbered stopping places, giving not only details of any forfeits and bonuses for the game, but also historical and geographical information for the instruction of the players. Within ten years the London publisher, Thomas Jefferys, Geographer to King George III, following the example of his namesake, produced a tour of Europe in 1768, and tours of the world and of England and Wales in 1770. These games were advertised as aids to the teaching of geography and became very popular. John Wallis of London, from the 1780s a prolific producer of all kinds of games and puzzles, published several map games in the early part of the 19th century; some of them he also made available as dissected maps (1.042), thus providing two educational games for the price of one.

Playing cards with maps as decoration have been popular since the 17th century. The earliest known set, by about 70 years, is a pack with maps of the counties of England and Wales, signed "W B inuent. 1590." The identity of "W B" is not known.

The next set of county map playing cards "The 52 Counties of England and Wales, Geographically described in a pack of Cards . . . ," undated, was published by Robert Morden in London in 1676, and later editions were issued in 1676 and 1680? and then nearly 100 years later. For many counties the playing card by Morden is the earliest separate printed county map to show any roads (Skelton, pp.16-18, 151-2). W Redmayne also brought out in 1676 (and again in 1677) a pack of playing cards engraved with English county maps, describing his publication as "Recreative Pastime by Cardplay" (Skelton, pp.153-4). Another London mapmaker Joseph Moxon was publishing sets of "Astronomical Playing Cards" and "Geographical Playing Cards, wherein is exactly described all the Kingdoms of the Earth, curiously engraved. Price plain 1s., coloured 2s., best coloured and gilt 5s., the Pack (as advertised in 1692; earlier advertisement in 1657)."

C GRAND-CARTERET, John (1896): *Vieux papiers, vielles images.* Paris, A Le Vasseur.

SKELTON, R A (1970): *County Atlases of the British Isles 1579-1850.* London, Carta Press.

WHITEHOUSE, F R B (1951): *Table Games of Georgian and Victorian Days.* London, Peter Garnett.

1.161 Perspective/Panoramic Map

A A perspective map shows features by means of bird's-eye views, while a panoramic map is one that consists entirely of a single bird's-eye view or, occasionally, of two views drawn with opposing horizons. Such maps range from the realistic to those that are foreshortened or are otherwise stylized. Herbert Hurst in 1899 used the term "a bird's flight view," in preference to "bird's eye view," for the drawing of a town as it would unfold itself to any one passing over it, as in a balloon, describing such a work more technically as a "view-plan in isometrical projection" (Hurst, pp.4-5).

B Plans of buildings and their grounds from Egypt in the 2nd millennium BC are near to being perspective maps in that they show individual features in elevation set on a ground plan; the same technique appears on a 6th century bas-relief from Nineveh showing the celebration of Assurbanipal's capture of Madaktu. The earliest truly perspective maps appear to be the representations of towns by oblique views of city walls, usually polygonal and sometimes showing stylized buildings inside. Athens and Corinth are thus represented on a cup from Tanagra (Boeotia), 2nd or 3rd century BC. This particular artistic convention can be traced, apparently as a continuous tradition, down to medieval and renaissance Europe. It appears strongly in the picture-maps from the Roman Empire, a tradition of perspective maps that was certainly in existence by the 2nd century AD.

The same convention for representing towns appears in China as part of a tradition of panoramic maps that clearly existed by the 16th century but is probably much older; 13th-century plans of Hangchow and Suchow show features in elevation on a

41

ground plan. The earliest topographical maps from Japan, 8th-century plans of estates, are panoramas, and panoramic maps played a continuing role in Japanese cartography, the town panoramas of the 17th to 19th centuries being noteworthy. Some 16th-century maps from Mexico seem to suggest that the perspective, even the panoramic, map may have been a part of Aztec cartography.

In medieval Europe the perspective/panoramic map developed earliest in Italy. Plans of Verona in the 10th century and of Rome in the 12th consist of stylized oblique views of city walls enclosing representations of each city's more distinctive monuments. Plans of other Italian cities, with increasing elaboration, were drawn in this form down to the late 15th century. By then the trend towards realism in Italian art had led to the production of fully detailed and realistic panoramic maps of cities: a view of Padua, as though from a slight elevation, by Giusto de' Menabuoi, 1382, may be seen as a precursor, but the late 15th-century city views by Francesco Rosselli mark the effective change. His views of Florence and Rome are known largely from derivatives, while those of Pisa and Constantinople are wholly lost.

Another significant development was the growth of a tradition of regional maps in north Italy. All were perspective maps, and many were panoramic, including the earliest known, which is of the area around Asti and Alba, 1291. By the 15th century panoramic maps had appeared among local maps in other parts of western Europe; examples show the villages of Talmay, Maxilly, and Heuilley by the Saône (1460), the lower Scheldt with Antwerp (1468), and a stylized view of Bristol with one feature, the High Cross, drawn from life (c. 1480).

A late re-appearance of the perspective town plan featured prominently in North American 19th century mapping. Plans of cities before the American Civil War in 1861 were drawn at low oblique angles and sometimes almost at ground level, with the street pattern indistinct. Other plan-views showed the cities as from a great height, a technique which appears to follow the invention of the balloon. Plans drawn after the Civil War (from 1865) were different in style in being generally more accurate and drawn from a higher angle (Hébert).

C ALMAGIÀ, Roberto (1951): "Un' antica carta del territorio di Asti," *Rivista Geografica Italiana, 58,* 43-4.

CAVALLARI, V, et al., ed. (1964): *Verona e il suo Territorio.* Verona, Istituto per gli Studi Storici Veronesi, *2,* 39-42, 232-3, 481-5, pl. opp. p.192.

DESTOMBES, M (1959): "A Panorama of the Sack of Rome by Pieter Bruegel the Elder," *Imago Mundi, 14,* 64-73.

HARVEY, Paul D A (1980): *The History of Topographical Maps.* London, Thames and Hudson.

HÉBERT, John R (1974): *Panoramic Maps of Anglo-American Cities. A Check List of Maps in the Collections of the Library of Congress.* Washington, DC, Library of Congress.

HURST, Herbert (1899): *Oxford Topography: an Essay.* Oxford, Oxford Historical Society at the Clarendon Press.

KRAELING, C H, ed. (1938): *Gerasa, City of the Decapolis.* New Haven, American Schools of Oriental Research, 341-51.

KURITA, Mototugu (1938): "Japanese Old Printed Maps of Cities," *Comptes Rendus du Congrès International de Géographie: Amsterdam 1938, 2,* Travaux des sections A-F, 362-80.

LAVEDAN, Pierre (1954): *Représentation des villes dans l'art du Moyen Âge.* Paris, Vanoest, 33-5, p.XVII.

LEVI, Annalina, and LEVI, M (1967): *Itineraria Picta: Contributo allo studio della Tabula Peutingeriana.* Rome, Museo dell' Impero Romano, 17-64.

LEVI, Annalina, and LEVI, M (1974): "The Mediaeval Map of Rome in the Ambrosian Library's Manuscript of Solinus," *Proceedings of the American Philosophical Society, 118,* 567-94.

RAVENHILL, William (1986): "Bird's-eye-view and Bird's-flight View", *Map Collector 35,* 36-7.

REPS, John W (1984): *Views and Viewmakers of Urban America.* Columbia, Mo, University of Missouri Press.

WILKINSON, John Gardner (1847): *The Manners and Customs of the Ancient Egyptians,* 3rd ed. London, John Murray, *2,* 5, 105, 129-48, pl. opp. p.94.

1.162 Pictorial Map

A A map in which topographical information is delineated by more or less realistic drawings, illustrations of features in elevation, and small bird's eye view sketches.

B A pictorial approach to the depiction of geographical phenomena has been commonly adopted by mapmakers from the earliest stages in mapmaking history. A pictorial plan or pictogram, incised in rock, of a village in the Val Camonica, province of Brescia in northern Italy, dates from *c.* 1500 BC. It features rudimentary illustrations of animals, animal pens, human figures, and stilt houses. The earliest pictorial regional map is of northern Mesopotamia and appears on a Babylonian clay tablet from Nuzi, variously dated from *c.* 2500 BC to *c.* 2300 BC. The first pictorial representations of the known world were also Babylonian – a map dated *c.* 600 BC shows Babylon encircled by a sea filled with fish, while various animals are seen to inhabit the seven islands of the world. Another pictorial map from the ancient world ranks as the oldest Hellenic map extant, if not qualifying to be the oldest map known showing more than a small area (Gottmann). It was excavated at Akro-Thera, on Thera (or Santorini) in the Aegean Sea, the ancient city buried under the ashes of a great volcanic eruption *c.* 1500 BC, and comprises wall paintings forming a frieze which decorates three sides of a room. Three towns are depicted linked by sea lanes, one of which clearly is Thera, while the other two may be in Egypt. The Madaba Map of the Holy Land, dated between 560 and 565, may also be described as a pictorial map (see Mosaic map, 6.132).

Roman mapmakers made extensive use of pictorial symbols. The route map known as the Peutinger map, an 11th or 12th century copy of a Roman map dating from the

4th or 5th century, probably from an earlier original (see Road map, 1.1810d), includes perspective drawings of buildings as well as ranges of hills drawn in elevation. From the first half of the third century, the Dura Europos shield has ships and fishes in the Black Sea and houses in perspective to denote settlements. The map miniatures in the Corpus Agrimensorum of the 6th to 9th centuries (copies of earlier originals) have ducks in estate ponds and a gnomon above centuriated land.

Some medieval mappae mundi (1.2320b) were encyclopaedic in their graphic depictions. The Hereford world map (Hereford Cathedral), c. 1285, for example, displays an exuberance of vignettes of classical as well as medieval origin, including the Cretan maze. The itineraries and maps of Palestine by Matthew Paris (after c. 1252) are notable for their depiction of architectural detail.

The portolan charts (1.0320d) drew some of their decorative character from a pictorial treatment of the land areas, particularly in the elaborate Catalan Atlas, c. 1375 (Paris, Bibliothèque Nationale, MS espagnol 30). The culmination of medieval map making came with Fra Mauro's world map, 1459 (Venice, Biblioteca Marciana), which carried an abundance of carefully annotated detail.

During the 16th century the wealth of information provided by the discoverers and explorers of new lands was expressed pictorially on renaissance world maps such as the map by Columbus's companion Juan de la Cosa, c. 1500 (Madrid, Museo Navale), and the Cantino world map, 1502 (Modena, Biblioteca Estense), representing Portuguese discoveries. It became accepted practice for Portuguese overseas expeditions to carry a "painter" to depict coastal scenes, which were later incorporated into the finished charts. The manuscript atlases and charts of Lope and Diego Homem well exemplify the peculiar blend of science and artistry displayed in works made for special presentation, for example, in Queen Mary's atlas, 1558 (British Library Add MS. 5415.A.). The hydrographers of Dieppe, who learnt much from the Portuguese, also employed artist/mapmakers ("le peintre"). The charts and atlases prepared at Dieppe and in its environs by Jean Rotz, 1542, Pierre Desceliers from 1546, and Guillaume Le Testu at Le Havre, 1555 [1556], were also masterpieces of iconographic detail (Wallis).

Of the printed maps published in this period one of the most remarkable for its wealth of pictures of sea creatures was the "Carta Marina," a woodcut map of northern Europe by Olaus Magnus (Venice, 1539). With the introduction of engraving and printing, the production of true pictorial maps gradually declined, the pictorial treatment being generally confined to the terrain and to ancillary items, such as the cartouche (6.031). During the 19th century, pictorial maps enjoyed something of a new lease of life, being extensively used as the basis of educational games (see Map games, 1.131) and as dissected maps (1.042).

C BAGROW, L (1964): *History of Cartography*. Revised and enlarged by R A Skelton. London, C A Watts & Co.; Cambridge, Mass., Harvard University Press.

BLUMER, Walter (1964): "The Oldest Known Plan of an Inhabited Site Dating from the Bronze Age about the Middle of the 2nd Millennium B.C.," *Imago Mundi, 18*, 9-11.

BRICKER, Charles, TOOLEY, R V, and CRONE, G R (1969): *A History of Cartography*. London, Thames & Hudson.

DESTOMBES, Marcel, ed. (1964): *Mappemondes: A.D. 1200-1500,* Vol. 1 of *Monumenta Cartographica Vetustions Aevi.* Amsterdam, N. Israel.

GEORGE, Wilma (1969): *Animals and Maps.* London, Secker and Warburg.

GOTTMANN, Jean (1984): *Orbits: The Ancient Mediterranean Tradition of Urban Networks.* The Twelfth J L Myres Memorial Lecture. London, Leopard's Head Press.

GROSJEAN, Georges, and KINAUER, Rudolf (1970): *Kartenkunst und Kartentechnik vom Altertum bis zum Barock.* Bern and Stuttgart, Verlag Hallwag.

HARVEY, Paul D A (1980): *The History of Topographical Maps.* London, Thames and Hudson.

LUGLI, Piero Maria (1967): *Storia e cultura della città italiana.* Bari, Editori Laterza.

UNGER, Eckhard (1935): "Ancient Babylonian Maps and Plans," *Antiquity, 9,* 311-22.

UNGER, Eckhard (1937): "From the Cosmos Picture to the World Map," *Imago Mundi, 2,* 1-7.

WALLIS, Helen (1985): "The Role of the Painter in Renaissance Marine Cartography," in *Imago et mensura mundi: Atti del IX Congresso internazionale di Storia della Cartografia.* Rome, Instituto della Enciclopedia Italiana, vol.2, pp.515-23.

1.1630 Plan

A A cartographic representation of a limited area normally known to the mapmaker through his own observation. See also Estate map, 2.0210c, and Property map, 2.0210e.

B Plans survive among the earliest cartographic works. In Europe, for example, plans of buildings seem to have existed in prehistoric times. A relief plan in terracotta of an apsed temple, dating from the late Neolithic and found in Malta, may have served as an architect's demonstration model or actual building plan (Trump). In Mesopotamia cuneiform writing on clay tablets, invented by the Sumerians, made possible the development of a surveying profession.

The Babylonians made maps to aid the practical purposes of daily life, for surveying, building, conducting business transactions, taxation, travelling, and waging campaigns. Theirs were the earliest scale maps. City plans of Babylon, Nippur, and Assur show temples, palaces, rivers, canals, walls, and fortifications and sometimes give distances and areas (Nemet-Nejat, p.5). Building plans are drawn with precision to the point sometimes of inscribing every brick, as in the plan of a building of *c.* 1800 BC from Babylon, drawn on a scale of about 1:60 (British Museum, Department of Western Asiatic Antiquities, no. 68641). A statue of Gudea, ruler of the Sumerian city-state of Lagash, *c.* 2200 BC (Gudea B, original in the Louvre), shows him holding a scale plan of the Ningirsu temple, with a measuring rule and a writing tool (Figs. 3 and 4). A second type of local map, the estate or field plan, occurs in the Neo-Sumerian and Old, Middle, and Neo-Babylonian periods. The earliest plans, comprising surveys of area measurements, were made to determine areas of fields within accepted boundaries (Nemet-Nejat, p.18). With the development (in the second millennium) of the *kudurrus* – carved stones, dealing with grants of public and private lands – cardinal points and details of boundaries were given. Late Babylonian field plans give contractual details of transactions and may be inscribed with the seeding capacity of the land and the number of date palms, showing also storehouses and canals, thus indicating the values of properties.

In ancient Egypt the annual flooding of the Nile necessitated the regular re-surveying of land to establish the boundaries of properties. Plans of circular and triangular plots and instructions on surveying are given in the Rhind Mathematical papyrus, *c.* 1600 BC (British Museum). The tomb of Amenhotpe-si-se, an official under King Tuthmosis IV, *c.* 1420 BC, shows a surveying team setting out to survey fields for the registration of the corn harvest. Accurate measurement was likewise demanded in major building works, such as the construction of the Great Pyramid, which was built with precise orientation. The most famous Egyptian plan extant is the Turin papyrus, on which is painted a map of the Nubian gold and silver mines, between the Nile and the Red Sea, *c.* 1200 BC, showing also hills, roads, houses, and shrines (Museo Egizio, Turin). The "Fields of the Dead," drawn on papyri partly in plan and partly in profile, may also be seen as idealized representations of real surveys. An example from the *Book of the Dead* of the priestess of Amun, Nesitanebtashru (from Upper Egypt, 21st dynasty, *c.* 1000 BC), shows her plot of land in the "Field of Rushes," one of the ancient Egyptian concepts of paradise. The plot is represented as a portion of fertile land surrounded by water and intersected by canals (British Museum, Department of Egyptian Antiquities, Greenfield Papyrus, 10554, sheet 81). The picture maps on painted coffins, such as those of *c.* 2000 BC from Dar el Bersha illustrating routes to the after life, may also be seen as an indication of what actual surveys were like. The accompanying text comprises the "Book of the Two Ways" which provides a guide for the departed soul's journey.

A well-known passage in Aristophanes's *The Clouds,* 423 BC, illustrates the familiarity of the Athenians with "allotment plans," as opposed to world maps. Although none on papyrus appears to survive, maps on stone are preserved, such as the plan of a 4th century BC mine at Thorikos, east of Attica. The Greeks used a system of squares or rectangles for surveying towns and of rectangles for rural areas, a type of survey said to originate with Hippodamus of Miletus (fl.473-43 BC). Their term for a plan was *ichnographia* (Dilke, p.87, 102). The Greek surveyors ("measurers of land") handed on their knowledge to the Romans, whose scale maps were among the most notable achievements of the ancient world. The surveyors were named *mensores* (measurers); the land surveyors *agrimensores* (land measurers). Their textbooks, the Corpus Argrimensorum, date from the first century AD onwards, and include plans used for teaching and demonstration. (The earliest are in MS. Arcerianus A, Wolfenbüttel, *c.* AD 500; those in the Vatican, Palatinus latinus 1564, are of the 9th century.) The authors recommended that the plans be engraved on bronze, an indication that the surveys were to be a permanent record (Harvey, 1980, p.126).

The plans which survive were made not on bronze but on stone. These are the marble tablets from Orange (Roman Arausio), showing plans of lands laid out by centuriation, AD 77, and the "Forma urbis Romae," the plan ("forma") of Rome. c. 3rd century, inscribed on marble tablets fixed to a wall. Plans of Jerusalem and of four buildings illustrate Adamnan's account of Arculf's journey to the Holy Land in 670 and may show traces of Roman traditions of mapping. A later example is the plan of a monastery drawn at Reichenau between 816 and 837 for the Abbot of St Gall. No later successor to the scale maps of the Romans is known.

In medieval Europe the functions of everyday life were performed without the use of maps. The earliest medieval plans date from the 12th century and come from Italy (Venice and Rome, both known only from 14th-century copies), the Holy Land (Jerusalem, from the 1140s), and England (Canterbury, between 1153 and 1161). The earliest recorded plans of localities in other areas are the Low Countries 1307 (Aardenburg and Beochoute, near Sluis), France 1422 (Vallée de Château Dauphin, near Grenoble), Germany 1441 (Wantzenau, near Strasbourg, but a plan of the Reichenau area was included on the Ebstorf world map, c. 1285). None is known from the Iberian Peninsula, the British Isles outside England, north Germany, Scandinavia, or eastern Europe. Very few plans are earlier than the mid-14th century, and by 1500 they were still quite exceptional productions.

Little research has been focussed on these plans as a whole to reveal a pattern. There seem, however, to be three significant local concentrations outside Italy: in eastern England around the Wash, in the Low Countries along a coastal strip, and in southwest Germany in Swabia and the Black Forest. Medieval Italian plans included two clear traditions: one of districts in north Italy from the late 13th century on, most 15th-century examples being of areas centred on one of the cities under Venetian control; the other of bird's-eye views of cities, especially Rome, which at first showed simply the city walls with a few principal monuments, but which by the late 15th century were detailed realistic views, the most notable being those by Francesco Rosselli in the 1480s (Florence, Rome) and Jacopo de' Barbari in 1500 (Venice). Of thirty local plans of places in England the earliest were all made for practical purposes, such as to provide a guide to concealed water pipes, as at Canterbury (1153-61) and Wormley (1220-30). By the 13th century maps were being made of disputed boundaries, such as the map of Wildmore Fen (1224-43). From the 14th century on, maps were also being made with an illustrative rather than an obvious practical purpose, as for example the plan of Thanet in Elmham's Chronicle (c. 1400) (Skelton and Harvey, pp.5-6).

It is also significant that most medieval maps are not drawn to a consistent scale. They are mainly either diagrams of ground-level features or pictures of landscape in which some or all of the features are shown as if seen from a height. Besides the St Gall monastic plan the only medieval scale plans known are of towns of the Holy Land by Pietro Vesconte, c. 1320, which are drawn on a measured grid, and a plan of Vienna with Bratislava about 1422, which has a scale bar (Historisches Museum, Vienna). This is the first known European local map since Roman times drawn explicitly to scale. An early 14th century plan of Venice is probably a copy of one drawn in the 12th century and appears to show a ground-plan drawn to scale. If so, this is the only scale-plan known from medieval Italy (Paolino Veneto, *Magna Chronologia, c.* 1320, Biblioteca Nazionale Marciana, Venice). Apart from this,

Fig. 3 – Statue of Gudea, ruler of the Sumerian city-state of Lagash, c. 2200 BC. He holds on his lap the plan of the Ningirsu temple. By permission of the Trustees of the British Museum, Department of Western Asiatic Antiquities, C. 27; cast of the original in the Louvre, Paris, photographed by permission of the Musée du Louvre. (Plan, 1.1630; Map surface, 6.1310)

Fig. 4 – Plan held on Gudea's lap (Fig. 3), comprising a drawing of the temple of the god Ningirsu, the patron deity of Lagash. A measuring rule with subdivisions is visible, and a graver of slate pencil lies alongside. This is the earliest known plan drawn to scale, c. 2200 BC.

Leonardo da Vinci's plan of Imola (*c.* 1502-3) ranks as the earliest scale-map of an Italian city since Roman times (Royal Library, Windsor Castle).

Military engineers were among the first to draw maps to scale, as illustrated by a plan of the new fortifications at Portsmouth, 1545 (British Library, Cotton MS. Augustus I.1.81; Harvey, 1980, pp. 80, 155, 160-1). Henceforth the scale-map of towns, fortifications, and estates became a common-place in Europe.

In China the tradition of local topographies was established early, and the earliest work dates from AD 52 (Needham, p. 517). The importance of geomancy, "the art of adapting residences of the living and the tombs of the dead so as to co-operate and harmonize with the local currents of cosmic breath" necessitated a careful appraisal of the topographical features of any locality and also encouraged local guides and mapping. The plan of an irrigation survey by Chhin Chiu-Shao, 1247 (Shu Shu Chiu Chang, ch.6), may be seen as comparable to the plans of waterworks in medieval Europe. It became traditional for the Emperor to have maps and plans made for all tours and military campaigns, and often very large pictorial plans in long scrolls were made for this purpose. A notable album is that by Ch'ien Wei-ch'eng of the fifth journey of the Emperor Ch'ien-lung, an 18th century manuscript (British Library, Or.12895.f.15).

Native Mexican cartography was also notable for its local mapping. A Mexican map of estates in the state of Pueblo, *c.* 1600, shows the division of nine pieces of land between the two chiefs of the towns of Itzcuintepec and Teteltepec. The plan, which is painted on amatl paper in black ink outline, was made for presentation to the Spanish court (Fig. 8).

C ALMAGIÀ, Roberto (1929): *Monumenta Italiae Cartographica.* Firenze, Istituto Geografico Militare, 1-13, pl. VII-XIII.

BONACKER, Wilhelm (1950): "The Egyptian Book of the Two Ways," *Imago Mundi, 7,* 5-17.

DAINVILLE, François de (1970): "Cartes et contestations au XVe siècle," *Imago Mundi, 24,* 99-121.

DILKE, O A W (1985): *Greek and Roman Maps.* London, Thames and Hudson.

HARVEY, Paul D A (1980): *The History of Topographical Maps.* London, Thames and Hudson.

HARVEY, Paul D A (1985): "The Spread of Mapping to Scale in Europe, 1500-1550," *Imago et Mensura Mundi. Atti del IX Congresso Internazionale di Storia della Cartographia,* edited by Carla Clivio Marzoli. Roma, Istituto della Enciclopedia Italiana, *2,* 473-477.

HEYDENREICH, L H (1965): "Ein Jerusalem-Plan aus der Zeit der Kreuzfahrer," in *Miscellanea Pro Arte,* Joseph Hoster and Peter Bloch, ed. Düsseldorf. Appendix to *Schriften des Pro Arte Medii Aevi, 1,* 83-90, pl. LXII-LXV.

HOFF, B van't (1962): "The Oldest Maps of the Netherlands: Dutch Map Fragments of about 1524," *Imago Mundi, 16,* 29-32.

HORN, W, and BORN, E (1974): "New Theses about the Plan of St Gall," in *Die Abtei Reichenau,* Helmut Maurer, ed. Sigmaringen, Thorbecke, 407-76.

NEEDHAM, Joseph (1959): *Science and Civilisation in China,* vol. 3. Cambridge, At the University Press.

NEMET-NEJAT, Karen Rhea (1982): *Late Babylonian Field Plans in the British Museum* (Studia Pohl: Series Maior: Dissertationes Scientificae de Rebus Orientis Antiqui). Rome, Biblical Institute Press.

REINHARDT, Hans (1952): *Der St Galler Klosterplan.* St Gallen, Verlag der Fehr'schen Buchhandlung.

RÖHRICHT, R (1898): "Marino Sanudo sen. als Kartograph Palästinas," *Zeitschrift des Deutschen Palästina-Vereins, 21,* 84-126, pl.2-11.

SCHULZ, Jurgen (1978): "Jacopo de' Barbari's View of Venice: Map Making, City Views and Moralized Geography before the Year 1500," *Art Bulletin* (College Art Association of America), *60,* 425-74.

SKELTON, R A, and HARVEY, P D A (1986): *Local Maps and Plans from Medieval England.* Oxford, Clarendon Press.

TRUMP, David H (1979): "I primi architetti: I costruttori dei Templi Maltese," φιλίας χάριν *Miscellanea in onore di Eugenio Manni.* Rome, Giorgio, Bretschneider, pp.2113-14 and Figs.

1.1630A Plat

A In 16th century England the term "plat" meant a plan or diagram, especially a ground plan of a building or of any part of the earth's surface (O.E.D.). In Tudor usage a plat generally, though not invariably, indicated a true ground plan (Skelton) of which details of features were often shown in perspective. The term was used in contradistinction to the prospect or view. In the 17th and 18th centuries plat, or platt, was a popular term among mariners, for a sea-chart. Some of the chartmakers of the Drapers' Company in London in the 17th century, for example, gave as their address "At the Signe of the Platt," referring to the signboards hanging outside their shops, which would display a chart. By the end of the 19th century plat was the technical term in the United States for a plan of land units for registration: *Scribner's Magazine* (New York, June 1893, 695) refers to plats from the State Land Offices showing lands subject to entry (quoted in O.E.D.).

B The first local map since Roman times to be drawn explicitly to scale is a plan of Vienna copied in the mid 15th century from one about 1422, with Bratislava drawn in the top left hand corner (Harvey, p.80). This would qualify as a plat in English 16th century usage. In Italy Leonardo da Vinci's plan of Imola, 1502-3, is the earliest undoubted scale map (or plat) of an Italian city since Roman times (Royal Library, Windsor Castle; Harvey, p.155). It is possible that Italian engineers recruited by Henry VIII in the 1530s and early 1540s brought the concept of the scale map to England. One of the earliest scale maps in England is a plat of Portsmouth, 1545, showing the town's fortifications (British Library, Cotton MS. Augustus I.1.81; Harvey, p.160).

In the United States, detailed manuscript or printed plats showing land boundaries, ownership, and land use date from the later years of the 19th century. They were bound into volumes as "plat books" in the early 20th century, replacing the more elaborate county atlases (see Cadastral map, 2.0210b, and County atlas, 8.0110a).

C HARVEY, Paul D A (1980): *The History of Topographical Maps.* London, Thames and Hudson.

MERRIMAN, Marcus (1983): "Italian Military Engineers in Britain in the 1540s," in Tyacke, Sarah, ed.: *English Map-Making 1500-1650.* London, British Library, pp.57-67.

SKELTON, R A (1965): "John Norden's View of London, 1600," *London Topographical Record, 22,* 14-25.

THROWER, Norman J W (1981): "The County Atlas of the United States," *Surveying and Mapping, 21,* 365-73.

1.1630B Town Plan

A A comprehensive large-scale map of a city or town delineating streets, important buildings, and other urban features.

B The idea of the town plan developed in ancient times. The earliest known is a wall-painting of a settlement found at Çatal Hüyük, Turkey, now in the museum at Konya, which dates from *c.* 6200 BC. Surveys in Mesopotamia on clay tablets included town plans. One of the city of Nippur, *c.* 1000 BC, shows the Euphrates, canals, the main temple, enclosures, the city wall, and gates. Hippodamus of Miletus is assumed to be the surveyor who laid out the plan of Miletus in 479 BC (Cicero, *De re publica*, i.22). The plans of the Greek colonies of Metapontum (Metaponto) and Heraclea (Policoro) in south Italy illustrate the competence of Greek "measurers of land" and show that they must have passed on their knowledge to the Etruscans and Romans (Dilke, p.88). No maps or plans, however, survive as evidence of their work. The most notable work from the Roman period is the large marble plan of Rome, Forma Urbis Romae, completed sometime between AD 203 and 208. It was probably not the first plan of Rome, as Vespasian and Titus are reputed to have had Rome measured in AD 73. Some mosaic plans of towns also survive from the classical period, including several preserved in the ruins of Rome's ancient port of Ostia, dating from *c.* AD 200. The most famous of the mosaic maps is the Madaba map of the early Byzantine period, evidently made between 560 and 565 for the floor of a church. It comprises a map of Palestine which includes pictorial plans of Jerusalem and other towns.

Maps of Jerusalem are pre-eminent among early medieval town plans. The most accurate and also one of the earliest is a plan dating from the 1140s, reflecting the local knowledge of the Crusaders who had held the city since 1099 (Bibliothèque Municipale Cambrai; Harvey, pp.70-72). Matthew Paris of St Albans in the 13th century and Pietro Vesconte and the Venetians Marino Sanudo and Paolino Veneto in the 14th also included in their works on Palestine detailed town plans of Jerusalem and Acre, which reflected Crusader influence.

The plans of Jerusalem together with a tradition of local topographic mapping appear to have influenced the development of Italian town plans, especially those of Rome. The earliest of the Roman plans, known from a 14th century copy, dates from the 12th century. These plans reflect a strong pictorial tradition and some are largely symbolic in style and content. In the late 15th century the more realistic bird's eye view became popular. Francesco Rosselli, an engraver of Florence (1445-1520), made large-scale views of this type, of which that of Florence, *c.* 1482, survived up to the Second World War. One or another of these plans is presumed to have influenced the most remarkable of all renaissance plan-views, the woodcut of Venice by Jacopo de' Barbari (*c.* 1450-*c.* 1511) composed on six large sheets and measuring in all 135 x 282 cm. It is believed to have been built up from a mosaic of sketches fitted into an outline plan of Venice (Schulz, pp.429-30). While derived presumably from the picture-map tradition, the plan may also reflect the separate tradition of the outline plan in which Italians were leading practitioners in the later Middle Ages. The earliest known town plan since Roman times drawn explicitly to scale is that of Vienna *c.* 1422 (see Plat, 1.1630a).

The renaissance plan-view may be regarded as one of the most accomplished cartographic devices of its day. It appealed to the civic pride of the citizens, as emphatically stated by Anton Kolb, Barbari's publisher, who wrote that the woodcut was issued "principally for the glory of this illustrious city of Venice." In technique the plan-view was intermediary between the plat (1.1630a) and the prospect or view (see Perspective/panoramic map, 1.161). In the plan-view the ground plan normally forms the basic map and the eye, in theory, is overhead, but important buildings are shown in elevation. The style represents a combination of techniques and is similar to what 17th century writers in optics called "scenographia," combining "ichnographia" (the ground plan) with "orthographia" (the profile or elevation).

The first printed city atlas, comprising the first uniform collection of plans of the cities of the world, was the *Civitates Orbis Terrarum,* published in five volumes at Cologne by Georg Braun (1541-1622) and Frans Hogenberg (1542-1600), who may have been the main instigator and whose work has been described as "a forerunner of modern scientific illustration." (Skelton, 1966, pp.v, xiii). A sixth volume completed the set in 1617-18. This atlas, produced as a complement to Abraham Ortelius's *Theatrum Orbis Terrarum* of 1570 (see Atlas, 8.0110), ranks as one of the monuments of renaissance cartography. Almost all contemporary techniques for representing towns are found within its pages, from plan to perspective, as Braun explicitly states: "tam geometrica, quam perspectiva pingendi ratione." It is clear from the editors' choice of materials that as geometrical techniques advanced, true plans were preferred to pictorial methods.

Nevertheless, perspective detail remained for many years a typical feature of the town plan. An unusual form is the strictly geometrical circular scale plan of Vienna by the Nuremberg cartographer Augustin Hirschvogel, 1547, flanked by a fish-eye view of the surrounding fortifications and countryside and inscribed with the cardinal points of the compass (City Archives of Vienna). This tradition of encircling perspective detail was a feature of the work of the Nuremberg school, as illustrated by Paul Pfinzing's plans of the city and environs, 1594 and 1596.

It is notable also how slowly the geometrical vertical plan without perspective detail came to supersede the popular semi-pictorial tradition. When John Ogilby and his wife's grandson William Morgan made their large map of post-fire London, 1676, at a scale of 100 feet to the inch, they chose an essentially two-dimensional and more mathematically accurate presentation as was appropriate to a city under reconstruction, referring to the plan as "ichnographically describing all the streets . . ." The survey represents very nearly the first linear ground plan or plot (plat) of any British town, and certainly is the first large multi-sheet plan of a British town so delineated (Hyde, section 8). The earlier plan-view with its bird's eye or vertical depiction of topographical detail gave a more vivid impression of the appearance of a town, and had proved an ideal means of representing the European renaissance city. At the same time the technique enabled the mapmaker to conceal his ignorance; as John Holwell remarked, "Those that are minded to draw the Map of any Town, City or Corporation, only with the Uprights of the Houses, will have no need to measure either the House, Courts or Allies thereof" (John Holwell, *A Sure Guide to the Practical Surveyor*, 1678, p.190).

After 1670 it became increasingly common for European surveyors to make geometrical town plans according to mathematically precise methods of survey. The city plan-view gained a new lease of life, however, in the United States of America and Canada where, before the American Civil War (1861-65), perspective plans became a popular genre and like the earlier renaissance city plans were designed to appeal to local pride.

In the 19th century national surveys ventured into urban mapping on an increasingly large scale, as in Great Britain between 1848 and 1852, when a scale of ten feet to a mile (1:528) was adopted for northern English towns to deal especially with sanitary problems. On the continent of Europe cadastral surveys provided the material for town plans indicating ownership of property. A notable early example is that of Vienna: "Grundriss der Kaiserl: Königl: Haupt und Residenz Stadt Wien, eingetheilt nach seinen Grundbüchern/Plan de la Ville de Vienne d'après le Cadastre 1799. Dessiné par Max de Grimm," hand-coloured, published by Artaria in Vienna (British Library, K. Top. XC.40.). The introduction of lithography and the development of thematic mapping from the middle years of the 19th century increased the facilities for, and the interest in, a more elaborate kind of urban survey. The firms of Charles E Goad from 1875 in Canada and D A Sanborn in the United States (from 1867) developed a major production of insurance plans of towns (Insurance map, 2.091). Charles Booth's poverty map of London (1889) marked a notable advance in sociological mapping (Poverty map, 2.162).

The early development of local mapping in China is illustrated by the fact that in the period of Hsi-Ning's reign, 1068 to 1077, ambassadors from Korea received maps on request from every county capital (*hsien*) and provincial capital visited on their travels (Needham, pp.549-50). A city map of Suchow, P'ing-chiang t'u, 1229, survives as an engraved stone, one of four drawn *c.* 1193 by the geographer Huang Shang and intended for the instruction of the future emperor, to whom Huang was tutor. The maps were engraved by Wang Chih-Yuan early in the 13th century and the steles were erected in the Confucian temple at Suchow. (A rubbing is preserved in the British Library, OMPB 15406.a.5.)

Large-scale plans of the principal cities of Japan became an important feature of 17th century Japanese cartography. A woodcut plan of Nagasaki Harbour, *Nagasaki Ezu,* 1680, was among the first Japanese written materials to reach northern Europe (from the Engelbert Kaempfer Collection, British Library, OMPB Or.75.g.25). Edo (now Tokyo), the centre of the Togugawa government, was well depicted in plans such as that by Ishikawa Ryūsen, *Edo no Ezu,* Edo, 1689 (woodcut; British Library, OMPB Or. 75.f.18.), in which special emphasis is given to the residences of the feudal lords (daimyō) with their family crests (*mon*). Japanese mapmakers were expert in combining aerial view and plane techniques, often drawing castles, shrines, and temples in exaggerated proportions (Nanba et al., p.154).

C DILKE, O A W (1985): *Greek and Roman Maps.* London, Thames and Hudson.

HARVEY, Paul D A (1980): *The History of Topographical Maps.* London, Thames and Hudson.

HYDE, Ralph (1976): *A Large and Accurate Map of the City of London . . . By John Ogilby.* Introductory Notes. Lympne Castle, Kent, Harry Margary.

NANBA Matsutaro, MUROGA Nobuo, and UNNO Kazutaka, eds. (1973): *Old Maps in Japan.* Osaka, Sogensha.

NEEDHAM, Joseph (1959): *Science and Civilisation in China,* vol. 3. Cambridge, At the University Press.

SCHULZ, Juergen (1978): "Jacopo de' Barbari's View of Venice: Map Making, City Views, and Moralized Geography before the Year 1500," *Art Bulletin* (College Art Association of America), *60,* 425-74.

SKELTON, R A (1965): "John Norden's View of London, 1600," *London Topographical Record, 22,* 14-25.

SKELTON, R A, ed. (1966): "Introduction," in Georg Braun and Frans Hogenberg: *Civitates Orbis Terrarum,* vol. 1. Cleveland and New York, The World Publishing Company.

WALLIS, Helen, and others (1974): *Chinese & Japanese Maps.* London, British Library Board.

1.164 Planisphere

A (MDTT 818.5) A continuous map of the celestial sphere or of the whole of the earth (the latter application is archaic). The meaning of this term has clearly changed over the years, with its application in cartography being broadened in early modern times. A precise definition suitable for international acceptance is therefore difficult to compose, but the following is suggested as embracing most aspects of its provenance: a map or chart formed by the projection of a sphere, or part of one, on a plane; originally a stereographic projection of the celestial sphere, later a name associated with the astrolabe and, more recently, used to denote a number of projections of the sphere on a plane.

B It is not possible to say who actually coined this hybrid of the two elements "plane" and "sphere." According to Marie Armand Pascal d'Avezac (pp.17-18), the new projection of the Greek astronomer Hipparchus of Rhodes (fl. 146-127 BC) was called by classical geographers a planisphere. It assumed the eye of the beholder to be opposite the concave hemisphere projected, and was thus a stereographic projection. Claudius Ptolemy (2nd century) adopted this concept and wrote a treatise entitled *Planisphaerium*. Knowledge of this work reached the West through Maslama ibn Ahmad al-Majriti, a Hispano-Muslim scientist of note, who translated it into Arabic with a commentary in the second half of the 10th century. This in turn was translated into Latin by Hermann the Dalmatian at Toulouse in 1143. A further treatise on the subject was composed by Jordanus de Nemore, a German mathematician and physicist, probably in Westphalia in the 12th century, who provided the first general demonstration of the fundamental property of the stereographic projection, namely, that all circles on a sphere are projected as circles. Ptolemy had proved it only in special cases.

The *Planisphaerium* became much more widely known when Hermann's translation appeared in print for the first time as an appendix to the Rome edition of Ptolemy's *Geography* in 1507. It was printed again in a collection of astronomical writings at a press in Basle during the year 1536, and both Jordanus's and Ptolemy's writings appeared under the editorship of Federici Commandini at Venice in 1558.

Since certain parts of many astrolabes are constructed on the principle of the stereographic projection, the term planisphere has frequently been linked with that instrument, and indeed the use of the stereographic projection was generally confined to astronomy up to the 16th century.

From the early years of the 16th century the stereographic form of projection began to be applied to terrestrial cartography, and the term planisphere was used by such notable mapmakers as Gemma Frisius (1508-1555), Oronce Fine (1494-1555), who described the projection in a treatise of 1544, and Rumold Mercator (1545-1599). Before the end of the century the term had broadened in its application in certain quarters, and it was being used not only with reference to the polar, equatorial, and oblique aspects of the stereographic but also to a number of other projections of the sphere on a plane, even those derived from a developed cylinder. A particularly interesting later use of the term was in the famous "Planisphère Terrestre," which was a large map on an azimuthal projection, laid out on the floor of the Paris Observatory in 1682 by Giovanni Domenico (Jean Dominique) Cassini, to incorporate the new determinations of longitude derived from observations taken of the eclipses of Jupiter's moons, a technique which he had done much to perfect.

C AVEZAC, Marie A P d' (1863): *Coup d'oeil historique sur la projection des cartes de géographie*. Paris, E. Martinet.

BLUNDEVILLE, Thomas (1597): *M. Blundeuile His Exercises, containing eight Treatises*. London Iohn Windet, 289, 338.

CARPENTER, Nathanael (1625): *Geography Delineated forth in two bookes*. Oxford, 1, vii, 174.

DIGGES, Leonard and Thomas (1591): *A Geometrical Practise, named Pantometria.* London, unpaginated, but see "The. 29 Chapter."

FINAEUS, Orontius (Oronce Fine) (1544): "Planisphaerium geographicum." Lutetia (Paris).

GREGORY, John (1649): *Gregorii Posthuma.* London, 309-11.

HEIBERG, Johan Ludvig (1970): *Ptolemaei opera astronomica minora.* Leipzig, 263-70.

NORDENSKIÖLD, A E (1889): *Facsimile-Atlas to the Early History of Cartography.* Stockholm, 16-17, 90-93. (Reprinted, Kraus Reprint, NY, 1961.)

SARTON, George A L (1927, I; 1931, II): *Introduction to the History of Science,* 3 vols. Baltimore, Carnegie Institution of Washington, Publ. 376. I, 668; II, Part I, 125, 173, 174, 177; II, Part II, 613, 614, 616.

ZIEGLER, Jacob (1536): *Sphaerae atque astrorum coelestium ratio, natura et motus; ad totius mundi fabricationis cognitionem fundamenta.* Basle.

1.1810   Route Map

A   (MDTT 822.8) "A small-scale map (sometimes in the form of a strip map, or schematic map) which shows routes between places by a selected mode of transport and which frequently indicates the distances . . . between them."

B   The route map is one of the basic types of map. People of every culture from the earliest times have shown an instinctive ability to express distance, often given in units of time such as a day's journey, and direction, established from observations of the sun and movements of the stars. They have no difficulty in making rough but accurate sketches of routes and the relative positions of places.

In ancient Egypt coffins were inscribed with the "Book of the Two Ways," comprising a guide to the departed soul's journey in the after-life. Three coffins found at Dar el Bersha, the graveyard of the princes of the Middle Empire (c. 2000-1750 BC), and transferred to museums in Berlin, Paris, and Cairo, bear these inscriptions which include picture-maps of itineraries. The plan on the Berlin coffin (Neues Museum, no. 14385), shows two routes, one a land-way, the other partly a waterway, the favourable one in red, the dangerous one in black; it appears to be based on maps of the Nile valley (Bonacker; Kamal). The Paris coffin resembles that of Berlin. The Cairo coffin (Cairo Museum, no. 28,083) shows a waterway leading along islands as a winding canal, and a land-way guarded by watchmen (Bonacker, pp. 11-12).

In China route maps of rivers from the 1st century BC reflected the great importance of rivers in Chinese economic life (Canal/river chart, 1.1810a). Long Chinese scrolls marking rivers and canals of the 18th century survive as some of the finest specimens of Chinese mapmaking.

The earliest extant maps of Romans roads date from c. 200-250; and the Peutinger map of the 4th or 5th century, comprising a road map of the whole Roman empire on a long narrow strip, is the most famous map of the ancient world. In the Middle Ages the 13th century English monk Matthew Paris was a pioneer in making itineraries in the form of strip maps for travellers to Italy and on the way to Palestine (itineraries from London to Apulia, of which four manuscript copies survive).

The first printed road map was made by Erhard Etzlaub at Nuremburg, *c.* 1500, to show routes to Rome. The first printed road atlas, Cologne, 1579-80, depicted the roads of the Christian world. The next major innovation was John Ogilby's *Britannia*, London, 1675, a road atlas of the British Isles showing roads on strip maps. The exact measurements made possible the production of a new type of map, the distance map (1.043), expressing topological relationships rather than topographical. Railway (1.1810c) and cycling (1.1810b) maps were produced in the 19th century in response to the needs of these new forms of travel.

Among non-European and non-Oriental peoples the American Indians have revealed special skills. Mexican maps of the 16th century now extant show a blend of Aztec and European conventions (Road maps, 1.1810d). Thomas Pownall in *A Topographical Description of the Dominions of the United States of America* (London, 1776), wrote that the North American Indian had a high degree of topographical knowledge: "The Habit of travelling mark to him the Distances, and he will express accurately from these distinct Impressions, by drawing on the Sand a Map which would shame many a Thing called a Survey" (De Vorsey, pp.46-7). A birchbark map found in 1841 attached to a tree between Lake Huron and the Ottawa River in Upper Canada depicts an Indian route by land and water (Fig.21). It was sent to London with a covering note: "Forwarded to the United Service Institution in the hope that it may shew young officers how small an effort is needed to acquire that most useful art of Military Sketching since even Savages can make an intelligible plan" (British Library, Map Library, R.U.S.I. Misc.1.; Coe, p.263; Leggett, p.52).

C BONACKER, Wilhelm (1950): "The Egyptian 'Book of the Two Ways," *Imago Mundi, 7,* 5-17.

COE, Ralph T (1976):*Sacred Circles, Two Thousand Years of North American Indian Art.* London, Arts Council of Great Britain.

DE VORSEY, Louis, (1966): *The Indian Boundary in the Southern Colonies, 1763-1775.* Chapel Hill, University of North Carolina Press.

HODGKISS, Alan G (1981): *Understanding Maps: A Systematic History of Their Use and Development.* Folkestone, Kent, Dawson.

KAMAL, Youssouf (1926): *Monumenta Cartographica Africae et Aegypti,* tom 1. Leiden, p.1.

LEGGETT, Robert (1975): *Ottawa Waterway – Gateway to a Continent.* Toronto and Buffalo, University of Toronto Press.

1.1810A Canal/River Chart

A A map designed primarily for the navigation of canals or rivers. A river or canal map may be made for other purposes, e.g. to illustrate a plan for the regulation of a river or for the building of a canal.

B Chinese geographers were concerned with hydrogaphy from the earliest times, since river systems were arteries for transport and communication and vital elements therefore in the Chinese economy. Sang Chhin of the 1st century BC wrote the *Shui Ching* (Waterways Classic) giving a description of 137 rivers. The work was greatly expanded by the geographer Li Tao-Yuan in about the 6th century AD and entitled *Shui Ching Chu* (A Commentary on the Waterways Classic). The titles of various books suggest that rivers were being mapped from the Chin dynasty (265–420) onwards. A chart of the river systems in west China is included in Fu Yin's *Yü Kung Shuo Tuan* (Discussions and conclusions regarding the Tribute of Yü), *c.* AD 1160 (Needham, pp.514-15). The *Hsing-shui chin-chien* (Golden Mirror of the Flowing Waters), by Fu Tsê-hung, 1725, is illustrated with panoramic maps of rivers and lakes (British Library, O.M.P.B. 15275.c.6.).

The Chinese scroll was an appropriate format for elaborate river and canal maps, although it necessitated considerable geometrical distortion. A manuscript scroll of the Grand Canal from Peking to Hangchou, 18th century, includes, for example, large sections of the Yellow River, the Huai, and the Yangtse (British Library, O.M.P.B. Or. 2362). A map of the Yellow River, "Wang Shi gu huang tu," signed by Wang Hui, 1704, but probably a copy *c.* 1900, drawn in ink and wash on a long scroll, straightens out the river and marks parallel to it the Grand Canal, which runs mainly north-south (British Library, O.M.P.B. Or.13990).

In western Europe the Flemings and Dutch were among the pioneers in river mapping since hydrographic control of rivers in the Low Countries was vital, and rivers were used to determine boundaries. An early example of such a boundary map was made in connection with the classification of students by "nations" in the medieval University of Paris. A small sketch map of 1357 in the registers of the university shows the course of the river Maas as the limit distinguishing the Dutchmen north of the Maas, known as "natio Anglicana," from those south, the "natio Picardia" (Keuning, pp.40-1). There were evidently maps of the river available at that time from which the sketch could be made. A large manuscript map of 1468 depicts the lower course of the Scheldt from Rupelmonde to its mouth in the North Sea (Algemeen Rijksarchief, Brussels). It is really more a bird's-eye view made for a lawsuit than a chart. The work of Pierre Pourbus, painter of Bruges, who made some of the best topographic maps of his day, is especially important for the depiction of waterways in the 16th century. His large manuscript map of the Franc de Bruges, begun in 1562, completed in 1571, shows waterways in detail (Musée Archéologique, Bruges; see also Map surface: cloth, 6.1310c). In 1630 Willem Jansz Blaeu published at Amsterdam a map of the Fossa Eugeniana, the canal between the Maas and the Rhine, "Fossa quae a Rheno ad Mosam duci coepta est Anno [16]XXVII."

In England one of the earliest local maps is a mid-13th century copy of a plan of *c.* 1220 showing the source of Waltham Abbey's water supply at Wormley, Hertfordshire. It marks springs (circles), tanks (rectangles), and water conduits. The plan is preserved in the cartulary of the abbey (British Library, Harley MS. 391.ff5ᵇ-6). An early panoramic manuscript chart of the Kent coast, believed to date from 1514, includes the river Swale from Faversham (see Coastal chart, 1.0320a). A manuscript chart more modern in style is that of the Tyne, *c.* 1570, showing the river's course from Newcastle to the sea. Soundings are carefully marked, and obstructions indicated by symbols (British Library, Cotton MS. Augustus 1.ii.5). In the later 17th century Andrew Yarranton was a pioneer in the planning of inland navigation. In his book *England's Improvement by Sea and Land*, London, 1677, he put forward projects for a deep-water dock in the Thames and foreshadowed the great canal-building schemes of the 18th century. Illustrating his book is "This Mapp Discovers to you how farr the great Rivers of England may be made Navigable . . ."

The first of the English canals planned in the 18th century was the Duke of Bridgewater's between Worsley and Manchester, begun in 1761 and constructed by James Brindley (1716-72). A map of the canal between Manchester and Runcorn, completed by John Brindley in 1773, was drawn by the Hanoverian engineer Johann Lüdewig Hogrewe in 1777 and dedicated to George III (British Library, King's MS.46.f.55, Pl.VII). Brindley's scheme for the Thames "A Plan of the River Thames and of the intended canal from Monkey Island to Isleworth. Survey'd by James Brindley 1770," engraved by Thomas Jefferys, Geographer to the King, was published in 1770. Thereafter numerous canal schemes were put forward, each accompanied by plans of the intended project. (Many are preserved in a three-volume work in the British Library, Maps C.25.d.2-4). The Scot John Rennie the elder (1761-1821), who settled in London in 1791, was one of the leading surveyors and engineers of his day. His plans for the Kennet and Avon canal, surveyed 1792, published 1794, for the drainage of the Bedford level in the Fens, 1810, and for the East India Docks, London, 1804, illustrate some of his major projects.

Italy was notable for its early hydrographic mapping. The artist Leonardo da Vinci, a military engineer by training, illustrated various projects, including an embanking scheme to divert the river Arno near Florence, 1502-3. His sketch reflects his interest in water movement as well as his pioneering skills in cartographic techniques (Royal Library, Windsor Castle, n.12680). Two hundred years later another scientific polymath, the cosmographer Father Vincenzo Coronelli of Venice, applied himself as an hydraulic engineer. His scheme for the canalization of the Danube at Vienna, 1717, with its maps (Biblioteca Marciana, Venice, MSS. Italiani, Classe IV, n.31), brought him a reward of a gold collar from the Emperor Charles VI. His most ambitious plan was for the diversion of part of the Adige into Lake Garda (his drawings are preserved in Padua, Centro Studi Coronelliani). In 1712 he presented the Provveditori with a large topographical map of the Adige-Garda (Venice, State Archives; *Miscellanea Francescana,* pp.92-107, 303-5, 310).

The major European canal project of the 17th century was the French Canal Royal of Languedoc, built between 1666 and 1681 to link the Atlantic Ocean and the Mediterranean Sea. Two manuscript surveys were made in 1665, "Plan géometrique du canal . . ." by François Andréossy and Jean Cavalier, and "Carte particulière de la coste de la mer . . . et le Canal," by Cavalier (Bibliothèque Nationale, MS cinq cent de Colbert, t.202). The first printed map was Pierre Duval's "Canal des eaux de la Montagne Noire . . . pour le nouveau canal de Languedoc," Paris, 1666. This was followed by Nicolas de Fer's "Karte du nouveau Kanal," Paris, 1669, a work notable also for its use of a new phonetic orthography. Andréossy, chief surveyor and cartographer of the canal, achieved the finest portrayal with his "Carte du canal royal . . . ," 1669, which represented France uniting the Atlantic, in the aspect of Neptune, and the Mediterranean, depicted as a water nymph.

The Italian naturalist Count Luigi di Marsigli was a pioneer in the geographical and hydrographic depiction of a river system. His six-volume *Danubius Panonico – Mysicus, observationibus geographicis, astronomicis, hydrographicis, historicis, physicis, perlustratus,* The Hague and Amsterdam, 1726, provides a wealth of cartographic material, including sectional charts of the Danube. A "Mappa Potamographia" depicts the rivers and lakes of the region (vol.1, Part III, tab.34), and others show the declivities of the river bed (vol.1, tab.40).

In the 19th century with the development of tourism, panoramic maps and charts of rivers became popular with the general public. Friedrich Wilhelm Delkeskamp (1794-1872) of Bielefeld was a pioneer in this genre. His "Panorama des Rheins," Frankfurt-am-Main, ran to various editions from 1825 to 1868 and was published in English at London, [1835]. William Tombleson produced a similar panoramic style "strip map" in *Tombleson's Thames,* edited by William Gray Fearnside, which was issued also in a German edition, both published in London, 1834.

C DAINVILLE, François de (1961): *Cartes anciennes du Languedoc XVIe-XVIIIes.* Montpellier, Société Languedocienne de Géographie.

HARVEY, Paul D A (1968): "A 13th-Century Plan from Waltham Abbey, Essex," *Imago Mundi, 22,* 10-12.

KEUNING, Johannes (1952): "XVIth-Century Cartography in the Netherlands," *Imago Mundi, 9,* 35-63.

MISCELLANEA FRANCESCANA (1951): *Il P. Vincenzo Coronelli dei Frati Minori Conventuali 1650-1718 nel III Centenario della nascita.* Rome.

NEEDHAM, Joseph (1959): *Science and Civilisation in China,* vol. 3, Cambridge, At the University Press.

REAL COMMISSIONE VINCIANA (1941): *I manoscritti e i disegni di Leonardo da Vinci . . . I disgni geografici,* Roma, La Libreria dello Stato.

SMET, Antoine de (1947): "A Note on the Cartographic Work of Pierre Pourbus, Painter of Bruges," *Imago Mundi, 4,* 33-36.

1.1810B Cycling Map (British)

A Any of a variety of small-scale, folding pocket road maps aimed at British cyclists. They were variously published as series of sectional maps, county maps, and district maps centred on towns or cities; or as one- or two-sheet whole-country maps; or as strip maps. Most had protective card covers attached.

B Except for the strip maps, few of the early maps were specially produced for cyclists; instead the majority were ordinary road or railway maps, usually given titles saying that they were intended for cyclists and other road users, such as walkers, horse-carriage drivers, and tourists in general. Most were derived from existing maps – for example, those of J and C Walker or G and J Cary (later G F Cruchley), or those issued with the newspaper the *Weekly Dispatch* and later re-published by G W Bacon and George Philip. By the end of the century, however, Bartholomew, W & A K Johnston, and others had published new maps.

The data likely to interest cyclists were frequently overprinted, for example, dangerous hills, classes of roads according to quality by colours, distances between towns, and towns with cycling club representatives or with hotels and repair shops offering special terms to cyclists. The strip map, in which Gall & Inglis and George Philip specialized from about 1893, opened out usually to show only one road, with its natural and man-made features depicted in considerable detail.

Soon after 1900 publishers began to appeal to motorists as well, and gradually the motorists took precedence. Typically, Bacon's "Half Inch District Cycling Maps" became, or were supplemented by, "Cycling and Motoring Maps," then by "Motorists' and Cyclists' Maps," and then by "Motoring Maps."

C BRITISH PARLIAMENTARY PAPERS (1896): Report of the Departmental Committee appointed by the Board of Agriculture to consider the arrangements to be made for the Sale of Ordnance Survey Maps. Vol. 68, C.8147 and 8148.

NICHOLSON, T R (1983): *Wheels on the Road: Road Maps of Britain, 1879-1940.* Norwich, Geo Books.

PIGGOTT, Charles Antony (1980): "When the Cycle Was King of the Road," *Map Collector, 13,* 14-18.

1.1810C Railway Map

A (MDTT 822.9) A strip or schematic map of any scale designed specifically to delineate a single railway, its connections, and/or a system, or an entire network. Such a map often used a topographic or general map as a base and included stations and junction points, as well as other features of concern to rail users. It sometimes contained names of specific lines, distances, and transit time between points.

Special types include (1) topographical strip survey maps depicting individual lines that show only a ribbon of land adjacent to the railway, (2) large scale legal rights-of-way plans accurately delineated from actual surveys, and (3) schematic maps distorted in scale and/or line manipulation to portray favourably a railway, (4) railway junction diagrams.

B The railway map had its beginnings in the first years of the 19th century, first in England and soon after in North America. Surveys for, and construction of, track for pioneer railways created demands for special mapping and also induced mapmakers to show the progress of surveys, the construction and completion of lines, both on separate and on general maps and also in atlases, itineraries, travellers guides, and timetables.

In England, lines began appearing regularly on maps after 1803 when the Surrey Iron Railway was constructed between Wandsworth and Croydon. Charles Smith's detailed map of Surrey in his *New English Atlas* (1804) shows the line by symbol and name. British railways were first depicted in print on a map in the *Atlas of Europe* by Philippe Marie Guillaume Vander Maelen, published at Brussels, 1829-33. The first specialist British railway map was George Bradshaw's map of rivers, canals, and railways in the Midlands, 1830 (Hodgkiss). Bradshaw's further publications included "Maps and sections of the railways of Great Britain," Manchester and London, 1839.

In North America the earliest survey map, which shows a commercial "tramway" in Pennsylvania, was drawn by John Thomson in October 1809 and was entitled "Draft Exhibiting the Railway Contemplated by John Leiper Esq. From His Stone Sawmill and Quarries to His Landing on Ridley Creek." Inauguration of more extensive railway projects in Europe and North America after 1830, coupled with the technological advances in papermaking and printing, greatly contributed to the volume and variety of railway map publications.

C FORDHAM, Herbert George (1914): *Studies in Carto-Bibliography, British and French, and in the Bibliography of Itineraries and Road-Books.* Oxford, Clarendon Press.

GARNETT, David (1984): "John Airey's Undated Early Railway Maps," *Map Collector,* 26, 28-31.

HANEY, Louis Henry (1908): *Congressional History of Railways in the United States,* 2 vols. Madison, Wisconsin, Democrat Printing Co., also University of Wisconsin (1910).

HODGKISS, A (1981): *Understanding Maps: A Systematic History of Their Use and Development.* Folkestone, Dawson.

MILLER, Sidney Lincoln (1933): *Inland Transportation.* New York and London, McGraw-Hill.

MODELSKI, Andrew M (1975): *Railroad Maps of the United States.* Washington, DC, Library of Congress.

NOCK, Oswald Stevens (1978): *World Atlas of Railways.* London, Mitchell Beazley.

POOR, Henry Varnum (1860): *History of the Railroads and Canals of the United States of America.* New York, J H Schulz & Co.

TANNER, Henry S (1829): *Memoir on the Recent Surveys, Observations, and Internal Improvements, in the United States.* Philadelphia.

1.1810D Road Map

A A medium- or small-scale map of a road network designed for route planning and route finding. Such maps usually distinguish between roads suitable for different categories of traffic. They may also indicate the distances between places (see Distance map, 1.043).

A special type is the strip map, depicting individual routes and not attempting overall topographic coverage. This is also called a route map.

B Itineraries and itinerary maps were in common use in the Roman Empire. The Romans used either lists of places with distances ("itineraria") or painted (i.e. coloured) maps ("itineraria picta") as guides to their road networks. The earliest surviving road map seems to be that drawn on a parchment covering of a shield dated *c.* 200-250 found during excavations at Dura Europos, Syria, in 1923. The map shows a Roman road, with stopping places, leading from Byzantium to the mouth of the Danube and a number of places beyond. The most notable example of a Roman strip-map showing routes is the Peutinger map (so-called after its 16th century owner, Konrad Peutinger), known only from a copy made in the 11th or 12th century from a Roman road map of the 4th or 5th century, probably revised from a Roman original of the 1st century. It comprises eleven sheets of parchment which together form a strip 6.75 m. long by 34 cm. high, and shows the main roads of the Roman Empire, many with distances, and the towns to which they lead. The westernmost part (the twelfth sheet), which contained most of Britain and most of Spain, is lost.

In medieval Europe route maps for travellers, especially pilgrims and merchants, continued to be one of the main forms of topographic mapping. In England Matthew Paris, a monk of St Albans, drew the earliest English road map *c.* 1250 which showed the route from London to Otranto in Apulia, by way of Boulogne, Paris, and Rome. Another of his innovations, the earliest historical map of Great Britain, was his "Scema Britannie," a sketch showing the four main Roman roads. From the 14th century there survives a large manuscript map of Great Britain drawn *c.* 1360 for government use and known as the "Gough map" (named after its 18th century owner). It marks the roads as single red lines, and numbers give distances between towns in customary miles, as they were later known.

The first printed road map was made by Erhard Etzlaub at Nuremberg, *c.* 1500. Centred on Nuremberg and entitled "Das ist der Romweg . . ." (The way to Rome . . .), it shows routes for pilgrims to Rome in the Holy Year 1500. Roads are indicated by dotted lines, each dot denoting one German mile (about 7 km) (See Fig. 18). Martin Waldseemüller of St Dié in Alsace produced in 1511 a large wall map of Europe, entitled "Carta Europae," and beneath, outside the frame, is the title "Carta Itineraria Europae." It is now known from an impression of 1520 preserved in the Museum Ferdinandeum, Innsbruck. Roads are marked as dotted lines.

The *Itinerarium Orbis Christiani* (Cologne, 1579-80), usually attributed to Johannes Metellus Sequanus (1520-98), is thought to be the first printed road atlas. It employed double parallel lines to indicate roads, a convention thereafter in general use. John Norden's map of Middlesex (1593) is the first English engraved county map to show roads. The next major innovation was achieved by John Ogilby with the publication of his celebrated road-book, the *Britannia* (London, 1675), comprising a road atlas of the British Isles, in which particular roads are depicted in sections by means of strip maps. The work was an instant success, and Ogilby's road maps were revised and reissued many times through the 17th and 18th centuries. John Adams used the roads in Ogilby's *Britannia* to compile a large-scale road and distance map, "Angliae totius tabula . . . 1684," published with his *Index Villaris,* 1690 (see Distance map, 1.043).

The earliest post-road map in France – ranking also as the earliest post-road map

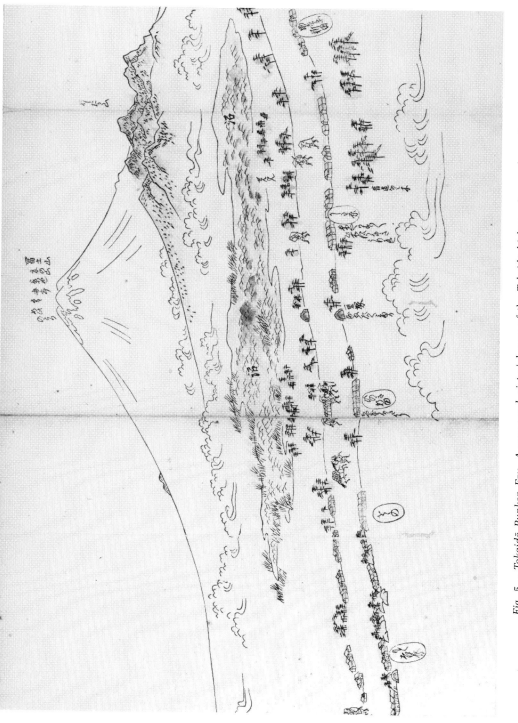

Fig. 5 – Tokaidō Bunken Ezu. A measured pictorial map of the Tokaidō highway, Japan, drawn by Hishikawa Moronobu, compiled by Ochikochi Dōin. Edo, published by Hangiya Shichirobē, 1690. Woodcut, hand coloured. Scale 1:12,000. In 5 vols. This section shows Mount Fuji in the background. Orientation is indicated by means of a compass box, drawn wherever the road changes direction. By permission of the British Library. (Road map, 1.1810d; Tourist map, 2.202; Orientation, 4.151)

known – is Melchior Tavernier's "Carte Géographique des Postes qui traversent la France," drawn by Nicolas Sanson, Paris, 1632. This shows distances by dashed lines. Road systems of Central Europe were mapped by Joh. Georg Jung and Georg Conrad Jung in "Totius Germaniae Novum Itinerarium," Nuremberg, 1641. The earliest map to mark distances in figures was Johann Ulrich Müller's "Tabula Geographica totius S. Imperii Romani," Ulm, 1690, which has been described aptly as a milestone in road mapping (Pabst, p.166). The first true post route map has been identified probably as Johann Peter Nell v. Nellenburg's "Postarum seu Veredariorum Stationes per Germaniam et Provincias adiacentes. Neu-vermehrte Post-Charte durch gantz Teutschland," which covers the whole network of Taxis posts. The edition published by J B Homann at Nuremberg in 1714, and subsequently much copied, is inscribed "Inventa Anno 1709". Although no issue of 1709 is known, versions published by Eugene Henry Fricx in Brussels, 1711, and by Schenk [1710] in Amsterdam were probably plagiarized from the earlier version.

Ogilby's style of strip map was copied in France by various map makers, notably by Michel in *L'Indicateur fidèle, ou Guide des Voyageurs* . . ., Paris, published by Louis-Charles Desnos, 1764. The strip map also became popular in the United States of America. Christopher Colles's *A Survey of the Roads of the United States of America* (New York, 1789), includes 73 small maps of major roads, strip map in style. It ranks as the earliest American road guide, and is described as the ancestor of the 200,000,000 road maps now distributed annually to American motorists (Ristow, in Colles).

In non-European cultures a number of road maps survive from an early period. After the Spanish conquest of Mexico in the early years of the 16th century, the native Mexicans made maps on which Mexican and European symbols are blended. The maps in the Codex Tepetlaoztoc (mid-16th century) show roads by double parallel lines in European style, and also with a row of footprints, following the Mexican tradition. Similarly in 17th century Japan, the highway from Edo (Tokyo) to Kyoto, known as as the Tokaidō and measuring about 300 miles, was surveyed in 1651, mapped at a scale of 1:12,000, and published in 1690 in five volumes (Fig. 5). In China a number of maps from the 18th century include roads. For example, the maps of the public granaries of Hopei Province, 1753 (*I-ts'ang t'u, c.* 1800), show the major roads (British Library, OMPB 15239.b.11.vol.1).

C BAGROW, Leo (1954): "Carta Itineraria Europae Martini Ilacomili, 1511," *Imago Mundi, 11,* 149-50.

BONACKER, Wilhelm (1973): *Bibliographie der Strassenkarte,* with an Introduction by R Kinauer. Bonn-Bad Godesberg, Kirschbaum.

BURLAND, C A (1960): "The Map as a Vehicle of Mexican History," *Imago Mundi, 15,* 11-18.

COLLES, Christopher (1961): *A Survey of the Roads of the United States of America, 1789,* ed. Walter W Ristow. Cambridge, Mass., Harvard University Press.

ELIAS, Werner (1981): "Road Maps for Europe's Early Post Routes 1630-1780," *The Map Collector, 16,* 30-4.

FORDHAM, Sir Herbert G (1924): *Road Books and Itineraries of Great Britain 1570-1850.* Cambridge.

GUZMÁN, Eulalia (1939): "The Art of Map-making among the Ancient Mexicans," *Imago Mundi, 3,* 1-6.

KRÜGER, Herbert (1951): "Erhard Etzlaub's *Romweg* Map and Its Dating in the Holy Year of 1500," *Imago Mundi, 8,* 17-26.

KRÜGER, Herbert (1958): "Des Nürnberger Meisters Erhard Etzlaub älteste Strassenkarten von Deutschland," *Jahrbuch für Fränkische Landesforschung, 18.*

PABST, Wilhelm D (1985): "Im Schatten der Grössen: Zeit und Werk des Ulmer Kartographen Joh. Ulrich Müller (1653-1715)," *International Yearbook of Cartography, 25,* 159-182.

VAUGHAN, Richard (1958): *Matthew Paris.* Cambridge, at the University Press.

WOLKENHAUER, August (1903): "Über die ältesten Reisekarten von Deutschland aus dem Ende des 15. und 16. Jahrhunderts," *Deutsche Geographische Blätter, 26,* 120-38.

ZÖGNER, Lothar (1975): *Strassenkarten im Wandel der Zeiten Ausstellung.* Berlin, Staatsbibliothek Preussischer Kulturbesitz.

1.191 Spherical Chart

A A section of a globe with a radius of about 1 meter representing without distortion a part of the surface of the earth. Named (in Dutch) by its inventor "de ronde ghebulte Pas-caert."

B The spherical chart was invented by the Dutch mathematical practitioner Adriaen Veen and patented in 1594. No spherical chart is known to have survived. The two spherical charts mentioned in Veen's patent covered the northern and southern parts of the west coasts of Europe and were inscribed with systems of loxodromes (4.123). The device was intended to be used at sea, together with an instrument called the "Zee-passer" (sea compasses), also invented by Veen. These were dividers with three legs, to be used for plotting and scaling-off distances on the chart. In his book *Napasser van de Westersche ende Oostersche Zee-vaert,* Amsterdam, 1597, Veen explains and illustrates (in a vignette on the title page) the use of the chart and the "Zee-passer."

C CRONE, E (1966): *Une traduction inédite d'un traité d'Adriaen Veen, Cartographe Hollandais.* Coimbra, Agrupamento de estudos de cartografia antiga, Junta de investigacões do ultramar.

KOEMAN, C. (1976): *The Application of a Pair of Dividers with Three Legs, as Explained by the Dutch Mathematical Practitioner Adriaen Veen, in 1597.* Rio de Janeiro, Ministério da Marinha, Serviço de Documentação Geral da Marinha. (Segunda Reunião Internacional de História de Náutica e da Hidrografia, Salvador, 1976.)

1.192 Symbolic Map

A A map of a real or imaginary place "with an overall significance other than, or additional to, the representation of topographical or thematic information"; cf imaginary map (1.091); satirical map (2.192); religious map (2.181).

B From early times people have drawn maps and diagrams to express their ideas about the world and the universe. In China relief maps were designed for liturgical purposes; for example, a pottery hill-censer (*po shan hsiang lu*) of the Han period, 1st century BC, represents one of the magic mountain islands of the Eastern Sea (Needham, p.581). A literary text in a Taoist work, 4th century, describes how the Emperor Han Wu Ti visited the legendary goddess of the West, Hsi Wang Mu, who showed him a "Map of the True Topography of the Five Sacred Mountains" (*Wu Yo Chen Hsing Thu*) (Needham, p.566). The map must have represented a Buddhist-Taoist view of the cosmos, centred on Mt Khun-Lun, the source of the Yellow River. Sino-Korean atlases as late as 1800 normally included as their frontispiece a world map representing the ancient sino-centred cosmography. In Japan Buddhist cosmography is depicted in maps of the shield-shaped continent of Jambu-dvīpa, with the sacred lake of Anavatapta at the centre (see Religious map, 2.181). Indian cosmographical maps are centred on the sacred Mount Meru.

The Christian equivalent is the medieval world map with Jerusalem at the centre (Mappa mundi, 1.2320b). The manuscript "Bible moralisée" included symbolic depictions of the creation of the world, with God or Christ measuring the earth, shown in cross section, as in French manuscripts of the 13th century (Vienna, Nationalbibliotek; Oxford, Bodleian Library). In mappae mundi the symbolic decoration was often in the borders, displaying Christ surmounting the earth, as in the Psalter map, late 13th century (British Library, Add. MS. 28681.f.9.). This peripheral symbolism gave way to a blending of symbolism with the cartographic detail itself when Opicinus de Canistris, a clerk from Pavia, showed on a portolan-style chart, 1335-36, the Mediterranean as a "sea of sin," its coasts portrayed in the figures of a man and woman (Satirical map, 2.192). Another example is found in a 13th century manuscript which depicts Rome in the shape of a lion (Dilke and Dilke); and it is probable that a number of similar maps were drawn at the time.

A late example of the circular cosmographical world map is that by the Maltese doctor of theology Antonio Saliba, "Nuova Figura Di Tutte Le Cose . . .," Naples, 1582, in which the northern hemisphere (in conventional form) encloses the subterranean world and infernal regions, and is encompassed by circles of storm and clouds, the heavens, and an outer zone of fire (Shirley, pp.168-9).

In the 16th century maps showing continents or countries as people or animals became a favourite genre. The idea was not new, as the maps of Rome and satires of Opicinus de Canistris were precursors; but with the enlarging market for printed maps and the great interest in animals of the old and new world such maps proved of great popular appeal. The earliest, designed in 1537 by Joannes Bucius (Putsch), of Salzburg and Innsbruck, depicted Europe as a woman. The map became well known through the version published in Sebastian Münster's *Cosmographia* between 1588 and 1628, and in Heinrich Bünting's *Itinerarium Sacrae Scripturae*, Magdeburg, 1581. Bünting made a companion piece showing Asia as a winged horse, "Asia secunda pars terrae in forma Pegasir" (1581). To accommodate the animal's shape the geographical outlines of the continent were grossly distorted. A third map by Bünting (1581) shows the old world as a clover leaf, with America, England and Scandinavia fitted round the edge.

Of all symbolic maps in animal form by far the most popular was "Leo Belgicus." The Austrian Baron Michael von Eytzinger originated the concept in his history of the Low Countries, *De Leone Belgico*, Cologne, 1583. He cleverly superimposed the shape of the lion on a geographically accurate base drawn at a much larger scale than in the earlier examples. The symbolism was apt in that many of the seventeen provinces of the Netherlands had a form of lion in their coats of arms. The work gave rise to a succession of variants, of which that by the Roman Jesuit Famiano Strada ran into some 90 editions. A version of Leo Belgicus was published at Leiden as late as 1807. Other forms of animal symbolism appeared elsewhere, such as Berne as a bear, Austria as an eagle, and France as a cock.

Maps based on the human form included the Baltic as Charon, found in Olof Rudbeck's *Nova Samolad, sive Laponia Illustrata,* Uppsala, 1701; and the Spanish empire as a queen, by Vincente de Memije, "Aspecto Symbolico del Mundo Hispanico," Manila, 1761.

Other forms of symbolic map are conventional in cartographic detail but present the map in a symbolic setting. The "fool's cap map," in which the world is depicted as the visor of a jester's head, is known in a woodcut of Jean de Gourmont, Paris, *c.* 1575, and in an anonymous Latin version, Antwerp?, *c.* 1590 (Shirley, pp.157-8, 189-90; also *Map Collector, 15,* p.47; *18,* pp.39-40).

The degree of symbolism varied from the single image, as with Leo, to complex allegorical maps in which the whole content of the map was symbolic or satirical. The "Mappemonde nouvelle papistique," probably by Jean-Baptiste Trento, published at Geneva in 1566, was filled with anti-papal vignettes (see Satirical map, 2.192). In 17th century France the fashion for "precieux" literary conceits prompted Madelaine de Scudéry's "Carte de Tendre," in *Clélie*, Paris, 1656-60, providing a model for the love map.

The symbolic map had a new lease of life as a political cartoon in the later years of the 19th century (Satirical map, 2.192).

C DILKE, O A W, and DILKE, Margaret S (1975): "The Eternal City Surveyed," *Geographical Magazine, 47,* 744-50.

HILL, Gillian (1978): *Cartographical Curiosities.* London, British Library Board.

NEEDHAM, Joseph (1959): *Science and Civilisation in China,* vol.3. Cambridge, At the University Press.

SHIRLEY, Rodney W (1983): *The Mapping of the World.* London, The Holland Press.

TOOLEY, Ronald Vere (1963): *Leo Belgicus. An Illustrated List.* London, Map Collector Series, No. 7.

WALLIS, Helen, and others (1974): *Chinese & Japanese Maps.* London, British Library Board.

1.201 Tactile Map

A A map which is designed and constructed so as to be read by the sense of touch, requiring that portions be elevated using the third dimension. This term is not used in referring to three-dimensional maps designed to be read visually, such as relief maps or models and primitive works such as the Marshall Island stick charts or the early clay tablets with recessed symbolism. If these were to be included, the oldest forms of mapping would indeed be tactual.

Tactile maps are also called "tactual maps," "maps for the blind," and "braille maps."

B Tactual systems for reading and education date at least to the 14th century, but make no direct references to graphic devices or maps. The earliest references to maps of this type are made in an essay on the instruction and amusements of the blind by Richard Phillips, London, 1819, and in *Lehrbuch zum Unterrichte der Blinden,* by Anton Strauss, Wien, 1819. Phillips refers to early maps made by Weissenbourg as not being very satisfactory and describes in detail more advanced maps produced at the school for the blind in Paris. This reference is undoubtedly to the first school for the blind, the Institution Royale des Jeunes Aveugles de Paris, opened by Valentin Haüy in Paris in 1784. Louis Braille, an innovative blind student and later an assistant master at the Institution Royale, was presented with a sound-writing system in 1821 by Captain Charles Marie Barbier de la Serre, a former Captain of Artillery. Braille modified the "night writing" system which Barbier had invented and by 1825 completed the system of writing for the blind that is universally adopted today.

The selected tactile maps listed below depict innovative ideas and/or stages in development. These maps originated in many different countries and indicate a widespread interest in helping the blind to comprehend spatial relationships. A relief globe of papier mâché, 41 cm. in diameter, was made in 1822, inspired by Johann August Zeune, the geographer-cartographer, who was associated with the first German school for the blind which opened in 1806. The Map of the Kingdom of Bohemia, *c.* 1830, was printed in Vienna, using the so-called "ectographical" printing method. Many books, charts, and other materials were embossed by this method, which is akin to the present so-called "thermographic" printing and quite like the English "solid dot" braille. The New England Institution for the Education of the Blind produced a "Printed Embossed Map of New England, 1833" (Report of the Trustees of the New England Institution for the Education of the Blind, covering the first two years of the school, Boston, Mass.), and the *Atlas of the United States for Use of the Blind,* 1837 (published by Samuel Gridley Howe in Boston at the New England Institution for the Eduation of the Blind at the expense of John C Cray). Soon afterwards an "Outline Map," 1844, was prepared at the Perkins School for the Blind, Boston, Mass., by Samuel P Ruggles, who also designed a large tactual globe in 1851.

By 1843 Dobbs & Co. in London had started making "relief or model" maps which the Royal Geographical Society considered useful for instructing the blind (Royal Geographical Society, archives, 28 June 1842, 17 April 1844). The maps combined "the English process" of embossing with ordinary engraving. The firm published *c.* 1844 a "Relievo" Geological Map of England and Wales, 25 miles to one inch, and

"Arabia Petraea & Idumaea" and "Palestine," both 15 miles to one inch (Murchison, pp. cxxiii-cxxiv). A decade or so later a leading authority was William Moon (1818-1894), who was blind from 1840 and the inventor of the Moon system of reading for the blind. He published at his establishment in Brighton various embossed atlases and maps, including Moon's Biblical Pocket Atlas, large maps of Palestine, England and Wales, Marching Order of the Israelites, and St. Paul's Travels, as well as astronomical maps (Rutherfurd, pp.178, 276-7). His maps of the Eastern Hemisphere and the Western Hemisphere were published in 1884 by the British and Foreign Blind Association, the organization which became the Royal National Institute for the Blind. A map of the United States of America, by John Bartholomew, modelled by G R Boyle, was published in London, Glasgow and Edinburgh by William Collins, [1895].

Other publications include Laas d'Augen's "Map of the World Showing the Two Hemispheres," 1847. D'Augen was blind from early childhood, and his work was much admired by many contemporary geographers. The "Map of North America" (embossed by Feuguiere in Rio de Janeiro, Brazil, 1858) is interesting in showing that the use of braille was then common in Brazil. The "Map of Australia," c. 1890, embossed by Ravenstein of Frankfurt, Germany, shows, in the lower corner, sketches of the steamships of that time and the tropical vegetation of the Pacific Islands.

C A CENTURY AND A QUARTER OF EMBOSSED MAPS FOR THE BLIND: A historical exhibition of maps by the Blindiana Museum of Perkins School for the Blind. Watertown, Massachusetts, 1956, Nelson Coon, Librarian.

CURTIS, John B (1928): "Braille Maps," *Teachers Forum, 1,* No. 3, 2-4.

GUILBEAU, E (1908): "De enseignement de la géographie aux aveugles," *L'Educateur Moderne, 3.*

LENDE, Helga (1953): *Books about the Blind, A Bibliographical Guide to Literature Relating to the Blind.* New York, American Foundation for the Blind.

MACKENZIE, Sir Clutha (1954): *World Braille Usage.* UNESCO.

"MAPS FOR THE BLIND" (1943): *New Beacon, 27,* 157-59, 185-86, 208-10, 224-25; *ibid.* (1944): *28,* 14-16, 32-33.

MURCHISON, Roderick Impey (1844): "Address to the Royal Geographical Society of London," *Journal of the Royal Geographical Society, 14,* cxxiii-cxxv.

RUTHERFURD, John (1898): *William Moon, LL.D., F.R.G.S., F.S.A., and His Work for the Blind.* London.

SCHLEUSSNER, K (1908): "Modelle zur Veranschaulichung Geographischen Grundbegriffe," *Blindenfreund, 28,* 279-81.

WIEDEL, Joseph W, and GROVES, Paul A (1972): *Tactual Mapping: Design, Reproduction, Reading and Interpretation.* University of Maryland, Occasional Papers in Georgaphy, No. 2.

1.202 Thematic Map

A (MDTT 823.1) An important class of maps, the objective of which is to portray the character of a particular distribution in contrast to general maps, e.g., topographical maps, which show the locations of a variety of phenomena. Formerly called geographic, special purpose, or applied maps, the name thematic apparently stems from N Creutzberg's use of the term in 1953 ("Zum Problem der thematischen Karten in Atlaswerken," *Kartographische Nachrichten, 3,* Heft 3/4, 11-12), after which thematic was quickly and generally adopted. For thematic maps the range of subject matter and the appropriate degree of generalization is unlimited.

B Although an occasional thematic map was produced prior to the latter part of the 17th century, thematic maps may be said to have their beginning with the maps of ocean currents (3.151) by Athanasius Kircher (*Mundus subterranus,* 1665) and Eberhard Werner Happel (*Mundus mirabilis,* 1687) and with the maps of the winds (1686) (3.232) and compass declinations (1701) (3.031) made by Edmond Halley. The map of compass declinations was widely used and regularly revised. Halley's "curve lines" (isogones; see 5.0910e) were ultimately put to many cartographic applications in thematic cartography. A few other thematic maps, mostly geological, were made during the 18th century but not until the early years of the 19th century did this class of cartography begin to flourish. Such developments as the great studies in natural science, the initiation of censuses, the industrial and social revolutions and epidemic cholera all called for thematic maps to portray physical and cultural geographical complexities. The development of lithography as a cheaper alternative to copper engraving was also a contributing factor. By mid-19th century thematic maps were commonplace.

Some notable events in the rapid development of thematic cartography in the 19th century that seem to have had wide influence are Carl Ritter's first thematic "atlas," *Sechs Karten von Europa* (1806); William Smith's "A Delineation of the Strata of England and Wales ..." (1815); Alexander von Humboldt's introduction of the isotherm (1817) (5.0910g); Baron Charles Dupin's publication of the first choropleth map (1827) (1.034); André Michel Guerry's mapping of moral statistics in 1829 (with Adrien Balbi) and 1833; Adolphe Quetelet's introduction of the concept of social physics in the 1830's; C J Minard's trade and transport maps of the 1840s to 1860s; and especially the publication from 1838 to 1848 of the various parts of Heinrich Berghaus's *Physikalischer Atlas,* 1845, 1848, a collection of 90 plates of thematic maps of physical geography, including the organic world. A variety of lesser mapmakers contributed to the surging development of thematic mapping, which by 1860 had matured in almost all respects.

C ARNBERGER, Erik (1966): *Handbuch der thematischen Kartographie.* Wien, Franz Deuticke, 79-182.

ECKERT, Max (1921, 1925): *Die Kartenwissenschaft,* 2 vols. Berlin and Leipzig, Walter de Gruyter.

ROBINSON, Arthur H (1982): *Early Thematic Mapping in the History of Cartography.* Chicago, University of Chicago Press.

THROWER, Norman J W (1972): *Maps and Man.* Englewood Cliffs, New Jersey, Prentice-Hall, Inc.

ZÖGNER, Lothar (1979): *Carl Ritter in seiner Zeit (1779-1859).* Staatsbibliothek Preussischer Kulturbesitz Ausstellungskataloge 11.

1.203 Topographical Map

A A map showing detailed features of landscape; in its origin it would be a map of a limited area known to the draughtsman through his own observation.

B Some of the maps from so-called primitive societies can be classed as topographical maps; examples are the carved spear throwers of the Bindibu of Western Australia, the carved coastal reliefs of Greenland Eskimos and maps of North American Indians.

The earliest maps from more complex societies are also topographical and thus include the earliest of all maps known. The map found at Çatal Hüyük, Anatolia, apparently shows a volcano and a street pattern and probably dates from about 6200 BC. From Mesopotamia the clay tablet from Nuzi is inscribed with a map of mountains and rivers, about 2500-2300 BC. From Egypt coffins painted with maps of the next world, possibly based on maps of parts of the Nile, date from about 2000 BC. From China references to maps date from the 3rd century BC and actual maps drawn on silk showing areas between Canton and Kweiling, from about 168 BC. From the Roman Empire maps carved on stone about 77 AD show centuriated fields near Orange. From Japan maps of estates, preserved in temple archives, date from the 8th century. All these are the earliest examples known of what were demonstrably continuous traditions of topographical mapping in these particular cultures. Only one topographical map survives from the Hellenic civilization; namely, on a coin of Zancle (i.e. Messina), 525-494 BC, there is a representation of the curving sandbank that forms the town's harbour. Doubt is now cast on the supposed relief map on coins from Ephesus, about 335 BC.

Islamic topographical maps of various sorts (in relief; with non-pictorial symbols; on a grid) are known or reported from the Middle Ages, but their origin and development have not been investigated. Peru of the Incas provides records of maps in relief from the mid-15th century onwards, though none survives. From Mexico 16th century topographical maps in Aztec style may take their origin in the picture-stories, with some topographical content, of which examples date from the 14th century (Codex Zouche-Nuttall, Codex Vindobonensis) and the 15th century (Codex Xolotl). Certain features of some North American Indian 18th century maps (both surviving and reported) suggest a pre-European tradition of topographical maps.

The earliest surviving examples of various types of topographical maps from medieval Italy are conventionalized bird's-eye views of 10th century Verona and 12th century Rome, a ground-plan of Venice from the first half of the 12th century and a regional map of the area around Alba and Asti in a manuscript of 1291. Maps of Jerusalem were being drawn from the first half of the 12th century and of the whole of

73

Palestine from the first half of the 13th. Elsewhere in medieval Europe topographical maps were produced in England from 1153-61 (Canterbury Cathedral and its surrounds) and in the Low Countries from 1307 (Aardenburg and Boechoute, near Sluis) and came to be fairly widespread (though far from common) in the 15th century.

Many regional and local maps were made by landscape painters or topographical draughtsmen in the 15th century. Often the work was undertaken by sketching from horseback, as when Antoine Actuhier and another painter Jean d'Ecosse mapped the counties of Valentinois and Dios in 1423. Their procedures illustrate the process of turning landscapes into maps.

In the 15th and early 16th centuries the development of astronomical and mathematical studies at the European universities, especially Vienna, Nuremberg, and Freiburg, was a major factor in the introduction of planimetric topographical mapping. Martin Waldseemüller of St Dié and his collaborators were responsible for the topographical map of Lorraine which is included in the Strassburg Ptolemy, 1513. The renowned Flemish mathematician Gemma Frisius (1508-55), who was a student and then lecturer at the University of Louvain, described and presumably invented in 1533 the technique of triangulation. He set out the principles in his *Libellus de locorum describendorum ratione,* provided as an appendix to his edition of Peter Apian's *Cosmographicus liber,* Antwerp, 1533. Although the technique became indispensible for accurate large-scale mapping, it came into general use rather slowly. It was used for the map of Bavaria by Philipp Apian (son of Peter), surveyed from 1554 to 1561, published at Ingolstadt in 1568; for a map of the island of Hven, Denmark, by the celebrated astronomer Tycho Brahe, 1596; and for a map of Baden *c.* 1600.

The development of the scale-map rather than the introduction of triangulation explains the major change in European topographical mapping in the 16th century, namely, the fact that the surveyor (*geometricus* or *landmeter*) took over from the artist as mapmaker. By the end of the century, moreover, maps were regarded as an essential part of a survey, rather than as a mere adjunct to the written description (Harvey, pp.164-5, 167-8). In the 17th century, estate plans were often entitled "topographical description," emphasizing the concept of the topographical map as one based on precise survey, by whatever method. The association of the topographical map with surveys of planimetric accuracy based on triangulation and geodetic controls is a more recent concept. Thus in 1973 MDTT 821.1 defines the topographic map as "a map whose principal purpose is to portray and identify the features of the Earth's surface as faithfully as possible within the limitations imposed by scale."

By the modern definitions (listed by M. L. Larsgaard, pp. 3-9), the development of a topographical survey over a wide area, a whole country, was achieved first by France. As a first step, the Académie Royale des Sciences, Paris, engaged David Vivier in 1668 to survey Paris and its environs under the direction of Jean Picard. Some ten years later the map was published in 9 sheets at a scale of 1:86,400 (1 ligne to 100 toises) and demonstrated the feasibility of such a topographical survey and map. The first major national production of a multi-sheet topographical map was begun in France in 1747 based on the 1733-44 national triangulation by Jacques Cassini and his son Cassini de Thury. Under the direction of Cassini de Thury all except Bretagne was mapped by 1789, and it was finally finished in 1818. Officially it is the "Carte Géométrique de la France," but it is more appropriately and better known as the "Carte de Cassini."

Surveys were undertaken in Denmark in 1761, the Netherlands in 1767, Norway in 1773, and the preparation of official topographical maps became standard practice in the early 19th century. In Great Britain the establishment of the Ordnance Survey is officially dated 1791, but its precursor was the manuscript survey of Scotland by William Roy, 1747 to 1755 (British Library, K.Top. XLVIII.25.). It is significant that because of the lack of planimetric accuracy due to inferior instruments Roy described his survey as "rather . . . a magnificent military sketch, than a very accurate map of a country" (O'Donoghue, p.16).

The early topographical survey maps were essentially planimetric maps but often with some sort of pictorial or symbolic representation of the land surface, not always very successful. The means for obtaining accurate horizontal position (triangulation, plane table, alidade, etc.) were at hand, but a technique was lacking for determining elevations (hypsometry) accurately. Neither the barometer nor the measurement of vertical angles was sufficiently exact and expeditious. The introduction of the contour (see 5.0910a) and the survey technique of levelling, which became standard in France *c.* 1815 and with the British Ordnance Survey *c.* 1840, allowed the effective representation of the land surface on topographical maps thereafter.

C ALMAGIÀ, Roberto (1951): "Un' antica carta del territorio di Asti," *Rivista Geografico Italiana, 58,* 43-4.

BERTHAUT, Henry Marie Auguste (1898-99): *La Carte de France 1750-1898: étude historique,* 2 vols. Paris, Service Géographique de l'armée.

BONACKER, Wilhelm (1950): "The Egyptian 'Book of the Two Ways,'" *Imago Mundi, 7,* 5-17.

BROWN, Lloyd A (1949): *The Story of Maps.* Boston, Little, Brown and Company, 241-79.

CAVALLARI, V, et al., ed. (1964): *Verona e il suo Territorio.* Verona, Istituto per gli Studi Storici Veronesi. *2,* 39-42, 232-3, 481-5, pl. opp. p.192.

CLOSE, Sir Charles (1926): *The Early Years of the Ordnance Survey.* Chatham, Kent, The Institute of Royal Engineers, 141-44. (Reprinted, Newton Abbot, Devon, David and Charles, 1969).

CRÉDIT COMMUNAL DE BELGIQUE (1978): *La cartographie au XVIIIe siècle et l'oeuvre du comte de Ferraris (1726-1814).* Colloque International, Spa, 8-11 Sept. 1976. *Actes.* Collection Histoire Pro Civitate.

DAINVILLE, François de (1958): "De la profondeur à l'altitude," in *Le Navire et l'économie maritime du moyen âge au XVIIIe siècle principalement en Méditerranée,* Travaux du Deuxième Colloque international d'histoire maritime, Paris, 195-213. (Reprinted in *International Yearbook of Cartography, 2* (1962), 151-62. Translation into English in *Surveying and Mapping, 30* (1970), 389-403.)

GOTTSCHALK, Maria Karoline Elisabeth (1955-58): *Historische Geografie van Westelijk Zeeuws-Vlaanderen.* Assen, Van Gorcum. *1,* 148-9.

HARVEY, Paul D A (1980): *The History of Topographical Maps*. London, Thames and Hudson.

HSU, Mei-Ling (1978): "The Han Maps and Early Chinese Cartography," *Annals of the Association of American Geographers, 68,* 45-60.

LARSGAARD, Mary Lynette (1984): *Topographic Mapping of the Americas, Australia, and New Zealand*. Littleton, Colorado, Libraries Unlimited.

LEVI, Annalina, and LEVI, M (1974): "The Mediaeval Map of Rome in the Ambrosian Library's Manuscript of Solinus," *Proceedings of the American Philosophical Society, 118,* 567-94.

MEEK, Theophile James (1935): *Excavations at Nuzi: Vol. 3, Old Akkadian, Sumerian and Cappadocian Texts from Nuzi*. Cambridge, Mass., Harvard Semitic Series No. 10, pp. xvii-xviii.

MELLAART, James (1964): "Excavations at Çatal Hüyük, 1963, Third Preliminary Report," *Anatolian Studies, 14,* 39-119.

MELLAART, James (1967): *Çatal Hüyük: A Neolithic Town in Anatolia*. London, Thames and Hudson.

NEEDHAM, Joseph (1959): *Science and Civilisation in China, vol. 3*. Cambridge, At the University Press, 534, 582.

O'DONOGHUE, Yolande (1977): *William Roy, 1726-1790, Pioneer of the Ordnance Survey*. London, British Library Board.

PIGANIOL, André (1962): *Les Documents cadastraux de la colonie romaine d'Orange*. Paris, Gallia, Supplement No. 16.

RAMMING, M (1937): "The Evolution of Cartography in Japan," *Imago Mundi, 2,* 17-21.

SCHULZ, J (1970): "The Printed Plans and Panoramic Views of Venice," *Saggi e Memorie di Storia dell' Arte, 7,* 16-17.

SKELTON, R A, and HARVEY, Paul D A, ed. (1986): *Local Maps and Plans from Medieval England*. Oxford, Clarendon Press.

STEWARD, Harry (1980): "The Çatal Hyük (sic) Map," *Mapline, 19.*

1.204 Typographic Map

A A map printed entirely from specially cast metal, movable elements, not only for the lettering, but also for symbols and line work. Also known as a typometric map.

B Early experiments in 1773 were made by August G Preuschen, a Karlsruhe publisher, and the technique was developed by Johann Gottlob Immanuel Breitkopf, a printer-typefounder of Leipzig, and Wilhelm Haas (father and son), typefounders of Basel, reaching its peak in the late 18th century. Other printers employing the technique included Periaux (1806), Firmin Didot (1823), Alexandre Douillier (1840), and Eugene Duverger (1830), all in France; Georg Michael Bauerkeller (1832), and Mahlau (1851) in Germany, and Raffelsperger (1839) in Vienna. It was revived largely as a curiosity by the Boston Type Foundry, with examples of their map elements in two standard sizes in their 1883 type specimen book. Except for the efforts of the Haas family, the method never progressed beyond the experimental stage, and finally had to compete with the many other relief printing processes (wax-engraving, photomechanical line block, etc.) developed in the 19th century.

C AMERICAN DICTIONARY OF PRINTING AND BOOKMAKING (1894): New York, Howard Lockwood, 367.

HARRIS, Elizabeth M (1975): "Miscellaneous Map Printing Processes in the Nineteenth Century," in David Woodward, ed., *Five Centuries of Map Printing.* Chicago, University of Chicago Press, 113-36.

HOFFMANN-FEER, Eduard (1969): "Die Typographie im Dienste der Landkarte," *Regio Basiliensis, Basler Zeitschrift für Geographie, 10,* No. 1. Separately published Basel, Helbing & Lichtenhahn.

MCMURTRIE, Douglas C (1925): *Printing Geographic Maps with Movable Types.* New York, privately printed.

1.231 Wall Map

A (MDTT 818.19) A map, usually consisting of several hand-painted or printed sheets, used for display or as wall decoration. Sometimes the map is painted on wood or directly on the wall. Roman surveys on stone were fixed to the walls of offices for public use.

B Probably the oldest known pictorial wall map is the frieze excavated on Thera (Santorini) in the ancient Minoan city of Akro-Thera which was buried in the ashes of the volcanic eruption of *c.* 1500 BC. The painting shows three seaside towns, one of which may have been on the Nile delta. It would appear to be the earliest map known of a large-scale area (Gottmann, pp.5-6). Three Roman cadastral surveys of the region of Arausio (Orange) made for taxation purposes and cut in marble in the late 1st century AD were kept in the local *tabularium* or record office of the town and are said to have been attached to three internal walls of the building (Dilke, pp.108-9). The large-scale plan of Rome, "Forma Urbis Romae," carved on marble tablets between AD 203 and 208 measuring 60 feet wide and 42 feet high (18 x 13 m), was fixed to the wall of the Forum Pacis, the Temple of Peace, of the Emperor Vespasian (Dilke, pp.104-5).

In the Middle Ages large mappae mundi (1.2320b) were made or used as wall maps. Henry III (1207-72) of England had a world map on the wall in the Painted Chamber at Westminster, the Chronicles of England were illustrated around the wall, and a star map was on the ceiling. Reduced copies of Henry's map (by Matthew Paris of St Albans) are preserved (British Library, Cotton MS. Nero D.V.f.1; Corpus Christi College, Cambridge). Henry also commissioned in 1236 a mural mappa mundi for the Great Hall at Winchester (Tudor-Craig). The two great world maps from the 13th century, the Ebstorf (c. 1235) and the Hereford (c. 1285)., were presumably made for display at the religious institutions where they were preserved. The Hereford map is believed by tradition to have served as an altarpiece in the cathedral (Richard Gough, *British Topography,* London, 1780, *I,* 71).

In Italy a mappa mundi decorated the public loggia at the rialto market of Venice in the 14th century (F Sansovino: *Venetia città nobilissima descritta in XIIII libri,* Venice, 1581). According to a report wall maps were made for the ducal audience chamber in the Palazzo Ducale of Venice under the Doge Francesco Dandolo (1329-39). The finest mappa mundi from the 15th century is the large map by Fra Mauro of Venice, 1459, now on the wall of the Biblioteca Marciana, made for the Venetian government ("a contemplatiō de questa illustrissima signoria"). It may have been a replica of the world map made by Fra Mauro between 1457 and 1459 for King Alfonso V of Portugal, one of the many lost maps known only from documentary records.

The great mural maps of the 16th century were also Italian. On the third floor of the Vatican were maps which depicted the oecumene. In the west wing thirteen regional maps, comprising England and Ireland, Spain, France, etc, were by "Stefano Francese," identified as Étienne du Pérac of Paris and Rome (1525-1604). The great Dominican mapmaker Egnazio Danti from Perugia (1536-86) designed for the north wing the mappa mundi in two hemispheres (now extant) and other maps which deteriorated and were replaced. Pope Gregory XIII had invited him to Rome in 1580 to depict the whole Papal State as murals on the second floor of the Vatican Palace, and these survive as monuments of 16th century mapmaking. He was also the author of the maps painted in the Guadaroba of the Palazzo Vecchio in Florence, 1563-75. Another famous work is the "Mural Atlas" of the palace of Caprarola, the country residence of Cardinal Alessandro Farnese, 1574. The "Sala del Mappamondo" has its walls painted with maps of the known parts of the earth and has the celestial sphere on the ceiling. It thus gives a complete view of the universe (Kish). The artist is identified as Giovanni Antonio de Varese, a north Italian who was active in Rome from 1562 to 1596 and was working on the Vatican's cosmographic loggia from 1562 to 1564.

The palaces of Northern Europe also had maps on their walls for information and decoration. In England the inventories of Henry VIII's possessions at the time of his death in January 1547 show that the maps hanging in the palace at Whitehall included "mappe mondes" and "plats" of his early victories, some painted on canvas and put on the wall framed. These works were evidently made for public display. In France the Galerie de Cerfs at Fontainebleau displays as frescoes mural maps of royal estates, c.1604.

The invention of printing and engraving techniques in the later years of the 15th century introduced a new artefact, the printed map which could be reproduced in numerous copies. To form a wall map the item was printed from a number of wood blocks, and the prints then mounted. The oldest extant printed wall map is the woodcut town view of Venice (135 x 282 cm) by Jacopo de' Barbari (c. 1440-c. 1511), dated 1500 and printed from six large wood blocks. Although the publisher, Anton Kolb claimed that no such view had ever been made before, large printed bird's-eye views of Italian cities were already known (see Town plan, 1.1630b). The oldest printed wall map of the world was that by Martin Waldseemüller (1470-1518) of St Dié in Lorraine, 1507, from twelve wood blocks. A later notable example was Giacomo Gastaldi's world map on nine woodcut sheets "Cosmographia Universalis," *c.* 1561, 820 x 900 mm (only known copy in the British Library).

In the middle of the 16th century the printing technique changed and wood blocks were joined by engraved copper plates. The Italians led the field with such work as Gastaldi's map of Africa, Venice, 1564. In Germany Gerard Mercator made wall maps of Europe, 1554, the British Isles, 1564, and the world, 1569, which were masterpieces of 16th century cartography. Sebastian Cabot's elliptical world map, published at Antwerp in 1544, engraved on eight sheets (Bibliothèque Nationale, Paris), was revised and re-issued in 1549 by Clement Adams in London. This new version hung on the walls of many merchants' houses and the most famous exemplar was on view in the privy gallery at the Palace of Whitehall, from at least *c.* 1566 to *c.* 1620, and treated as a document of major political importance as evidence of England's claims to North America. Next to it hung, from some time after 1580, the manuscript world chart recording the circumnavigation of Sir Francis Drake, 1577-80.

In the Low Countries Abraham Ortelius made a large eight sheet world map on a cordiform projection, published at Antwerp in 1564, but shortly to be eclipsed by Mercator's of 1569. Peter Plancius followed with the first large world map as a wall map in the Netherlands, mounted from eighteen sheets and published at Amsterdam and/or Antwerp, with measurements 2330 x 1460 mm. The Netherlands now became a major centre for the production of wall maps by the family firms of Hondius, Blaeu, Danckerts, Frederik de Wit, and Hugo Allard. Many identifiable wall maps are depicted in the paintings of Dutch interiors by 17th century artists such as Vermeer.

Wall maps in consequence of their large size and as a result of the continuously varying conditions of light, temperature, and humidity are especially vulnerable to deterioration. For this reason only a few copies of mounted wall maps still exist. In the 1660s a number of Dutch wall maps were bound together in three giant atlases for presentation to royal and noble persons (Klencke Atlas, British Library, London; Atlas of the Great Elector, Berlin; Rostock-Atlas, Rostock). Consequently, these wall maps were well protected and are now preserved in good condition.

At the end of the 17th century the prominent part of the Dutch in making wall maps was relinquished to French cartographers, such as Guillaume de l'Isle, Jean Baptiste Nolin, and others.

C ALMAGIÀ Roberto (1929): *Monumenta Italiae Cartographica.* Firenze, Istituto Geografico Militare.

ALMAGIÀ, Roberto (1952-55): *Le pitture murali della galleria delle carte geografiche. Le pitture geografiche murali della terza loggia e di altre sale Vaticane: Monumenta Cartographica Vaticana,* vols. 3 and 4. Città del Vaticano, Biblioteca Apostolica Vaticana.

ALPERS, Svetlana (1983): *The Art of Describing: Dutch Art in the Seventeenth Century.* Chicago, University of Chicago Press.

BANFI, Florio (1952): "The Cosmographic Loggia of the Vatican Palace," *Imago Mundi, 9,* 23-24.

CRONE, Gerald Roe (1954): *The World Map by Richard of Haldingham in Hereford Cathedral, circa A.D. 1285.* London, Royal Geographical Society.

DILKE, O A W (1985): *Greek and Roman Maps.* London, Thames and Hudson.

FISCHER, Joseph, and WIESER, Franz von, eds. (1903): *Die älteste Karte mit dem Namen Amerika aus dem Jahre 1507 und die Carta Marina aus dem Jahre 1516 des M. Waldseemüller (Ilacomilus).* (Also in English.) Innsbruck, Verlag der Wagner'schen Universitäts-buchhandlung.

GOTTMANN, Jean (1984): *Orbits: The Ancient Mediterranean Tradition of Urban Networks.* The Twelfth J L Myres Memorial Lecture. London, Leopard's Head Press.

KISH. George (1953): "The 'Mural Atlas' of Caprarola," *Imago Mundi, 10,* 51-6.

KLEMP, Egon (1971): *Kommentar zum Atlas des Grossen Kurfürsten: Commentary on the Atlas of the Great Elector.* Stuttgart, Berlin, Zürich, Belser Verlag.

SCHULZ, Juergen (1978): "Jacopo de' Barbari's View of Venice: Map Making, City Views, and Moralized Geography before the Year 1500," *Art Bulletin (College Art Association of America), 60,* 425-74.

TUDOR-CRAIG, Pamela (1957): "The Painted Chamber at Westminster," *Archaeological Journal, 114,* 92-105.

WELU, James A (1978): "The Map in Vermeer's 'Art of Painting'," *Imago Mundi, 30,* 9-30.

1.2320 World Map

A A map representing the whole surface of the Earth (or the greater part of it).

B World maps were made by the ancients in Babylonia, Greece, and China (see World map, ancient, 1.2320a). The earliest extant is a map of the earth and surrounding cosmos inscribed on a Babylonian tablet, 7th or 6th century BC. The Greeks inherited from the Babylonians an idea of the earth as a flat circular disc surrounded by the primordial ocean, and then developed in the 6th century BC the concept of a spherical earth. Following the capture of Egypt by Alexander the Great, the library at Alexandria was established during the reign of Ptolemy II Philadelphus (283/2 BC) as the centre of applied mathematics and scientific learning. Eratosthenes (c. 275-194 BC), director of the library, made the first scientific map. Marinus of Tyre (fl. AD 100-110) wrote a work called by Claudius Ptolemy "Correction of the World Map" and this was used by Ptolemy and, later, by Arab geographers. Ptolemy (c. AD

90-*c.* 168) worked at Alexandria. His Geography, written *c.* 150, provided tables of co-ordinates for other places in the world and instructions for making world maps according to three projections (see Map projection, 4.134). Whether Ptolemy drew maps himself is a matter of dispute, but his influence on cartography itself lasted for more than 1400 years. Through this posthumous fame he may be accounted the most important of ancient geographers.

Marinus's map and Ptolemy's Geography, perhaps in Syrian translation, were used by Arab geographers in the 8th century. The Academy of Science set up by al-Ma'mun, Caliph of Baghdad, 813-833, produced a map of the world, now lost (Dilke, pp. 155-56). In the late 13th century a Byzantine monk Maximus Planudes (*c.* 1260-1310) revised a manuscript text of Ptolemy's Geography and supplied it with maps from which (in Leo Bagrow's view) the later maps derive (Fig. 14). At the request of the emperor Andronicus II Palaeologus (1281-1328), a copy of the Geography was made with maps, including a world map on Ptolemy's first (simple conic) projection (Vatican Library, Vaticanus Urbinas graecus 82). When contacts between Byzantine and western European scholars were more fully established in the later 14th century, arrangements were made for Latin translations of Ptolemy. In 1415 two Florentines, Francesco Lapaccino and Domenico Boninsegni, latinized the maps of the manuscript Urbinas graecus 82. The first printed Ptolemy world map was published in 1477 in the Bologna edition of the Geography, the first printed Ptolemy with maps. Other versions followed in the Rome edition 1477-78, and the Ulm, 1482 (see Atlas, 8.0110).

Prior to the rediscovery of Ptolemy, Byzantine scholars were making world maps which combined Greek and Christian cosmology. The mosaic map at Nicopolis in Epiras, 6th century AD, gives a general picture of the world surrounded by the ocean. The most curious blend of the two traditions is found in the work of Cosmas Indicopleustes (fl. AD 540), a well-travelled Alexandrian merchant. He showed the universe as a vaulted box like the Ark of the Covenant, with the world depicted in its lower section. His oecumene (in another work) is a flat rectangle with surrounding ocean.

The Arabs modified the traditions of world mapping inherited from the Greeks. Whereas the Greek and Roman world maps normally had north at the top, most Arabic maps were oriented to the south, perhaps because south was sacred to the Zoroastrians or because Mecca lay south of Baghdad and other Islamic centres. Alternatively, the preference may have been already established in dynastic Egypt (Dilke, p.177). World maps schematic in style were included in the sets of maps known as the Atlas of Islam, made by various authors in the 10th century.

Until the 12th century Arabic map making remained separate from the main stream of western European cartography. A connection was then established when the Norman King Roger II of Sicily summoned the Arab geographer al-Idrisi (born 1099) to Palermo and commissioned a description of the world illustrated by maps. Al-Idrisi completed in 1154 his large world map and a small circular one. Copies of the smaller map (such as that in the Bodleian Library, Oxford, Uri MS. 887) are marked with curved *Klimata.* The influence of al-Idrisi's world maps persisted to as late as the 16th century.

Whereas Ptolemy's Geography was lost to European knowledge in the Middle Ages, the classical traditions continued in the form of mappae mundi, appropriately adapted to Christian knowledge and iconography. A separate tradition of world maps evolved in the 14th century when European sea charts were extended, notably by the Catalan school of cartographers, to encompass discoveries in Africa and Asia. A number of 15th century maps, although mappae mundi in style, represent the beginning of renaissance cartography (see Mappa mundi, 1.2320b). With the revival of Ptolemy's Geography in the 15th century in western Europe, mapmakers introduced a new type of world map, Ptolemaic in form (e.g. with north at the top), but corrected in the light of the recent geographical knowledge. Henricus Martellus Germanus included such a map in his "Isolario," made at Florence c. 1490 (British Library, Add. MS, 15760.ff.68b-69). A much larger MS map by Martellus, similar but slightly later, is in Yale University Library. These are the last maps of the oecumene. The discovery of the American continents (the New World) opened a new era, but Ptolemy's influence persisted.

The earliest chart to show both America and the old world is that of the Spanish navigator Juan de la Cosa, c. 1500 (Museo Naval, Madrid). The Portuguese world map known as the "Cantino," 1502 (Biblioteca Estense, Modena), ranks as the earliest Portuguese chart to display the New World. These works herald the development of the planisphere as a new form of world map. Those of Diego Ribeiro, 1525, 1527, and 1529, are notable as the first to show the whole circuit of the globe between the polar circles (see Nautical Chart, 1.0320c). These manuscript charts, being specially commissioned and of restricted circulation, were of limited influence. The printed map of the world was a much more potent document of intelligence and propaganda.

From the Italian workshop of Francesco Rosselli came the first printed world map to show the mainland of America, by Giovanni Matteo Contarini, engraved at Florence in 1506. (The Map Library of the British Library has the only known copy.) A somewhat similar map by Johann Ruysch in the Rome Ptolemy of 1507 features as one of the "tabulae modernae." Martin Waldseemüller (1470-1518) of St Dié, Lorraine, provided a new prototype for world maps with his two large woodcut maps, Strassburg, 1507 and 1516. In the later years of the century Gerard Mercator (1512-1594) of Rupelmonde and Abraham Ortelius (1527-1598) of Antwerp commanded the field: Mercator with his world chart published at Duisburg, 1569, drawn on the projection named after him, Ortelius with his "Typus Orbis Terrarum," in the *Theatrum Orbis Terrarum,* Antwerp, 1570, the most popular world map of the century.

From 1470 to the end of the 17th century some 650 European world maps were published (Shirley, p.IX). They come in many shapes and sizes. They change with the development of geographical knowledge and of new techniques of projection and production. They reflect the varied interests, patriotic, scientific, decorative, commercial, as may be, of a host of mapmakers from many walks of life.

In the orient the most celebrated "medieval" world map is the so-called Korean world map by Li Hui and Chhüan Chin, entitled *Hun-I Chiang-Li Li-Tai Kuo Tu chih Thu* (Map of the territories of the one world and the capitals of the countries in successive ages), manuscript, 1402. It is based on a compilation from two Chinese maps, and goes back to the time of Chu Ssu-Pên (1273-1337). The version now known (in a copy of c. 1500 in Japan) shows Europe, Africa, and Asia and is much in advance of contemporary European or Arabic world maps (Needham, pp.555-56).

When Father Matteo Ricci (1552-1610) went to China in 1582 as the first Jesuit missionary, he made and published woodcut world maps to inform Chinese officials and scholars about the true size and configuration of the world. His first map of 1584 and the second, at twice the size, are both lost. The third version, published at Peking in 1602, is the "K'un-yü wan-kuo ch'üan-t'u" (Map of the ten thousand countries of the earth). It was made to fit a folding screen twelve feet by six on six panels. The map is based on that of Ortelius, 1570, but rearranged to place China near the centre. It is reported that the edition ran to "many thousands" of copies to meet the large demand. A derivative is the woodcut Shōhō map, "Bankoku Sōzu" (Complete map of the world), published at Nagasaki Harbour in 1645. This ranks as the earliest Japanese printed map of the world. It was accompanied by forty pictures of the peoples of the world as the second sheet of a two-fold screen. Various other versions were published in Japan in response to the interest in decorative maps.

C CRONE, Gerald R (1978): *Maps and Their Makers: An Introduction to the History of Cartography.* Folkestone, Kent, Dawson; Hamden, Connecticut, Archon Books.

DESTOMBES, Marcel, ed. (1964): *Mappemondes: A.D. 1200-1500,* vol. 1 of *Monumenta Cartographica Vetustioris Aevi.* Amsterdam, N. Israel.

DILKE, O A W (1985): *Greek and Roman Maps.* London, Thames and Hudson.

NEEDHAM, Joseph (1959): *Science and Civilisation in China,* vol. 3. Cambridge. At the University Press.

SHIRLEY, Rodney W (1983): *The Mapping of the World: Early Printed World Maps, 1472-1700.* London, Holland Press.

WALLIS, Helen (1965): "The Influence of Father Ricci on Far Eastern Cartography," *Imago Mundi, 19,* pp.38-45.

1.2320A World Map, Ancient

A A map representing the whole surface of the Earth (or a greater part of it), made before the fall of the western Roman Empire in the later years of the 5th century.

B The Babylonians made some of the earliest known maps, although few are extant. A clay tablet bearing a cuneiform inscription from Sippar, central Iraq (British Museum, Department of Western Asiatic Antiquities, no.92687), shows a world map identified by the inscription as representing the four "shores" of the world, and dates from the 7th or 6th century BC, but a note at the end of the tablet states that it was copied from an earlier tablet (Fig. 6). It ranks as the only surviving world map of the pre-Christian era. Babylon is near the centre, and a river, probably the Euphrates, descends from the northern mountains. A rectangular area across the bottom may represent the Persian Gulf (as it is known today). East of Babylon lies the land of Assyria. The disc of the earth is encircled by the Bitter River (the Cosmic Ocean). Outside lie the "Eight Regions," described in detail on the reverse of the tablet.

Fig. 6 – Babylonian map of the world, 7th or 6th century BC, on a clay tablet from Sippar (central Iraq); the only surviving world map of the pre-Christian era. By permission of the Trustees of the British Museum, Department of Western Asiatic Antiquities, 92687. (World map, ancient, 1.2320a; Map surface, 6.1310)

The description of the world by the Greek geographer Strabo (*c.* 67 BC-AD 23), with its report of remote islands "at the ends of the Ocean," may follow the Babylonian concept of the cosmos. A much earlier design of the cosmos, dating from about 3500 BC, appears in the remains of a large fresco painting at Teleilat Ghassul, Jordan, and is believed to belong to the Babylonian tradition.

Greek world maps before Ptolemy's time (2nd century AD) are not extant, but some are described in the Illiad (xviii.478-617). Homer relates that Achilles' shield included on one of the bronze plates an engraving of the earth, sky, and sea, and round this whole was the river ocean. Whether such an object was ever made is not known. The Dura Europos shield — the only shield map known — dates from much later, from the period immediately preceding AD 260 (see Road map, 1.1810d). Homer is said to have lived on the west coast of Asia Minor, and the city of Miletus on that coast was the birthplace of Greek mapmaking. It would follow therefore that the Greeks might well see Homer as the father of Greek geography (Dilke, p.20). According to the geographer Eratosthenes, the earliest Greek map of the world was made by Anaximander of Miletus (early 6th century BC). Strabo and Agathemerus (date unknown) described it as a *pinax,* i.e. a painted (or sometimes a bronze) panel. The progress of further mapmaking at Miletus is indicated by the story in Herodotus that in 499-8 BC Aristagorus, tyrant of Miletus, visited mainland Greece carrying with him a *pinax* (bronze tablet) engraved with a map of the world.

Eratosthenes (*c.* 275-194 BC), was the first to make a scientifically drawn world map. It was based on a method of measuring the size of the earth and displayed parallels and meridians, although the grid was irregular. The inhabited world (oecumene) was divided into areas called "sphragides." This map influenced the development of world maps up to the renaissance. The mathematician and astronomer Hipparchus of Rhodes (fl.146-127 BC), voicing various criticisms of Eratosthenes, proposed a revision to a more regular system of divisions. Claudius Ptolemy (*c.* AD 90-168) explained in his Geography, *c.* AD 150, how to make a map of the globe according to three mathematical projections. Whether the world maps contained in manuscripts of Ptolemy were copied at several removes from a map drawn for Ptolemy's Geography by Agathodaemon, an Alexandrian cartographer of uncertain date (perhaps sixth century, or he may have lived soon after Ptolemy's death), or whether they are based on Byzantine reconstructions is disputed. Manuscripts of the astronomical work, Ptolemy's "Handy Tables," dating from the 13th century but descended from an archetype of the Roman Empire, contain a miniature comprising a combined world map and wind rose (see Wind map, 3.232). The mapmaker may have been a Greek living in Upper Egypt in the 5th or 6th century (Dilke, p.170).

The concept of the earth as a sphere is said to have derived from Pythagoras, an Ionian who flourished in south Italy in 530 BC, although even in antiquity this was a matter of dispute (Kahn, pp. 115-18). The idea was developed by Parmenides of Elea in southern Italy (born *c.* 515 BC), who is reputed to have been the first to divide the spherical earth into five zones, one hot, two temperate, and two cold. It is probable that he illustrated these divisions on a map or globe (Dilke, p.25; see Zone map, 3.261). The problem of how to show the oecumene was a matter for discussion among geographers. The Homeric scholar Crates of Mallos (*c.* 150 BC) designed the oecumene in the form of a hemisphere (Crates' Orb). He divided the world into four symmetrical land-masses: Europe, Asia and the known part of Africa; to the south the Antoikoi, "dwellers opposite"; to the west the Perioikoi, "dwellers round"; and to the southwest the Antipodes (Dilke, p.36). Macrobius developed the system, which was adopted by some medieval mapmakers (see Mappa mundi, 1.2320b).

The Romans required maps for practical purposes, such as road maps for travel and military use. These were compiled into a map of the oecumene, planned by Marcus Vipsanius Agrippa (64 or 63-12 BC), acting emperor for Augustus, to foster Roman imperial expansion (Dilke, p.41). It was completed by Augustus after Agrippa's death in 12 BC and was displayed in Rome on the wall of the portico named after Agrippa. The map does not survive, but the elder Pliny (AD 23/24-79) made frequent references to it in his Natural History. Derivatives can be traced in medieval world maps as late as the 13th century.

In ancient China the concept of spherical movements of the celestial sphere centred on the earth dates from the 4th century BC. In the 1st and 2nd centuries AD the universe was likened to a hen's egg; the earth was "as round as a cross bow bullet (*yuan ju tan wan*; Needham, pp.216, 498-9). A tradition of religious cosmology — Buddhist and Jain — in the form of "wheel maps" developed in eastern Asia and may be compared to the Christian cosmology of medieval Europe (see Symbolic map, 1.192). Their origin may have been Indian or perhaps even Babylonian (Needham, pp.567-8). The legendary Mt Khun-Lun of Central Asia was usually the hub of the map. The cosmology of Tsou Yen in the 4th century BC postulated an encircling ocean without the central mountain. The scientific tradition remained predominant, however, in Chinese mapmaking.

C AUJAC, Germaine (1966): *Strabon et la science de son temps.* Paris, Société d'Édition "Les Belles Lettres."

BAGROW, Leo (1943): "The Origin of Pholemy's [sic] Geographia," *Geografiska Annaler, 3-4*, 318-87.

BRINCKEN, Anna-Dorothee von den (1973): "Die Klimatenkarte in der Chronik des Johann von Wallingford — ein Werk des Matthaeus Parisiensis?" *Sonderdruck aus der Zeitschrift "Westfalen,"* Bd 51, Hf 1-4, pp. 47-56.

BRINKEN, Anna-Dorothee von den (1985): "Mundus figura rotunda," in Ornamenta Ecclesiae. Kunst und Künstler der Romanik. *Sonderdruck aus dem Katalog zur Ausstellung des Schnütgen-Museums in der Josef-Haubrich-Kunsthalle,* Köln, pp.98-106.

DILKE, O A W (1985): *Greek and Roman Maps.* London, Thames and Hudson.

DILKE, O A W, and DILKE, M S (1985): "The Imprint of Ptolemy," *Geographical Magazine, 57,* 544-49.

KAHN, Charles H (1960): *Anaximander and the Origins of Greek Cosmology.* New York, Columbia University Press.

KLOTZ, A (1931): *Die geographischen Commentarii des Agrippa und ihre Überreste.* Leipzig, Dieterich (reprinted from *Klio, 24* (1931), 38-58, 386-466).

NEEDHAM, Joseph (1959): *Science and Civilisation in China,* vol.3. Cambridge, At the University Press.

NEUGEBAUER, O (1975): "A Greek World Map," in *Le monde grec: . . . Hommages à Claire Préaux,* ed. Jean Bingen *et al.* Bruxelles, pp.312-17 and Pl. III.2.

PAULY, August Friedrich von, and WISSOWA, Georg (1965): *Realencyclopädie classischen Altertumswissenschaft.* Stuttgart, Druckmüller. "Ptolemaios als Geograph," Supplement X, cols. 680-833.

UNGER, Eckhard (1937): "From the Cosmos Picture to the World Map," *Imago Mundi, 2,* 1-7.

1.2320B Mappa Mundi

A The term mappa mundi was used for medieval (usually circular) world maps, made between the fall of the western Roman empire in the later years of the 5th century AD and the end of the 15th century. The Latin word "mappa" means patch, napkin, or cloth.

B The two main types of mappae mundi made in the period from the 5th to the 10th centuries were derivatives of maps from classical Greece and Rome. One type, defined by M C Andrews as "hemispherical," shows the earth as two circular hemispheres which are divided into zones and presupposes a spherical earth (see Zone Map, 3.261). These diagrammatic maps were used to illustrate the text of Ambrosius Aurelius Theodosius Macrobius (395-436). In the Commentary on Scipio's Dream, a work in Cicero's *De re publica* (51 BC), Macrobius follows Crates and Cicero in their theory of an equatorial belt of water (Dilke, p.174). The six-zone maps are associated with this work of Macrobius or with the *De nuptiis Mercurii et Philologiae* of Martianus Capella (fl.410-439; Dilke, p.174). Early examples show two hemispheres, later examples only one comprising that of the known world.

A specialized type of hemispheric map is the climate map, as distinct from the zone map (3.261). The earth is divided into seven climates, belts of the sphere, as interpreted by Eratosthenes. Knowledge of the Greek climate maps was preserved and interpreted by Arabic geographers, whereas no early medieval Latin examples are known. With the medieval renaissance of the 12th century, Christians aided by Moslems and Jews translated the Arabic scientific texts into Latin, and the four extant climate maps derive from this period of the late Middle Ages. The earliest (12th century), by Petrus Alphonsus, a converted Jew from Huesca, Spain, shows the climate zones without further definition and follows Arabic models with Arin (or Arim) as the mythical world centre. Another of this type is found in the early editions of the works of Joannes de Sacrobosco (John of Holywood, fl.1220-1256), for example, *Opusculum spericum cum figuris optimis et novis* (Leipzig, *c.* 1500), but is somewhat altered from the manuscript original *c.* 1250. The maps of Petrus Alphonsus and Sacrobosco have south at the top. The third, by Pierre d'Ailly, 1410, with north at the top, shows the influence of the rediscovery of Ptolemy. The fourth and most unusual map is that in the Chronicle of John of Wallingford, infirmarer of St Albans (died 1258; British Library, Cotton MS. Jul.D.VII). He was a friend of Matthew Paris, fellow monk of St Albans, who was the author of much material in the chronicle and presumably also of this map. The map displays the climates and is inscribed with legends recording Christian and topical information. Although east is marked at the top of the page, the map is drawn sideways across the sheet, with south at the top of the map itself. This orientation, together with the marking of "Aren civitas," shows Arabic influence. A total of eight climates appears in the inhabited hemisphere. The cosmographical information given in the legends corresponds with the teachings of al-Idrisi the Arab geographer (1099-1164) and was known to Guillaume de Conches (*c.* 1080-*c.* 1154). The concept herein depicted of a *chlamys extensa* (chlamys is a short Greek mantle) as half of a hemisphere was the basis of Ptolemy's world map and became known in the Middle Ages through Macrobius.

The second main type of mappa mundi is the "T-O" map, oriented with east at the top. This belongs to what Andrews describes as the "oecumenical family." The O represents the oecumene, the T the axis of the Mediterranean, crossed by the meridian from the Don to the Nile. In its usual form it derived from the world map of Agrippa (after 12 BC), elaborated in the light of later Christian knowledge. The prototype of the Christian version was based on the scientific work of the Spanish scholar Isidore of Seville (560-636). The printed edition of Isidore's *Origines* (Augsburg, 1472), contains a T-O map which is the earliest known printed map. Some of the T-O maps deviate from the circular shape. Henry of Mainz's, *c.* 1110, is lenticular, perhaps to fit the page of the manuscript (Corpus Christi College, Cambridge). The map drawn in the year 1109 by the converted Moors at the monastery of Silos, in the diocese of Burgos, Old Castile, to illustrate a Commentary on the Apocalypse represents the earth as quadrangular.

The T-O maps were elaborated into works of increasing complexity. One of the earliest of these is the Anglo-Saxon world map, made at the end of the 11th century (British Library, Cotton MS. Tib. B.v.f.59) (see also Fig. 7). The largest, the Ebstorf and Hereford world maps, are encyclopaedic in content. The Ebstorf, *c.* 1235 (destroyed in 1943), is believed to be by Gervase of Tilbury, a 13th century provost of Ebstorf. The Hereford world map (preserved in Hereford Cathedral), made by Richard of Haldingham *c.* 1285, is the largest extant mappa mundi. G R Crone has shown it to be the last in a series of copies of a Roman original in the tradition of Agrippa (Crone, 1978, p.6). Many Christian features have been added in the course of the centuries. Jerusalem is at the centre, and the Garden of Eden in the orient is at the top.

The medieval style of world mapping with its emphasis on tradition remained separate from the evolution of the mariners' chart from the late 13th century onwards. The Catalan school of cartographers, mainly Majorcan, introduced a new element incorporating the results of Mediterranean charting and of explorations in Asia. The circular medieval world map was thus expanded as in the Catalan Atlas, *c.* 1375 (Bibliothèque Nationale, Paris, MS. espagnol 30), entitled "Mappa mondi . . . " (mappa-mundi, that is to say, the image of the world and of the regions which are on the earth and of the various peoples who inhabit it). The map incorporates data from the account by the Venetian traveller Marco Polo of his and his uncles' journeys across Asia to China, 1255-95.

A number of notable 15th century world maps illustrate the transition between the medieval mappae mundi and renaissance cartography: the Borgia world map *c.* 1430 (Vatican, Biblioteca Apostolica Vaticana, Borgia XVI), engraved on two plates of copper; three maps by Giovanni Leardo made at Venice in 1422, 1448, and 1452-53

Fig. 7 – Medieval world map, in a 13th century psalter, drawn on vellum. Christ, flanked by angels, is depicted embracing the earth, shown in T-O form. Descriptive text fills the three continents. This and the better known "Psalter map" on the recto date from about 1275 and were probably added to the manuscript at a later date. By permission of the British Library, Add. MS. 28681, f.9ᵛ. (Mappa mundi, 1.2320b; Religious map, 2.181; Map surface, vellum, 6.1310h)

89

(Verona, Biblioteca Communale; Vicenza, Museo Civico; Milwaukee, Wisconsin, American Geographical Society); the circular map of Andreas Walsperger made at Constance in 1448 (Vatican, Biblioteca Apostolica Vaticana), which has some similar delineations to the maps of Leardo and that of Andrea Bianco of Venice, 1436. The "Genoese" world map of 1457 (Florence, Biblioteca Nazionale Centrale) is lenticular and augments a medieval mappa mundi with material from nautical charts and from Ptolemy. The world map of Fra Mauro, a monk of Murano near Venice, which was completed in 1459 or 1460 and may be regarded as the final flowering of the mappa mundi, is elaborated with a host of new data and is revised with cognisance of, but without undue deference to, Ptolemy. Jerusalem is displaced westward from its usual central location and the longitudinal extents of Europe and Asia are now roughly correct. These maps mark the end of an era in the mapping of the oecumene.

C ANDREWS, Michael C (1926): The Study and Classification of Medieval Mappae Mundi," *Archaeologia, 75,* 61-76.

BRINCKEN, Anna-Dorothee von den (1983): "Weltbild der Lateinischen Universalhistoriker und – Kartographen," *Settimane di studio del Centro italiano di studi sull'alto medioevo* (Spoleto), *29,* 377-421.

BRINCKEN, Anna-Dorothee von den (1985): "Mundus figura rotunda," in Ornamenta Ecclesiae. Kunst und Künstler der Romanik. *Sonderdruck aus dem Katalog zur Ausstellung des Schnütgen-Museums in der Josef-Haubrich-Kunsthalle,* Köln, pp. 98-106.

BROWN, Lloyd A (1949): *The Story of Maps.* Boston, Little, Brown and Co.

CRONE, Gerald R (1954): *The World Map of Richard of Haldingham.* London, Royal Geographical Society.

CRONE, Gerald R (1978): *Maps and Their Makers: An Introduction to the History of Cartography,* 5th ed. Folkestone, Kent, Dawson; Hamden, Connecticut, Archon Books.

DESTOMBES, Marcel, ed. (1964): *Mappemondes: A.D. 1200-1500,* vol. 1 of *Monumenta Cartographica Vetustioris Aevi.* Amsterdam, N. Israel.

DILKE, O A W (1985): *Greek and Roman Maps.* London, Thames and Hudson.

KIMBLE, George H T (1938): *Geography in the Middle Ages.* London, Methuen & Co.

MILLER, Konrad (1895-1898): *Mappaemundi, die ältesten Weltkarten,* 6 vols. Stuttgart.

SHIRLEY, Rodney W (1983): *The Mapping of the World: Early Printed World Maps, 1472-1700.* London, Holland Press.

UHDEN, Richard (1931): "Zur Herkunft und Systematik der mittelalterlichen Weltkarten," *Geographische Zeitschrift* (Leipzig), *37,* 321-40.

WOODWARD, David (1985): "Reality, Symbolism, Time, and Space in Medieval World Maps," *Annals of the Association of American Geographers, 75,* 510-21.

WRIGHT, John K (1925): *The Geographical Lore of the Time of the Crusades.* New York, American Geographical Society. (Reprinted, Dover Publications. NY, 1965.)

GROUP 2

Maps of Human Occupation
and Activities

2.011 Archaeological Map

A A map drawn to locate archaeological sites and monuments so as to relate them to one another and to the contiguous landscapes.

B The seed of this cartographic concept probably first formed in maps produced to illustrate the Holy Places in Jerusalem and Palestine at the time of the Crusades, as for example, in Matthew Paris's map of Palestine showing Jerusalem and Acre (*c.* 1250). When a secular antiquarianism developed among the archaeologically-minded humanists of the 15th century the seed germinated, and an early expression of this is to be found in the bird's-eye view map drawn to accompany Flavius Biondo's account of Rome's antiquities *De Roma instaurata* which was written in 1444-6. It is known only in copies drawn later, one by Alessandro Strozzi in 1474. Other copies were incorporated into Pietro del Massaio's manuscripts of Ptolemy's *Geography* in 1469, 1472 and the 1490s. A wider archaeological scene is depicted in Eufrosino della Volpaia's engraved map of the Roman Compagna of 1547.

In Renaissance Britain archaeological monuments excited the interest of topographical writers, as for example, William Camden, and actually appeared on maps, including John Norden's maps of Cornwall. A significant step in the evolution of the archaeological map was Timothy Pont's inclusion of the Antonine Wall on one of his maps of Scotland in the last decade of the 16th century.

A surge of interest in British and Roman antiquities arose after the appearance of Gibson's enlarged and annotated English edition of Camden's *Britannia* in 1695. This

91

volume placed in the hands of the country gentry, doctors and clergymen a basic summary against which they could set the results of their own local enquiries in archaeology. Into this ferment of interest entered a number of eminent archaeologists, and leading the way in the 1720s was William Stukeley whose percipient eye for the geographical aspects of archaeology set the cartographic pattern to be followed by Alexander Gordon in 1726 and John Horsley in 1732. In their works the appreciation of antiquities in relation to one another and to the landscape expressed in map form is an essential element. With these archaeological field surveyors emerged the true archaeological map, articulated in words by their successor, William Roy, who declared that during the Scottish Survey the Antonine Wall "was observed in the ordinary way but without the wall itself becoming the principal object . . . it was therefore judged proper, in 1755, to survey accurately the line of this old intrenchment."

Incorporated within the idea of the archaeological map are the two concepts of location and space relationships, and it is these with particular reference to sites which in the 19th century were formulated into an interpretative tool for the archaeologist. One of the first maps specifically constructed to express distribution of sites was that of inscribed British coins by John Yonge Akerman published in 1849. Since then the distribution map has become an essential tool for the study and exposition of problems in archaeology and the graphic representation of areal distribution a canon of its methodology.

C AKERMAN, John Yonge (1849): "On the Condition of Britain from the Descent of Caesar to the Coming of Claudius, Accompanied by a Map . . . Shewing the Finding of Indigenous Coins," *Archaeologia, 33,* 177-190.

CAMDEN, William (1586, 1587, 1590, 1594, 1600): *Britannia.* London.

GIBSON, Edmund (1695): *Camden's Britannia.* London, 959.

HORSLEY, John (1732): *Britannia Romana: or the Roman Antiquities of Britain.* London, 158, N1-N10; 176, N1-N5.

MACDONALD, Sir George (1933): "John Horsley, Scholar and Gentleman," *Archaeologia Aeliana* (Newcastle-upon-Tyne), 4th series, *X,* 1-57.

NORDEN, John (1604): "Speculi Britanniae Pars, or A Topographicall and Historical Description of Cornwall." Library of Trinity College, Cambridge, and British Library Harleian MS.6252.

PARIS, Matthew (*c.* 1250): Map of Palestine showing Jerusalem and Acre. British Library, Royal MS. 14.C.vii, f.4^v and f.5.

PIGGOTT, Stuart (1950): *William Stukeley, An Eighteenth-Century Antiquary.* Oxford, Clarendon Press.

PONT, Timothy (*c.* 1590): Sterlingshyre. National Library of Scotland, Pont Maps 32.

ROY, William (1773): *The Military Antiquities of the Romans in North Britain.* IId, 157. British Library, Kings MS.248, f.43.

SCAGLIA, G (1964): "The Origin of an Archaeological Plan of Rome by Alessandro Strozzi," *Journal of the Warburg and Courtauld Institutes, 27,* 137-63.

STUKELEY, William (1743): *Abury, A Temple of the British Druids.* London.

VOLPAIA, Eufrosino della (1914): *Mappa della Compagne Romana del 1547 di Eufrosino della Volpaia, riprodotta dall'unico esemplare esistente nella Biblioteca Vaticana, con introduzione di Thomas Ashby.* Roma.

WEISS, Roberto (1969): *The Renaissance Discovery of Classical Antiquity.* Oxford, Blackwell, 90-94.

2.0210 Boundary Map

A A map showing the limits of properties or territories; or one specifically produced for the purpose of delineating a boundary line or frontier.

B Surveys made for official or private purposes to delimit the boundaries of estates and other properties date from the ancient civilizations of Egypt, Mesopotamia and China (see Estate map, 2.0210c, and Property map, 2.0210e). Administrative maps (2.0210a) recording the boundaries of subdivisions of a state are equally early, and became important with the rise of the nation state in Europe in the 15th century. Jurisdictional maps (2.0210d), which are similar to administrative, express territorial limits with special reference to ecclesiastical, governmental and military activities. Cadastral maps (2.0210b), also early in origin, derive their name from the Roman cadastre, and emerged as a major element in the national mapping of European states in the 17th and 18th centuries. They represent the most accurate depiction of territorial ownership designed for fiscal control.

Maps made to record actual or proposed boundaries and frontiers were well known in Roman times, and Roman land surveyors were often required to solve boundary disputes. What appears to be the earliest boundary map in England illustrates the determination of a boundary in Sherwood Forest, *c.* 1150, and was probably drawn *c.* 1300 (Map in the Kirkstead Psalter, MS., Beaumont College, Old Windsor). An unusual boundary is depicted in a map of the Isle of Thanet in the 15th century chronicle of Thomas of Elmham, a monk of St Augustine's Abbey at Canterbury (Trinity Hall, Cambridge, MS 1, f.28v.). The map illustrates the "run of the deer," the pet deer of Queen Domneva of Mercia, who was given as much land as it encompassed in a single run (Harvey).

One of the earliest geopolitical boundary maps is the manuscript planisphere drawn by an unknown Portuguese cartographer in 1502 for Alberto Cantino, secret agent of Ercole d'Este, Duke of Ferrara, Italy. It shows the division of the world between the Spanish and Portuguese empires as determined by the Treaty of Tordesillas, 1494 (Biblioteca Estense, Modena, Italy). In North America probably the most famous of the boundary surveys was that made by the two English mathematicians Charles Mason and Jeremiah Dixon between 1763 and 1767. Known henceforward as the Mason-Dixon Line, the boundary determined the frontier between the provinces of Maryland and Pennsylvania. The line was recorded in 1768 on a long strip map entitled "A Plan of the Boundary Lines between the Province of Maryland and the Three Lower Counties on Delaware . . ." Another map "A Plan of the West

Line or Parallel of Latitude, which is the Boundary between the Provinces of Maryland and Pennsylvania," 1768, completed the line (Schwartz and Ehrenberg; Christie's New York sale, Thursday, 1st April 1982).

In the 18th century maps were accepted as essential instruments in the arbitration of territorial disputes. The most important political map in North American history is the "Red-lined Map," known also as "King George III's Map," which comprises "A Map of the British Colonies in North America," by John Mitchell, London, 1755 [1775], with manuscript additions by Richard Oswald, c. 1782. The map records the official British interpretation of the negotiations at Paris, 1782, for the establishment of the United States of America in the Treaty of Paris, 1783 (British Library, Map Library, K.118.d.26; K.Top. CXVIII.49b).

C BRITISH LIBRARY (1975): *The American War of Independence 1775-83.* London, British Museum Publications.

DILKE, O A W (1971): *The Roman Land Surveyors: An Introduction to the Agrimensores.* Newton Abbot, David & Charles.

HARVEY, Paul D A (1980): *The History of Topographical Maps.* London, Thames and Hudson.

HYDE, Charles Cheney (1933): "Maps as Evidence in International Boundary Disputes," *American Journal of International Law, 27,* 311-16.

PRICE, Derek J (1955): "Medieval Land Surveying and Topographical Maps," *Geographical Journal, 121,* 1-10.

SCHWARTZ, Seymour I, and EHRENBERG, Ralph E (1980): *The Mapping of America.* New York, Harry N Abrams.

WALLIS, Helen (1968): "The Use of Early Maps in International Disputes over North American Territories," 21st International Geographical Congress, *Abstracts of Papers.* Calcutta, National Committee for Geography, 309.

WEISSBERG, Guenter (1963): "Maps as Evidence in International Boundary Disputes: A Reappraisal," *American Journal of International Law, 57,* 781-803.

2.0210A Administrative Map

A A map made or used by government employees for administrative purposes. More particularly, a map placing emphasis on the names and boundaries of the administrative subdivisions of a state.

B The first of these definitions would doubtless apply to the map of the Changsha area of China, showing military posts and numbers of households (c. 168 BC). It would also apply to a large proportion of the maps made since that time, many of which have nothing in their appearance to distinguish a public from a private origin.

Political boundaries were certainly shown on Roman maps. Their existence has been deduced for the lost map of 20 BC by Marcus Vipsanius Agrippa, and they are probably represented by the straight-line boundaries marked on certain medieval world maps, such as those of Richard of Haldingham (Hereford map, *c.* 1285) and Ranulphus Higden (*c.* 1350).

Modern administrative maps begin with the rise of the nation state. An administrative framework is implicit in the order of 1460 to the governors of cities, territories and castles in the Venetian Republic to supply maps of the areas under their jurisdiction. An early British example is a letter to Henry VIII in 1526 enumerating the counties of Ireland "as by the platt may appear" (*Letters and Papers, Foreign and Domestic, of the Reign of Henry VIII,* Vol. 4, Pt. 2, 1077). Some of Christopher Saxton's topographical maps of the English counties show the boundaries of minor territorial divisions; but as topographical maps became increasingly crowded with the results of the more detailed surveys made in the middle 16th century, the value of skeleton maps emphasizing political boundaries became clearer. An example is the manuscript map of 1579 showing the territories from which the new Irish counties of Wicklow and Ferns were proposed to be formed.

Published official administrative diagrams appear in the early 19th century and seem to have originated as small-scale indexes to the sheet lines of large-scale maps, which, like many index maps, were soon found to have a value of their own. In England and Wales the redistribution of Parliamentary representation and political power proposed in the Reform Bill of 1831 and made law in 1832 involved the drawing of the boundaries of the new county divisions and boroughs which were to return members to Parliament. The maps produced by the Boundaries Commission were prepared by Lieut. Robert K Dawson and are among the earliest civil maps to be printed by lithography in Great Britain.

C DAWSON, R K (1831-32): Maps to accompany "Reports from Commissioners on Proposed Division of Counties and Boundaries of Boroughs," British Parliamentary Papers, *Accounts and Papers, 11-14,* Vols. 38-41.

DUNLOP, R (1905): "Sixteenth-Century Maps of Ireland," *English Historical Review, 20,* 309-37.

GALLO, Rodolfo (1955): "A 15th Century Military Map of the Venetian Territory of Terrafirma," *Imago Mundi, 12,* 55-7.

HSU, Mei-Ling (1978): "The Han Maps and Early Chinese Cartography," *Annals of the Association of American Geographers, 68,* 45-60.

2.0210B Cadastral Map

A A map which delineates property boundaries (MDTT 822.2(a)); more specifically a map which shows the boundaries of land parcels in a given administrative unit (e.g. parish, county, etc.) and usually provides information about their location, size and value for fiscal or other government purposes (see also Property map, 2.0210e).

B The earliest cadastral plans are thought to have been made in ancient Babylon and Egypt during the 2nd and 3rd millennia BC. Possibly the earliest surviving example is a plan drawn on a clay tablet *c.* 2500-2300 BC excavated at Nuzi in central

Mesopotamia (now Iraq) in 1930-31. The plan shows the location of an estate measuring 354 *iku* (about 120 hectares) of cultivated land which belonged to a landowner named Azala. The Romans used cadastral surveys for determining land ownership or tenancy and for assessing tax liability. At Orange in France substantial fragments of a Roman cadastral map on stone, the Arausio Cadasters, were discovered (mainly 1949-51) relating to the Roman colony of Arausio. A monumental inscription of AD 77, also found at Orange, records that the Emperor Vespasian ordered a map to be set up with a record of the annual rental due on each "century" or land unit. Whether all three cadasters are of that period is disputed.

In Japan following the Imperial Edict of 646 cadastral surveys were carried out in the years 742, 755, 773, and 786, and maps with considerable detail – the so-called "Denzu" – were made. However, none survives from this period (Ramming, 1937).

Although the practice of attaching estate or manorial plans to written descriptions of private estates (*terriers*), became widespread in 16th- and 17th-century Europe, the use of cadastral plans (i.e. plans for fiscal and government purposes) does not seem to reemerge until the 18th century. One of the earliest cadastral surveys was that compiled on a scale of 1:2372 by the Intendant of the Duchy of Savoy in 1728-38. Each parcel of land is numbered according to an accompanying register which gives the name and status of the owner, along with the area, type, and quality of the land. Such surveys culminated in the national cadastral map of France (*cadastre napoléonien*) ordered by Napoleon I in 1807.

In North America, on the passing of the Land Ordinance of 1785, extensive cadastral mapping followed the opening of the public domain of the United States for private ownership. A rectilinear method of survey was used in the western three-quarters of the United States and Canada to facilitate settlement and mapping. The maps came to form an important basis for local administration in both urban and rural areas.

C BINNS, Sir Bernard (1953): *Cadastral Survey and Records of Rights in Land.* Rome, FAO Agricultural Studies No 18.

DALE, P F (1976): *Cadastral Surveys within the Commonwealth.* London, HMSO.

DILKE, O A W (1971): *The Roman Land Surveyors: an Introduction to the Agrimensores.* Newton Abbot, David & Charles, 157-77.

DILKE, O A W (1973): "The Arausio Cadasters," in *Vestigia: Beiträge zur alten Geschichte, 17* (Akten des VI. Internationalen Kongress für greich. und lat. Epigraphik, München, 1972), 455-57.

GUICHONNET, P (1955): "Le Cadastre savoyard de 1738 et son utilisation pour les recherches d'histoire et de géographie sociales," *Revue de Géographie Alpine, 43,* 255-98.

HERBIN, R, and PEBEREAU, A (1953): *Le Cadastre Français.* Paris, Les Éditions Francis Lefebvre.

MEEK, Theophile James (1935): *Excavations at Nuzi: Vol. 3, Old Akkadian, Sumerian, and Cappadocian Texts from Nuzi.* Cambridge, Massachusetts, Harvard Semitic Series No. 10, pp. xvii-xviii.

PIGANOIL, André (1962): *Les Documents cadastraux de la colonie romaine d'Orange.* Paris, Gallia, Supplement No. 16.

RAMMING, M (1937): "The Evolution of Cartography in Japan," *Imago Mundi, 2,* 17-21.

SALVIAT, F (1977): "Orientation, extension et chronologie des plans cadastraux d'Orange," *Revue Archéologique de Narbonnaise, 10,* 107-18.

THROWER, Norman J W (1966): *Original Survey and Land Subdivision: A Comparative Study of the Form and Effect of Contrasting Cadastral Surveys.* Association of American Geographers, Monograph No. 4. Chicago, Rand McNally.

2.0210C Estate Map

A A map of a land property.

B The oldest known map, a clay tablet *c.* 2500-2300 BC found at Nuzi in central Mesopotamia (now Iraq), might be called an estate map, for its purpose seems to have been to record the position and size of an estate. Other Mesopotamian maps from the late third to early first millennium BC show boundaries of estates and fields; some are clay tablets that served as title deeds, while by the second millennium some were carved on *kudurru*, stones set up as a public statement of ownership. An estate plan of the Ptolemic period in Egypt survives (Jouguet). It dates from 259-258 BC, has west at the top, and is rectangular with measured plots subdivided by ditches. From the Roman empire plans of centuriated fields are the only known sort of estate map: at Orange in France are preserved fragments of three carved scale plans of the 1st and 2nd centuries of the sort that were drawn up by official surveyors, while the illustrations to surveyors' manuals, compiled in the 5th century, also show schemes of centuriation in the form of pictorial diagrams not drawn to scale.

Early records of maps in China do not refer to estate maps, which lie outside the fields of military use and imperial administration for which mapping seems to have been primarily developed there. However, maps of the estates of particular temples are among the earliest known Japanese maps, dating from the 8th century. Many of the maps drawn in Aztec style in 16th-century Mexico were in effect estate maps, showing the areas claimed by particular chiefs together with their genealogies; these were occasioned by repartitions of lands after the Spanish conquest, but earlier Aztec cartography may well have included estate maps.

When estate maps were drawn in medieval Europe, this was usually to illustrate some point that was in dispute, such as that of areas in IJzendijke and Oostburg in Zeeland, drawn for St Peter's Abbey, Ghent, following an agreement on tithes in 1358, or that of Inclesmoor in Yorkshire, drawn for the Duchy of Lancaster in 1405-8 to show claimed rights of peat cutting. This probably remained true up to the mid-16th century; the second half of the century, however, saw maps coming into use in the normal running of landed estates, particularly in replacing the written descriptions that had hitherto been the end-product of an estate survey. The maps drawn for All Souls College, Oxford, in the 1580s and 1590s and the 68 maps of the estates of the former abbey of Rijnsburg prepared by Simon van Buningen in 1598 are notable early examples from England and the Netherlands.

C BURLAND, C A (1960): "The Map as a Vehicle of Mexican History," *Imago Mundi, 15*, 11-18.

DILKE, O A W (1971): *The Roman Land Surveyors: an Introduction to the Agrimensores.* Newton Abbot, David & Charles.

HARVEY, Paul D A (1980): *The History of Topographical Maps: Symbols, Pictures and Surveys.* London, Thames and Hudson.

JOUGUET, Pierre, ed. (1928): *Papyrus Grecs,* Tome I, Fasc. IV, Pl. I. Paris, Institut Papyrologique de l'Université de Lille.

KING, Leonard William, ed. (1912): *Babylonian Boundary-Stones and Memorial-Tablets in the British Museum.* London, British Museum, 81-2, 85-6, 99-101, pl. 14, 16, 17.

LANGDON, S H (1916): "An Ancient Babylonian Map," *Museums Journal* (Philadelphia), *7*, 263-8.

MEEK, Theophile James (1935): *Excavations at Nuzi: Vol. 3, Old Akkadian, Sumerian, and Cappadocian Texts from Nuzi.* Cambridge, Massachusetts, Harvard Semitic Series No. 10, pp. xvii-xviii.

RAMMING, M (1937): "The Evolution of Cartography in Japan," *Imago Mundi, 2*, 17-21.

2.0210D Jurisdictional Map

A Those maps which express territorial limits specifically pertaining to ecclesiastical, governmental, and military activities. Similar to Administrative map (2.0210a).

B The maps in the Notitia Dignitatum, a list of Roman military and civil administrations and their spheres, originally compiled in the late 4th or early 5th century, were intended to show headquarters and jurisdictions in pictorial form. For example, the map of Britain shows five walled cities as the five subdivisions of Britain.

Ecclesiastical divisions are evident on several T-O maps from the 11th through the 14th centuries. Spheres of denominational influence are recognized by vague divisions with supplemental illustrations of religious buildings. Maps in this classification will show divergent interpretations of enclosures, which reinforces the idea that they are not scientific delineations but artificial zones of influence based on patriotic or religious points of view.

The woodcut map of the environs of Nuremberg, printed by Glogkendon in

Nuremberg in 1492 and identified as the work of Erhard Etzlaub, shows territorial boundaries in red and appears to be the first European political map of clearly delimited areas of dynastic jurisdiction (Krüger). Abraham Ortelius in his *Theatrum Orbis Terrarum* (Antwerp, 1570), included a map "Salisburgensis iurisdictionis ... descriptio," by Marco Secznagel of Salzburg, with accompanying text on the history of the diocese.

A series of charts of English coastal jurisdictions date from 1604. The earliest, by T.H. [Thomas Hood] and evidently made for Trinity House, is endorsed "A feare Carde of y[e] Coastes of England wth the kinges armes." It defines the "Kings Chambers," within which James claimed territorial sovereignty. The "chambers" numbered twenty-six and were defined by custom as formed by lines joining headlands (Hatfield House, Hertfordshire, MS. CPM I.76). The derived engraved chart was printed with an accompanying broadside on 4 March 1604/5.

In the 18th century international territorial disputes in North America led to the mapping of rival claims to lands and fisheries. For example, "A New Map of Nova Scotia and Cape Breton Island" by John Green, which appeared anonymously with accompanying text over the imprint of Thos. Jeffreys from May 1755 and ran to many editions up to 1794, depicted in detail the various claims in that region.

In China the long tradition of mapmaking and the needs of a centralized bureaucracy stimulated the production of a host of official maps in local topographies or district gazetteers as well as maps of larger regions. The "Ch'ien-k'un wan-kuo ch'üan-t'u ku-chin jen-wu shih-chi" by Liang Chou published at Nanking in 1593, although purporting to be a world map, was primarily an administrative map of China made for mandarin officials. A woodcut map of Canton province *c.* 1739 by an anonymous compiler by order of the emperor shows administrative divisions with information derived according to the author from 88 maps of separate parts of the province (Philip Robinson collection).

C CLEMENTE, G (1968): *La "Notitia Dignitatum."* Cagliari, Sarda Fossataro.

FEBVRE, Lucien P (1925): *A Geographical Introduction to History.* New York. (Reprinted Westport, Connecticut, 1974.)

GOODBURN, Roger, and BARTHOLOMEW, P (1976): *Aspects of the Notitia Dignitatum.* Oxford, British Archaeological Reports: supplementary series No. 15.

KRÜGER, Herbert (1951): "Erhard Etzlaub's *Romweg* Map and Its Dating in the Holy Year of 1500," *Imago Mundi, 8,* 17-26.

SANTAREM, M F de Barros, Vicomte de (1842-53): *Atlas composé de mappemondes* ... Paris.

SEECK, O, ed. (1876): *Notitia Dignitatum.* Frankfurt am Main, Minerva (reprinted in 1962).

SKELTON, R A, and SUMMERSON, John (1971): *A Description of Maps and Architectural Drawings in the Collection Made by William Cecil First Baron Burghley, Now at Hatfield House.* Oxford, Roxburghe Club, cat. no. 2, pl.15.

2.0210E Property Map

A A map, usually on a large scale, which shows the location, size, and boundaries of a plot or plots of land and distinguishes ownership (see also Cadastal map, 2.0210b; County atlas, 8.0110a).

B Basic to the addition of the adjective "property" to the word "map" must be the sense of ownership and possession of a circumscribed plot or plots of land. Such a map must therefore either have the owner's name inscribed on the individual plot or, by appropriate symbolism, accredit plots to names appearing in a legend or separate list. Thus defined, property maps are among some of the oldest maps known; for example, those extant on Mesopotamian clay tablets (their usual form of stationery), dating from the third millennium to the 8th century BC. It is clear also that much of the surveying undertaken by the Roman *agrimensores* had as its main aim the production of property maps (see Estate map, 2.0210c).

Local maps in medieval Europe, although not numerous, were made to provide supporting evidence in legal disputes over the ownership of land. In the middle of the 15th century European owners of property began to employ surveyors; examples are the plan of the Benedictine Abbey of Chertsey, Middlesex, 1432, from the Cartulary of the Abbey (London, Public Record Office E.164/25, m 222) and the anonymous map of the lands of the Benedictine Abbey of Honau-Wanzenau lying on both banks of the Rhine, *c.* 1450, indicating ownership and land use (Grenacher). In the 16th century property surveys became much more common. An example in the Netherlands is the plan of the environs of Alkmaar by Pieter Jacobsz and Symon Meeuwiz of Edam and Maerten Cornelisz, 1532, recording properties (The Hague, Algemeen Rijksarchief).

In London two meticulous surveys of the properties of the Clothworkers Company in 1612 and of Christs Hospital, the city's orphanage, in 1612-13 were undertaken by Ralph Treswell the elder (Schofield).

Destruction by fire demanded a special type of official survey to record the ownership of destroyed properties and to settle the claims of owners. After the great Fire of London in 1666 John Leake and others made "An Exact Surveigh of the streets lanes and churches contained within the ruines of the City of London ... by the order of the Lord Mayor, Aldermen, and Common Councell of the said city." The reduced version on one sheet bearing this title was engraved by Wenceslaus Hollar in 1667. A notable plan of properties in Vienna, based on the cadastral survey, is "Grundriss der Kaiserl: Königl: Haupt und Residenz Stadt Wien," by Maximilian von Grimm, 1799 (Vienna, Artaria et Compagnie). This gives the numbers of properties and distinguishes buildings and houses in colour as "Kaiserl. Königlich," "Freye Häuser," "Magistratisch."

When such European ideas of personal property were subsequently carried to the New World a similar need for a cartographic record of land ownership emerged. In the United States, during the 19th century, this record was maintained in the land office or the office of the county surveyor. These data became embodied in more elaborate fashion in county land ownership wall maps and then (from the 1860s) in county atlases (see 8.0110a) which record property location, size, limits, and ownership (Figs. 24 and 25).

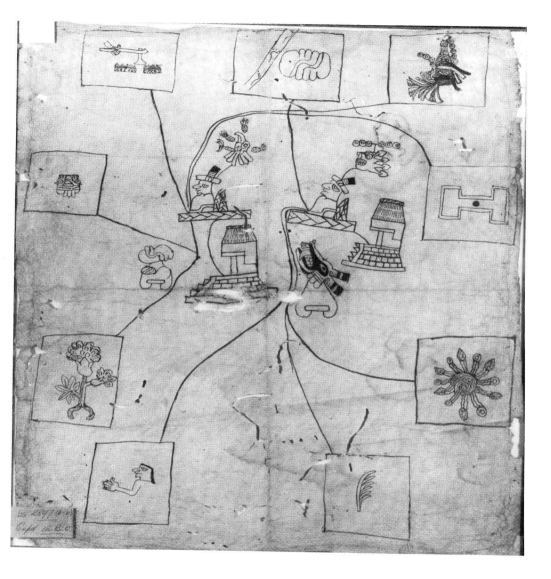

Fig. 8 – *Aztec map of estates in the state of Pueblo, Mexico, c. 1600, showing the division of nine pieces of land between the two chiefs of the towns of Itzcuintepec and Teteltepec. The row of footprints at the top was the Aztec conventional sign for a road, and may also represent the passage of time. Painted on amatl paper in black ink outline. By permission of the Trustees of the British Museum, Department of Ethnography, Egerton MS. 2897.(1). (Property map, 2.0210e; see also Plan, 1.1630; Conventional sign, 5.033; Map surface, 6.1310)*

The demands of commerce in the middle years of the nineteenth century led to the growth of another kind of property map, the insurance plan (2.091). An early example for London was Loveday's London Waterside Surveys for the use of Fire Insurance companies, merchants, brokers, agents, wharfingers, granary-helpers, etc., made for the Phoenix Fire Office, London, 1857 (British Library, Maps 4.b.1.). The firm of Charles E Goad which started in Canada and moved to Hatfield, Hertfordshire, in England then took over the field, producing insurance plans showing property ownership and land-use in Canada, Great Britain, Constantinople, Smyrna, and Egypt. Its counterpart in the United States of America was the Sanborn Map Company, whose fire insurance maps covered 12,000 cities and towns in the United States in the second half of the 19th century and early in the 20th century. The maps are on a scale of 50, 100, or 200 feet to one inch and denote individual buildings, their type of construction, and their use.

In England, from varying dates after the middle of the 19th century, local authorities have required the submission of maps for the construction of new buildings and for alterations to existing buildings.

An early property map from Mexico now in the Geography and Map Division of the Library of Congress is the "Oztoticpac Lands Map" of Texcoco, dated c. 1540. This Aztec pictorial document, with glosses in Nahuatl and Spanish, includes an estate plan and the subdivision of an agricultural area into individual plots with the names of the Indian families who farmed them (Cline). A more schematic Aztec map, c. 1600, shows the division of nine pieces of land between the two chiefs of the towns of Itzcuintepec and Teteltepec (British Museum, Department of Ethnography, Egerton MS 2897.(1.)) (Fig.8).

C CLINE, Howard F (1972): "The Oztoticpac Land Map of Texcoco, 1540," in *A la Carte: Selected Papers on Maps and Atlases* (Walter W Ristow, compiler). Washington, DC, USGPO, pp. 5-33.

DILKE, O A W (1971): *The Roman Land Surveyors: an Introduction to the Agrimensores.* Newton Abbot, David & Charles.

GRENACHER, Franz (1964): "Current Knowledge of Alsatian Cartography," *Imago Mundi, 18,* 60-77.

SCHOFIELD, John (1983): "Ralph Treswell's Surveys of London Houses c. 1612," in Tyacke, Sarah, ed., *English Map-Making 1500-1650.* London, British Library, pp. 85-92.

THROWER, Norman J W (1961): "The County Atlas of the United States," *Surveying and Mapping, 21,* 365-73.

THROWER, Norman J W (1966): *Original Survey and Land Subdivision: A Comparative Study of the Form and Effect of Contrasting Cadastral Surveys.* Association of American Geographers, Monograph, no. 4. Chicago, Rand McNally.

UNGER, Eckhard (1935): "Ancient Babylonian Maps and Plans," *Antiquity, 9,* 311-22.

2.041 Disease Map

A A map showing the incidence or spread of a disease. See also Poverty map (2.162) and Sanitary map (2.191).

B The earliest maps of the incidence of disease seem to be those concerning the occurrence of yellow fever on the eastern seaboard of the United States at the end of the 18th century. Lloyd G Stevenson (1965) lists several so-called "spot" maps prepared to bolster arguments of both the contagionists, who thought yellow fever an imported disease, and the anticontagionists, who thought it endemic, arising from local filth. The earliest were two maps by Valentine Seaman published in 1798 showing by dots and circles individual occurrences of yellow fever in waterfront areas of New York. The next such map appears to be the one by Felix Pascalis in 1819. Similar maps, as cited by Stevenson, mostly of yellow fever, are by the New York Board of Health in 1820; of Natchez, Mississippi, by Samuel A Cartwright in 1826; and of Gibraltar and the village of Catalan Bay by Nicolas Chervin, Pierre Charles Alexandre Louis, and Armand Trousseau in 1830. Maps of the incidence of yellow fever continued to appear in the United States and were followed in both the United States and Europe by similar maps of the incidence of cholera.

The first pandemic of cholera began in India in 1817, and several maps of that area appeared during the following decade showing the occurrence and spread of the disease. In the next several years numerous maps were published, ranging in coverage from individual cities to Eurasia and North America; Jarcho lists thirty-two such maps made by 1833. He points out that all the early maps were of epidemic diseases, particularly yellow fever and cholera, and observes that endemic diseases, such as tuberculosis, offered no comparable stimulus to mapping.

The most famous cholera map – a spot map – was that of Dr. John Snow, published in his book *On the Mode of Communication of Cholera.* It showed by means of black dots the distribution of deaths from cholera in the neighbourhood of Broad Street and Golden Square, London, from 19 August to 30 September 1854, and indicated the relationship of the deaths to use of the Broad Street pump, which proved to be the source of the outbreak through its contaminated water. Snow's mapping of the distribution of deaths on this and other maps was a major factor in his discovery and demonstrated that cholera was a water-borne disease.

The early disease maps were in general of three types: "spot" maps showing the occurrence of individual cases, shaded maps showing frequency of incidence, and route maps showing the spread of a disease. Beginning in the early 1830s disease maps became more elaborate, with many of them accompanying investigations into the probable environmental causes of epidemics, leading to a class called sanitary maps (2.191). By the 1840s disease maps had become quite sophisticated. Examples are Petermann's 1848 cholera map of the British Isles and Berghaus's 1848 world map of diseases in his *Physikalischer Atlas.*

C BERGHAUS, Heinrich (1845, 1848): *Physikalischer Atlas oder Sammlung von Karten . . .* Gotha, Justus Perthes.

GILBERT, E W (1958): "Pioneer Maps of Health and Disease in England," *Geographical Journal, 124,* 172-83.

JARCHO, Saul (1970): "Yellow Fever, Cholera, and the Beginnings of Medical Cartography," *Journal of the History of Medicine and Allied Sciences, 25,* 131-41.

PASCALIS, Felix (1819): *A statement of the occurrences during a malignant yellow fever, in the city of New York, in the summer and autumnal months of 1819; . . .* New York. (Extract, with map, in *The Medical Repository* (1820), new series, *5,* 229-56.)

PETERMANN, August (1848): "Cholera Map of the British Isles, Showing the Districts Attacked in 1831, 1832, and 1833." London.

SEAMAN, Valentine (1798): "An Inquiry into the Cause of the Prevalence of the Yellow Fever in New York," *The Medical Repository, 1,* No.3.

SNOW, John (1855): *On the Mode of Communication of Cholera,* 2d ed. London.

STEVENSON, Lloyd G (1965): "Putting Disease on the Map: The Early Use of Spot Maps in the Study of Yellow Fever," *Journal of the History of Medicine and Allied Sciences, 20,* 226-61.

2.051 Enclosure Map

A In a general sense an enclosure map is a commonplace and universal cartographic record of land reform, whereby land formerly unfenced, openly accessible, and managed by the community at large was converted into private ownership and physically enclosed. Its more specific meaning in England and Wales denotes a large-scale map depicting land described in an accompanying enclosure award whereby that land was allotted to private individuals under an Act of Parliament. An enclosure map is a cartographic statement of intent, not a survey, and the arrangement of land and roads which it portrays may not have been implemented in all its details.

Less frequently the term "enclosure map" applies to maps which accompany private agreements to enclose land, usually in the late 17th or early 18th century. Some early enclosure maps are sketches which describe retrospectively enclosures achieved in the 16th and 17th centuries by Chancery decree or order in Star Chamber.

B In Britain, from *c.* 1720 to *c.* 1870, private and local acts of Parliament and the general enclosure acts, especially those passed in 1801, 1836 and 1845, provided the usual means of converting common arable, meadow and heathlands into private allotments. Six earlier enclosure acts had also been passed in the 17th century. The commissioners appointed under such acts usually directed a surveyor to make a map showing the new arrangement of fields, footpaths, private and public roads, land ownership, and any other features described in the accompanying award. In this period about 5,500 such awards were made in England and Wales, 85 percent of them in the years 1775-1830. The maps and awards had to be enrolled with the Clerk of the Peace for the County and thus constitute official documents which may be required as evidence in disputes over land or highway rights. The maps vary considerably in content, style and quality, since surveyors were often commissioned to survey several enclosures in an area and this tended to produce local styles and conventions. A map may show only the area being enclosed (sometimes no more than a few hectares) or the whole parish or township, distinguishing old enclosures from the land in question. A small proportion of awards may not have been accompanied by maps, but not all maps for 18th century awards have survived or been deposited in official repositories.

Most employed the scales common to local land surveys, e.g., 1:4,752 (6 chains to 1 in.; 21.04 cm. to 1 km.; 13.33 in. to 1 mi.), or, as in most tithe maps (see 2.201), 1:2,376 (3 chains to 1 in.; 42.09 cm. to 1 km.; 26.67 in. to 1 mi.), the earlier maps generally being of both smaller and less standardized scale.

Mid-19th century enclosure maps are sometimes accompanied by another map showing the arrangement of land prior to enclosure, frequently distinguishing the strips or selions of the open arable fields. Such preliminary surveys are sometimes referred to as draft enclosure maps. Such maps may not have been made by the enclosure surveyor or enrolled with the award, but may have been placed with it for convenience.

Enclosure maps can only be understood with reference to the accompanying documents, especially the award. Many, especially in the mid-19th century, were based on estate maps (see 2.0210c), tithe maps (see 2.201) or other local surveys, sometimes by the same surveyor.

C BREWER, James Gordon (1972): *Enclosures and the Open Fields: a Bibliography.* Reading, British Agricultural History Society.

HARLEY, John Brian (1972): "Enclosure and Tithe Maps," in Harley, John Brian, *Maps for the Local Historian: a Guide to the British Sources.* London, 29-39.

SLATER, Gilbert (1907): *The English Peasantry and the Enclosure of Common Fields.* London.

TATE, William Edward (1967): *The English Village Community and the Enclosure Movements.* London.

TATE, William Edward (1978): *A Domesday of English Enclosure Acts and Awards* (M E Turner, ed.). Reading, University of Reading Library Publication 3.

YELLING, J A (1977): *Common Fields and Enclosure in England 1450-1850.* London.

2.052 Ethnographic Map

A Maps which show, or purport to show, ethnographic distributions in terms of territory occupied by a specific group, or groups, of people having common ethnic affinities—hence the German description of such maps as "Völkerkarten." Inherent in the production of many of these maps is the presumption that the ethnic groups so identified have a right to a separate or independent cultural and political identity.

B The idea of depicting the location of real or imagined races and ethnic groups on maps may be traced back to medieval times. Thus Maximus Planudes (*c.* 1260-1310) on his maps of Western Europe for Ptolemy's *Geography* (late 13th century; Vaticanus Urbinas graecus 82) delineated the holdings of native tribes (see Letter symbol, 5.122). Later antiquarian studies inspired the tribal maps of Philipp Clüver (1580-1623), whose *Germaniae Antiquæ Libri tres*, Leiden, 1616, included maps to show how German tribes had changed their locations over time. The earliest and simplest form of ethnographic map, however, shows the tripartite world of Isidore of Seville (died 636), described in *De Natura Rerum* or *Liber Rotarum* (written between 612 and 615) and the *Etymologiae* (622-633). The world maps, T-O in type (see Mappa mundi, 1.2320b), found in these manuscripts illustrate the Biblical tradition that the three sons of Noah peopled the earth. On a 12th century version the names of Sem, Iafet, and Chaos (Chem) are inscribed on the continents of Asia, Europe and Africa respectively (Paris, Bibliothèque Nationale, MS Latin 7672, f.137ᵛ). The first Western printed map is a simple woodcut T-O map by Günther Zainer of Augsburg, similarly inscribed, in the Augsburg edition of Isidore's *Etymologiae,* 1472. A more elaborate version of the T-O map appears in "La Fleur des Histoires" of Jean Mansel, *c.* 1455, illustrated by Simon Marmion (Brussels, Bibliothèque Royale, MS 9231), and has the Caucasian figures of Sem, Japhet, and Cham depicted on the continents. A very late T-O map made, probably in Bruges, in 1482 for Edward IV is notable in showing two black men in Africa (British Library, Royal MS 15.E.III. f.67ᵛ).

When overseas voyages in the late 15th and 16th centuries brought Europeans into direct contact with foreign lands, mapmakers illustrated their works with vivid portrayals of peoples. Notable examples are the "Miller" atlas, probably by Lope Homem and Pedro and Jorge Reinel, *c.* 1519 and later (Paris, Bibliothèque Nationale, Rés. Ge.DD.683 and Rès. Ge.AA.640), and the "Boke of Idrography," 1542, by Jean Rotz, hydrographer of Dieppe, with its eye-witness scenes including a Tupinámba village in Brazil (British Library, Royal MS. 20.E.IX.).

In the early 17th century the Dutch cartographer Willem J Blaeu on his wall map of the world, 1606-7, introduced and popularized the device of illustrating the peoples of the world by couples or groups of people in pictures round the border. His source for the peoples, as for the accompanying pictures of towns, was the *Civitates Orbis Terrarum* of Georg Braun and Frans Hogenberg (from 1572). The device was taken up not only by later Dutch and other European cartographers but also by the Japanese on their painted world maps on screens, such as that in the Kyoto Palace, Kyoto, made before 1614, and the engraved Shōhō world map, 1645, and its derivatives (Schilder).

Although such general maps included much dispersed and selective information, ethnographical mapping proper dates from the late 17th century; and language, as a test of ethnic affinity, assumed great initial importance. In 1696 H Jaillot [Alexis Hubert Jaillot] produced his "Nova Transilvaniæ principatus Tabula, ad Usum Serenissimi Burgundiæ Ducis: Principauté de Transilvanie divisée en cinq nations, subdivisée en quartiers et comtez" (Paris), in which the five nations were Hungarians, Saxons, Sicules, Moldaves, and Valaques. Gottfried Hensel of Nuremberg published four maps of the continents showing the distribution of languages in 1741 (see Linguistic map, 2.122). The Romantic Movement of the late 18th century in Europe quickened interest in vernacular languages and paved the way for the belief expressed by the nationalist leaders of the 19th century that only through its language could a nation fulfill itself and make its unique cultural contribution to the rest of the world. It was thus in the 19th century with the advent of nationalism as a powerful political force that the detailed delineation of ethnic distributions emerged as a serious branch of mapmaking.

Most of the ethnographic maps published in the early decades of the century such as those of Europe by F A O'Etzel ("Völkerkarte von Europa," 1821), and M A Denaix (1829) were based on language. Particularly influential in this respect were Ami Boué's map of Turkey-in-Europe (1847) and the monumental work of Karl von Czoernig on the Austro-Hungarian Empire (1855).

In spite of Darwinian notions and the application of the Linnaean method to the study of man, race in the biological sense was not widely regarded as a criterion in the compilation of ethnographic maps, although Gustav Kombst's ethnographic map of Europe in A K Johnston's *The National Atlas . . .* (Edinburg, 1846) was based on racial criteria. The powerful and politically conscious religious institutions and the continued existence of community loyalties based on religious as opposed to linguistic ties played an important role in the compilation of ethnographic maps in eastern Europe in the latter half of the 19th century, and later in the Middle East and India. The dilemma of choosing significant criteria as a measure of national sentiment is well expressed by Carl Sax's map of Turkey-in-Europe (1878) which portrayed no fewer than twenty-eight ethnic groups based on a cross-tabulation of language and religion.

C CZOERNIG, Karl von (1855): "Ethnographische Karte der Österreichischen Monarchie," scale 1:1,584,000, Wien.

DENAIX, Maxime Auguste (1829): "Carte de l'Europe, présentant le tableau des peuples qui l'habitent, classés par rapport à la dépendance relative des nations et à l'affiliation des langues," *Atlas physique, politique et historique de l'Europe.* Paris, Map 7.

HENSEL, Gottfried (1741): *Synopsis universae philologiae, in qua miranda unitas et harmonia linguarum totius orbis terrarum . . . eruitur.* Nuremberg.

SCHILDER, Günter (1979): "Willem Jansz. Blaeu's Wall Map of the World, on Mercator's Projection, 1606-07 and Its Influence," *Imago Mundi, 31,* 36-54.

WILKINSON, H R (1951): *Maps and Politics. A Review of the Ethnographic Cartography of Macedonia.* Liverpool, University Press.

WILKINSON, H R (1952): "Ethnographic Maps," *Proceedings, Eighth General Assembly and Seventeenth International Congress,* International Geographical Union, Washington, DC, 547-55.

2.081 Historical Map

A Every map ultimately becomes old and therefore an historical document. In contradistinction to old maps, an historical map portrays facts derived from the critical interpretation of source materials concerning cultural and physical elements which did not exist when the map was made. Three classes of historical maps may be recognized: (1) *cross-sectional* maps in which the subject is restricted to a narrow chronological period, such as a map of Europe in 1648 after the Thirty Years' War; (2) *development* maps concerning cultural-historical-political changes which took place over a period, such as a map of the territorial development of Prussia from 1415 to 1870, or a map of Poland in the period of partitions, 1772-1795; and (3) *thematic* maps concerning particular subjects, such as a map of Benedictine monasteries in England, 1066-1350.

B Historical maps made by medieval chroniclers and historians qualify as examples of the cross-sectional variety, type (1). Matthew Paris, monk of St Albans and historian (*c.* 1200-1259), drew *c.* 1250 an outline sketch of England and Scotland (entitled "Scema Britannie"), showing the main Roman roads. This is identified as the earliest English historical map (see also Road map, 1.1810d). Matthew Paris's four maps of Palestine (*c.* 1250) include biblical features and in this respect are historical. In the 15th century the *Geography* of Claudius Ptolemy, published with maps from 1477 onwards, comprised in effect an historical atlas. The revival of interest in the ancient and classical world, which had begun in the later Middle Ages and was a major feature in the European renaissance, was reflected in the works of leading 16th century cartographers such as Abraham Ortelius, who included in his world atlas *Theatrum Orbis Terrarum* (1570) a map of Palestine by the German Tilmann Stoltz showing the route of the wanderings of the Children of Israel from Egypt to the Land of Israel. During the period from 1579 to his death in 1598 Ortelius himself drew 38 historical maps for his *Parergon,* which comprised an historical atlas of ancient geography appended to the *Theatrum* and in 1624 published as a separate book. Ortelius's *Parergon* encouraged the developing relationship between the two scholarly disciplines of geography and history, and their interrelationship led in turn to the creation of a new discipline, historical geography. Philipp Clüver (1580-1623) of Gdansk, was among the first to draw attention to it in his *Introductio in universam geographiam* (1629).

The production of historical maps based on the analysis of sources began in the mid-18th century with France taking a leading part, especially through the work of the pioneer Jean Baptiste d'Anville (1697-1782). After the second quarter of the 19th century leadership in the production of historical maps shifted to Germany. The Pole, Joachim Lelewel (1786-1861), was also active.

As the field of historical geography developed in the 19th century, new methodologies for historical mapping were advanced. These derived in part from the scholarly planning and organization involved in the preparation of the *Geschichtlicher Atlas der Rheinprovinz* (Hermann Behrendt, Bonn, 1894) and the *Historischer Atlas der österreichischen Alpenländer* (Österreichischen Akademie der Wissenschaften, Wien, 1895–). The most useful technique was the "retrogressive method" which involved beginning with well-documented later times and then going backward step by step exposing earlier cultural and political landscape "layers."

There are three basic classes of phenomena dealt with in historical maps. One has to do with the natural landscape in which one works backward from the present state and assumes that the changes were relatively minor and local. This involves the reconstruction of floristic relationships, mainly having to do with forests and meadows and changes in hydrography such as new river courses and the disappearance of lakes and marshes caused by reclamation. A second element concerns the cultural landscape resulting from the economic activities of man, involving roads and the shapes, sizes, names, and arrangements of settlements. The third class, with which the older historical maps were largely concerned, has to do with the political landscape. This involves the reconstruction of territorial and political frontiers, state, church, and judicial boundaries, and the jurisdiction over, or ownership of, estates, production centres, towns and villages.

Historical maps may be made at any scale. When historical maps are grouped and the collection published as an entity, it constitutes an historical atlas.

C ARNBERGER, Erik (1966): *Handbuch der thematischen Kartographie.* Wien, Franz Deuticke.

ARNOLD, S (1951): *Geografia historyczna Polski.* Warszawa.

CLÜVER, Philipp (1629): *Introductio in universam geographiam.* Leiden.

EAST, W Gordon (1966): *An Historical Geography of Europe,* 5th ed. London, Methuen & Co.

FRANZ, Günther (1962): *Historische Kartographie: Forschung und Bibliographie,* 2d ed. Hannover.

GOBLET, Y M (1932): "La géographie historique et l'histoire de la géographie," *Revue historique, 170.*

JACUNSKI, W A (1955): *Istoricheskoye geografiya i yey wozniknowii ney i razvitiya v XIV-XVIII vekakh.* Moskva.

KOEMAN, C (1969): *Atlantes Neerlandici.* Amsterdam, Theatrum Orbis Terrarum.

LELEWEL, Joachim (1852-57): *Géographie du moyen-âge,* 4 vols. and atlas. Bruxelles.

VAUGHAN, Richard (1958): *Matthew Paris.* Cambridge, At the University Press.

2.091 Insurance Map

A Insurance maps or plans, prepared primarily for the use of fire insurance underwriters, show detailed, accurate, and up-to-date information on potential fire hazards and risks for individual residences and commercial and industrial installations in urban and suburban areas. Information shown includes the type of interior and exterior construction material, dimensions and heights of buildings, block, lot and building numbers, occupancy of buildings, and such fire protection facilities as watermains and lines, fire hydrants, and fire alarm boxes.

B Fire insurance maps developed in response to the need of underwriters to ascertain the potential fire risk of structures in various cities, without incurring the expense of personal inspection. The Phoenix Assurance Company, which still maintains headquarters in London, is generally credited with having introduced this specialized cartographic format. Between 1785 and 1820, Phoenix extended its fire insurance coverage to Canada, the United States, and the West Indies. Manuscript maps of certain cities in these regions were prepared for the head office to locate the insured properties and to document the nature of the risks. The only such plan that did not remain in manuscript appears to be one of Charleston, South Carolina, which was published in 1790 from a survey made for Phoenix by Edmund Petrie in August 1788. The single sheet plan, at the scale of one inch to 400 feet, identifies by numbers or letters public buildings, churches, wharves, business establishments, streets, and public and private wells. A more detailed insurance map of the Cities of London and Westminster at the scale of 26 inches to a mile, was prepared for the Phoenix Company, between 1792 and 1799, by Richard Horwood. The 32-sheet map identifies by street number every commercial and dwelling structure then standing.

Although insurance maps originated in England, they had their greatest development and use in Canada and the United States. Prior to about 1820, buildings in American cities were, with some few exceptions, insured by English underwriters. During the next several decades a number of fire insurance companies were organized in the United States and Canada. At first the companies were small and limited their risk coverages to local properties, but following a major conflagration in 1835 in New York City, which wiped out many small firms, the industry was reorganized with larger companies and more dispersed risk coverage. Inspection of properties by company personnel was no longer possible, and there developed a need for detailed insurance maps. George T Hope, an insurance company officer, and William Perris, an engineer, are credited with having produced the first large-scale insurance plan, of a portion of New York City, around 1850.

During the next half century a number of map companies published insurance plans of various urban and suburban areas utilizing the recently introduced reproduction technique of lithography. D A Sanborn, who prepared insurance plans for the Aetna Company in 1866, established his own company in New York City in 1867. The Sanborn Map Company extended its insurance map coverage throughout the United States, and gradually absorbed a number of smaller insurance map companies. Ultimately Sanborn obtained a monopoly of insurance plan production in the United States, and more than 12,000 United States towns and cities were mapped by Sanborn, many of them in multiple editions.

Although Sanborn prepared a few plans for Canadian cities in the later decades of the 19th century, Charles E Goad was responsible for most of the insurance map coverage for Canada. Goad established his company in Montreal in 1875. Ten years later he opened a branch in London, which subsequently became the head office. Goad published insurance plans for more than 1,300 Canadian cities, and for cities in England, Egypt, Turkey, and Venezuela.

Collection of insurance maps and atlases preserved in various institutions comprise valuable historical records of the development and growth of cities during the period 1850 to 1960. Several catalogs and checklists of insurance map holdings have been published.

C *Fire Insurance Maps in the Library of Congress, Plans of North American Cities and Towns Produced by the Sanborn Map Company* (1981). Washington DC, U.S. Government Printing Office.

GETTY, R P (1910): "Insurance Surveying and Map Making," *Cassier's Magazine, 39* (Nov.), 19-25.

HAYWARD, Robert J (1974): "Chas. E. Goad and Fire Insurance Cartography," *Proceedings,* Association of Canadian Map Libraries, Eighth Annual Conference, Toronto (June 9-13), 51-72.

HAYWARD, Robert J (1977): *Fire Insurance Plans in the National Map Collection.* Ottawa, Public Archives of Canada.

HOEHN, R Philip (1976): *Union List of Sanborn Fire Insurance Maps Held by Institutions in the United States and Canada.* Volume 1 (Alabama to Missouri). Santa Cruz, California, Western Association of Map Libraries.

LAMB, Robert B (1961): "The Sanborn Map: A Tool for the Geographer," *California Geographer, 2,* 19-22.

O'NEILL, Patrick B (1985): *A Checklist of Canadian Copyright Deposits in the British Museum 1895-1923. Volume 2 — Insurance Plans.* Halifax, Nova Scotia, Dalhousie University, School of Library Service.

RISTOW, Walter W (1968): "United States Fire Insurance and Underwriters Maps, 1852-1968," *Quarterly Journal of the Library of Congress, 25,* 194-217. (Reprinted *Surveying and Mapping, 30* (1970), 19-41.)

ROWLEY, Gwyn (1984): "An Introduction to British Fire Insurance Plans, *The Map Collector,* No.29, 14-19.

ROWLEY, Gwyn, and SHEPHERD, Peter McL (1976): "A Source of Elementary Spatial Data for Town Centre Research in Britain," *Area, 8,* 201-8.

2.121 Land-Use Map

A A map portraying the spatial distribution of the various forms of vegetation and occupation of the land.

B Many maps from ancient times onward show types of land use, although this was not the specific purpose of such maps. Of these early maps the closest to the land-utilization map was probably the estate survey (see Estate map, 2.0210c). John Blagrave's manuscript map of the Forest and Manor of Feckenham, Worcestershire, 1591, now known in a copy of 1744 (British Library, Maps M.T.6.b.1.(12.)), indicates by colours woodland, meadow, pasture and "corn ground" of the best, worst, and moderate quality. The earliest land-use maps, properly speaking, date from the end of the 18th century. Simple small-scale land-use maps of counties were first published in England in the 1790s, when the Board of Agriculture initiated statistical enquiries into the state of agriculture in the counties. Some of the reports included land-use maps coloured by hand. In John Middleton's *General View of the Agriculture of Middlesex* (1794), the map of Middlesex distinguishes by use of three colours arable, pasture, and nursery grounds.

These advances may have prepared the way for the production of the first true land-utilization map, 1800 in date, by Thomas Milne, an estate and county surveyor. Entitled "Milne's Plan of the Cities of London and Westminster, circumjacent Towns and Parishes as laid down from a trigonometrical survey taken in the years 1795-1799," and drawn on a scale of two inches to one mile (1:31,680), the map distinguishes twelve different types of land-use by key-letters (eg. arable land, meadows and pastures, woodland, market gardens, orchards), and hand colouring is applied to each enclosure. Only one example of the complete 6 sheet map is known (British Library, K.Top. VI.95); another example comprises one sheet only, coloured by parishes (British Library, Crace Maps XIX.28).

In the 19th century land-use information was incorporated into some of the English tithe maps of the 1830s and 1840s (see Tithe map, 2.201) and into the larger scale plans (1:2,500) of the Ordnance Survey. From 1853 to 1880 Books of Reference ("Area Books") for each parish stated the land-use of each parcel on the O.S. plans. In the handbook of instructions, 1875, the purpose of the land-use survey and its procedures were explained. Studies of the human environment and man's influence upon its transformation published in the last years of the century inspired a fuller mapping of the environment. Thus Hugh Robert Mill proposed on 6 March 1896 at the Royal Geographical Society a complete "geographical description" of the British Isles based upon the Ordnance Survey maps and to be called the "New Ordnance Survey." Taking southwest Sussex as an example, he showed how such a work would be made. The specification included "Vegetation and Agriculture." This project, for which he proposed support from the government, scientific societies, or a "patriotic millionaire," would have anticipated the first land utilization survey of Great Britain undertaken by Professor Dudley Stamp in the years 1930-38.

Various maps and surveys which included land-use were made in continental Europe and North America in the 18th century. In the cadastral survey of the Duchy of Savoy in 1728-38, the register to parcels of land recorded area, type, and quality (see Cadastral map, 2.0210b). The map of Hungary by Johann Matthias Korabinsky, published at Pressburg in a German edition of 1786 and in an English edition in London in 1797, includes natural productions and is partly a land-use map. The survey of the St Lawrence River made by order of Brigadier-General James Murray from 1763 gives land-use information. The surveyors were required "to take particular notice . . . how much of the land surveyed is plantable, and how much of it is barren and unfit for cultivation, and accordingly to insert in the survey . . . the true quantity of each kind of land . . ." These instructions have been described as "containing features of what was in essence, the first land-use plan to be applied to Canada" (Thomson, p.99); the maps are preserved in the British Library (K.Top CXIX.24-27), and also, as the "Murray Atlas," in the Public Archives of Canada.

A map of the environs of Vienna, "Umgebungen von Wien im 1:14,000 der Natur der Wiener Zoll gleich 200 Wiener Klafter," begun in 1830 by the Austrian Quartermaster General's Office is notable as the first Austrian land-use map. The map distinguishes by different printed colours areas of varying land use (see Colour printing, 7.031).

The demands of commerce in the middle years of the 19th century led to the development of urban land-use mapping in the form of the insurance plan. The firm of Charles E Goad, first established in Canada in 1875, published volumes of plans for cities in Canada, England, and elsewhere in the world, while the Sanborn Map Company from 1867 mapped towns and cities in the United States (see Insurance map, 2.091).

These surveys and maps mainly covered selected or restricted areas and regions. Only in the 20th century have their number, scope, and form been developed to any extent. This century should therefore be recognized as the period of the actual expansion of land-use researches.

C BULL, G B G (1956): "Thomas Milne's Land Utilization Map of the London Area in 1800," *Geographical Journal, 122,* 25-30.

DARBY, H C (1970): "Domesday Book–The First Land Utilization Survey," *Geographical Magazine, 42,* 416-23.

HARLEY, J Brian (1979): *The Ordnance Survey and Land-Use Mapping.* Exeter, University of Exeter, Historical Geography Research Series No.2.

MILL, Hugh Robert (1896): "Proposed Geographical Description of the British Islands Based on the Ordnance Survey," *Geographical Journal, 7,* 345-56.

MILNE, Thomas (1800): Milne's Plan of the Cities of London and Westminster, circumjacent Towns and Parishes, etc., laid down from a Trigonometrical Survey taken in the Years 1795-1799 (Map), 1:31,680. (Reprinted, with an introduction by G E G Bull, by London Topographical Society, Publications Nos. 118 and 119, 1975-6.)

THOMSON, Don W (1966): *Men and Meridians: The History of Surveyors and Mapping in Canada, Vol.I.* Ottawa, Department of Mines and Technical Surveys.

WALLIS, Helen (1981): "The History of Land Use Mapping," *Cartographic Journal, 18,* 45-8.

2.122 Linguistic Map

A A map showing the distribution of languages.

B The Nuremberg cartographer Erhard Etzlaub on the later states of his "Romweg" map, Nuremberg, *c.* 1500, attached a printed sheet which he called a Register in which he explained the various special features of his map. One of these features was to give colours to the countries and language areas adjacent to Germany. Not all the extant copies of the "Romweg" are coloured, apparently because the addition of linguistic information was decided upon only after the printing of the maps commenced (Krüger).

The first proper linguistic maps seem to be a set of four maps of continents by Gottfried Hensel published at the Officina Homanniana, Nuremberg, in 1741. Hensel showed the distribution by examples of the written languages of each people placed in correct geographical position. The maps illustrated Hensel's thesis that world languages may be classified in terms of their history and origin. His note on the map of Africa, for example, explained that the colourings distinguished the descendants of the three sons of Noah: Japheth, Shem, and Ham. Many ethnographical maps (see

2.052) were based on language. No other linguistic map appears to have been made until the German Julius Klaproth, a resident in Paris from 1815, published in 1823 a map in which he outlined and coloured twenty-three language areas of Asia in a *Sprachatlas* accompanying his scholarly study of the languages of that continent.

C HENSEL, Gottfried (1741): *Synopsis universae philologiae, in qua miranda unitas et harmonia linguarum totius orbis terrarum . . . eruitur.* Nuremberg.

KLAPROTH, Heinrich Julius von (1823): *Asia Polygotta nebst Sprachatlas.* Paris.

KRÜGER, Herbert (1951): "Erhard Etzlaub's *Romweg* Map and Its Dating in the Holy Year of 1500," *Imago Mundi, 8,* 17-26.

2.1310 Military Map

A Any map on either a large or a small scale made for military use or to show the results of military actions. Such maps can be for current or future use, or are historic in that they show previous military events. Battle maps (2.1310a) are generally retrospective, comprising detailed, large-scale plans showing how a battle progressed in time and space. Reconnaissance maps (2.1310b) result from preliminary or exploratory surveys that often precede a battle or are in anticipation of war. Fortification maps depict the defences of a single place. Plans of encampments and plans of the formations adopted by troops when manoeuvring or in camp are specific classes.

Military maps can also be divided into (a) officially sponsored, forward looking maps of a country or province, often entitled "Maps of the Seat of War . . . ," which provide physical and cultural information useful for future combat, and (b) historic, news, and educational maps made after the event.

B Official military maps existed in the ancient world. The earliest such maps are two from Changsha, China, which were made before 168 BC. The relief map shows landforms, drainage features, and roads. The garrison map shows relief with great skill and marks villages by numbers of households and loyalty, garrisons by their areal extent and their relationship to headquarters. Clearly highly sophisticated techniques of surveying and mapping were employed. It is reported of the Romans that "A Map of the conquer'd Country was always carried in the Victor's Triumph" (Cave, p.264). No Roman military map is known today, unless the Dura Europos shield map, dating probably from the first half of the 3rd century, was made for a military purpose (see Route map, 1.1810).

The Italians, building perhaps on a tradition of large-scale local mapping that had developed in Northern Italy, produced military maps from *c.* 1430 onwards, showing fortifications and towns (sometimes misplaced), in the course of wars between the cities, such as Venice against Milan, and between the Italians and the Turks. Two Italian military maps, both with Turkish connections, survive. The earlier, *c.* 1452, is an Italian map of the Balkans (Paris, BN, ms. cod. lat. 7239) once owned by Sultan Mehmed II (Babinger). The other map, of the entire Venetian territory of Terraferma

(Topkapi Saray Library, Istanbul), appears to have been sketched from personal knowledge and memory by a non-cartographer in 1469 or 1470 in the midst of a Venetian-Turkish war. No distances are shown, cities are mislocated, but city walls are carefully delineated and bridges identified. The principal names are in Latin characters that are repeated in the Cyrillic script of the period (Gallo). A crude military map of Scotland showing castles, c. 1440, accompanies the Chronicle of John Hardyng (British Library, Lansdowne MS 204, f.225).

Over the next two centuries, many military maps were compiled in manuscript for areas of current and potential conflict. In the 1530s and 1540s Henry VIII employed English and foreign surveyors (mainly Italians) to map and modernize the coastal defences and the fortification of ports. John Rogers's military map of the surroundings of Boulogne, c. 1546, drawn to a scale of one inch to 500 feet (British Library, Cotton MS. Aug. I.ii.75) and the plan of the defences of Portsmouth, 1545, 1:1200, one of the first scale-maps for England (Cotton MS. Aug. I.i.81), are examples from the numerous 16th century military maps in the Cotton collection of the British Library (Tyacke and Huddy; Merriman).

The charts of Robert Adams, 1588, to illustrate the Spanish armada were the most famous English military maps of their day. That of the Thames estuary, "Thamesis Descriptio. Anno 1588," manuscript on vellum, illustrated the military defences of London (British Library, K.Top. VI.17), while another copy marked Queen Elizabeth I's progress to the camp at Tilbury (British Library, Add. MS 44,839). A series of charts, engraved by Augustine Ryther, illustrated the Armada Campaign in *Expeditionis Hispanorum in Angliam vera descriptio Anno D. MDLXXXVIII*, London, 1590. These maps became famous through their reproduction on tapestries which hung in the House of Lords from about 1616 to 1834 (see Map surface, Cloth, 6.1310c).

Military maps of extensive regions in the 17th century may be seen as forerunners of the national surveys of the 18th century. In France the Ingenieurs du Roy, c. 1606-9, undertook for Henri IV the survey of Dauphiné (British Library, Add. MS 21,117; Buisseret, Dainville). Henri Sangré compiled in the 1670s and 1680s for Prince Louis II of Bourbon (Condé) maps of the Upper Rhenish area, which he was able to publish only in 1692 (Grenacher). During the War of the Spanish Succession (1704–1713/14), both French and German engineers were actively mapping the South Netherlands, and the maps of the Theatre of War were published in the Netherlands and elsewhere by Carel Allard, Pieter Schenk, R and J Ottens of Amsterdam, and others.

At this period French military surveys were based on geometric observations with graphometers, and the maps assumed the "flat-earth concept." No geographic co-ordinates are given though orientation is northerly. Villages are outlined and named; roads and paths are marked; arable land, meadows, and woodland are indicated by colour and symbol; primitive shading is used to indicate valleys. The development of engineering and military education in the 18th century resulted in the preparation of many large-scale maps based on exact surveys. Such maps served both for officers' training and for planning and carrying out military operations. Plan variants, the differences between the plan and the result, and even the succeeding stages of operations were often shown by overlays affixed in such a way that they could be lifted or removed so as not to cover permanently the basic map. At that time also additions were made to the stock of conventional signs. They distinguish detachments of troops of various kinds, the directions of their operations, and the engineering works carried out in the area. In particular, Johann Georg Lehmann's system for topographical mapping (1799) received wide approval in the military schools of Prussia, Saxony, Austria, and Russia, and was translated for use by the British in 1822 (see Characteristic sheet, 6.032).

Historical, news, and educational maps of a military nature date back to early times, with sometimes several different mapmakers producing views of the same event. From the 15th century onwards the maps were usually printed and sometimes carefully coloured. These maps were often designed to show pictorally where the armies were encamped, how and where the fire power (cannon, rifle) was directed, cavalry or ships in full encounter. While topographic accuracy is not ignored, it is of lesser significance than for battle maps. Military maps of a more specific kind, besides battle maps (2.1310a) and reconnaissance maps (2.1310b), are maps of sieges, lines of battle plans, route of march plans, and plans of forts and encampments. Of these various types, plans of forts and encampments date back at least to Roman times.

C BABINGER, Franz (1951): "An Italian Map of the Balkans, Presumably Owned by Mehmed II, the Conqueror (1452-53)," *Imago Mundi, 8*, 8-15.

BUISSERET, D J (1964): "Les Ingénieurs du roi au temps de Henri IV," *Bulletin de la Section de Géographie* (Bibliothèque Nationale), *77*, 13-84.

CAVE, Edward (1739): *Geography Reform'd: or, a New System of General Geography*. London.

CHANG, Kuei-sheng (1979): "The Han Maps: New Light on Cartography in Classic China," *Imago Mundi, 31*, 9-17.

DAINVILLE, François de (1968): *Le Dauphiné et ses contins vus par l'ingénieur d'Henri IV, Jean de Beins*. Geneva, Librairie Droz; Paris, Librairie Minard.

DILKE, O A W (1971): *The Roman Land Surveyors: An Introduction to the Agrimensores*. Newton Abbot, David & Charles.

GALLO, Rodolfo (1955): "A Fifteenth Century Military Map of the Venetian Territory of *Terraferma*," *Imago Mundi, 12*, 55-7.

GRENACHER, Franz (1964): "Current Knowledge of Alsatian Cartography," *Imago Mundi, 18*, 60-77.

LEMOINE-ISABEAU, Claire (1984): *Les militaires et la cartographie des Pays-Bas méridionaux et de la Principauté de Liège à la fin du XVIIᵉ et au XVIIIᵉ siècle*. Brussels, Musée Royal de l'Armée, Centre d'Histoire Militaire, Travaux 19.

MERRIMAN, Marcus (1983): "Italian Military Engineers in Britain in the 1540s," in Tyacke, Sarah, ed., *English Map-Making, 1500-1650*. London, The British Library, 57-67.

NISCHER-FALKENHOF, Ernst von (1937): "The Survey by the Austrian General Staff under the Empress Maria Theresa and the Emperor Joseph II., and the Subsequent Initial Surveys of Neighbouring Territories during the Years 1749-1854," *Imago Mundi, 2*, 83-88.

O'DONOGHUE, Yolande (1977): *William Roy, 1726-1790, Pioneer of the Ordnance Survey.* London, British Museum Publications for the British Library.

TYACKE, Sarah, and HUDDY, John (1980): *Christopher Saxton and Tudor Map-Making.* London, British Library Series No. 2.

2.1310A Battle Map

A A map illustrating tactical military operations, such as battles in the field and operations connected with the siege and defence of fortified points. Battle maps intended for direct operational use were drawn up either before operations were started in order to assist planning or during the operations in order to take into account the current situation.

Battle maps drawn up after the end of operations by military historians were aimed at (a) providing examples of operational situations for commanders training in the domain of tactics, or (b) illustrating descriptions in scientific and popular works of historians, or (c) disseminating news to an interested public, sometimes for the purpose of propaganda.

The groundwork for battle maps may be more or less exact topographical drawings. They include general data, such as relief, hydrography, forest cover, settlements, and networks of roads, as well as data significant for combatants, such as the distribution of military installations, especially fixed defenses and field works.

The essential problem of a battle map is the representation of the distribution and shift of detachments so as to illustrate two, three and sometimes four dimensional spatial elements. Consequently, there are static battle maps illustrating positions of detachments at a particular moment; dynamic maps illustrating movements of particular detachments in a stated time period; and static-dynamic maps portraying both initial and ultimate positions, as well as the routes of movements. In some cases, especially when repeated shifts of detachments in the same area took place, "phasic" battle maps have been made. These divide the events into several periods and devote a special map to each.

B The bas-relief picturing the battle fought by Ramesses II's Egyptian soldiers in 1294 BC at the walls of the Hittite south Syrian fortress Kadesh is thought to be the oldest extant battle map. Several different versions of the bas-relief can be seen at Abu-Simbel, Luxor, and Karnak (in the sanctuary of Amon and in the Ramesseum). Part of the map is composed of elements illustrated in orthogonal projection (hydrography, defense girth of Kadesh fortress) and part is in side view (walls of the fortress, its defenders, fights on the fortress's foreground with use of infantry and war-wagons). Moreover, several stages of the battle are portrayed on the one picture-map.

This type of cartographic representation of military operations and tactics, combining an orthogonal projection with the side view, was also used in the medieval plan-views. Only during the renaissance was it replaced by a drawing in central perspective or axonometric projection, which differs from the modern map mainly by the fact that it employs a diagonal projection, not orthogonal. Early examples of such a type of battle map are a miniature illustrating the besieging of Constantinople by the Turks in 1453 and the woodcut depicting the Turkish siege of the island of Rhodes in 1480.

The above mentioned types of battle maps were dominant in the 16th and 17th centuries, such as the copperplates illustrating the besieging of Metz in 1552; the naval battle at Lepanto in 1571; the battle at Bílá Hora (Ger: Weisser Berg), Bohemia, in 1620 in four stages; the besieging of La Rochelle in 1627; the battle of Rocroy in 1643; and the siege of Vienna in 1683. A large collection of such maps was published by the Matthaeus Merians (elder and younger) in "Theatrum Europeum" (30 vols., issued in the years 1642-1688). Two other collections are the "ordres de batailles" of the years 1600-1678 gathered by the Swedish military engineer Erik J Dahlbergh and deposited in the Krigsarkivet in Stockholm (partly published in the work of Samuel von Pufendorf, *De rebus a Carolo Gustavo Sueciae Rege gestis . . . ,* Nuremberg, 1696), and the later large compilation of plans of European battles, sieges, and the like, printed by Gabriel Bodenehr in his *Force d'Europe . . . ,* Augsburg, 1720-26. Maps drawn in orthogonal projection are often found next to maps of the perspective type. On these maps conventional signs begin to appear in the shape of quadrangles illustrating detachments and lines indicating direction of movement.

As a result of the extension of military cartographic services in the 19th century, many countries were completely covered with maps of tactical scale. These provided good foundations for battle maps, and partly released staff officers from carrying out map surveys. In the 19th century many writers tried to recreate in detail the course of former battles and sieges. These works included text as well as detailed base maps at the tactical scale drawn by military cartographic institutes. In addition, whole atlases came into existence to illustrate battles and sieges (e.g. William Siborne, *An Atlas to Accompany the History of the War in France and Belgium in 1815,* Field of Waterloo, 1844 (Fig.9); Louis Adolphe Thiers, *Atlas de l'histoire du Consulat et de l'Empire,* dressé et dessiné sous la direction de M T, Paris, 1853; and M Jahns, *Atlas zur Geschichte des Kriegswesens von der Urzeit bis zum Ende des XVI. Jahrhunderts,* Berlin, 1879).

An early example of a news map of a battle is Horatio de Marii Tigrino's map of Lepanto, entitled "Disegno dell' armate Christiana et Turchesca . . . 1571 . . . del golfo di Lepanto," 1572, which displays extensive text. Many Italian maps of the 16th century engraved by Lafreri and others were designed to illustrate recent battles for the interest of nobility and the general public. The speed with which news maps were produced in the 18th and early 19th centuries was remarkable. In the course of the American War of Independence, the first published plan of the Battle of Bunker Hill, "A Sketch of the Action between the British Forces and the American Provincials, on the Heights of the Peninsula of Charlestown, the 17th June 1775," was published in London on 1 August 1775 by Jefferys and Faden, and was printed only five days after news of the event reached London. The map has to be described as crude and topographically inaccurate (British Library, *The American War of Independence.* London, British Library, 1975, item 51).

A series of battle plans of the Peninsular War was printed by lithography by the Quarter Master General's Office at the Horse Guards in London from 1808 to 1815, as a cheap means of providing quick multiple copies for general use (British Library,

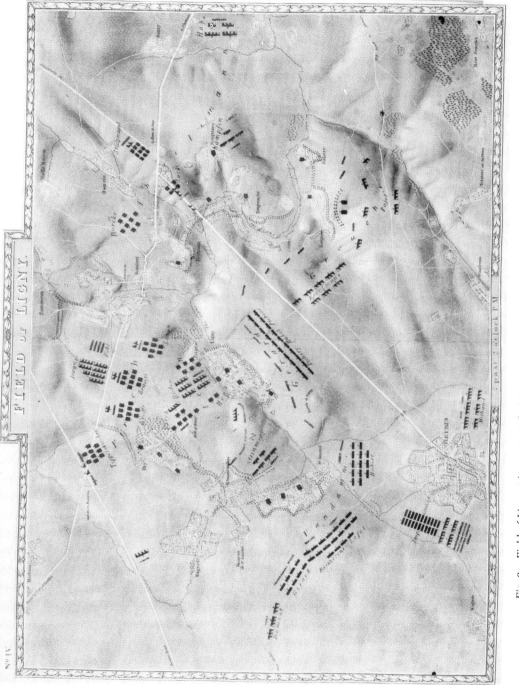

Fig. 9 – Field of Ligny, 1/4 past 2 o'clock P.M. Modelled by W Siborne. In William Siborne. An atlas to accompany the History of the War in France and Belgium, in 1815, London, 1844, pl. IV. Engraved by Freebairn. Bate's Patent Anaglyptograph. By permission of the British Library. (Battle map, 2.1310a; Anaglyptography, 7.011)

Maps C.18.l.1., and Maps C.18.m.1.). A "Sketch of the Battle of 21st August 1808 near Vimiera in Portugal," for example, was printed with an explanatory text at Whitehall on 5th September 1808 (see Autography, 7.1210a).

Plans of sieges may be classed as a special type of battle map. A number of striking maps illustrate the siege of Vienna by the Turks, 1683, such as "Grundriss der Kayserlichen Residenz-Stadt Wien, mit der türkischen Belagerung . . . 1683," published by J Hoffman, Nuremberg [1683?].

C DIEMER-WILLRODA, Ewald (1939): *Schwert und Zirkel: Gedanken über alte und neue Kriegskarten*. Potsdam.

EASTON, William W (1977): "A History of Military Mapping," *Bulletin, Geography & Map Division, Special Libraries Association, 109*, 40-4.

KOEMAN, Cornelis (1973): "Krijgsgeschiedkundige kaarten," *Armamentaria, 8*, 27-43.

OLSZEWICZ, Bolesław (1921): *Polska kartografia wojskowa* (Polish Military Cartography). Warszawa.

POGNON, E (1968): "Les plus anciens plans de villes gravés et les événements militaires," *Imago Mundi, 22*, 13-19.

STEBNOWSKI, Jan (1960): *Rozwoj kartografii wobec zagadnien wojskowych w starozytnosci* (Development of Cartography in Relation to Military Problems during Antiquity), *Studia i Materiały do Historii Wojskowosci* (Warszawa), *2*, part 2, 3-56.

2.1310B Reconnaissance Map

A A map resulting from a preliminary or exploratory survey, such as to obtain military information or to aid in planning a triangulation survey.

B Maps for military reconnaissance (and written reports) were the final stage of intelligence gathering. Most were not intended to survive. Such maps must date back to the ancient world. As Edward Cave reports (p. 22), Moses sent spies to discover the condition of Canaan (Numbers, 13.3.17); Augustus sent Dionysius of Charax, in Persia, into the East to make discoveries on behalf of his eldest son, then planning his Armenian expedition to fight the Parthians and Arabs (Pliny, Natural History, 1.6.c 17).

In renaissance Europe reconnaissance mapping became an accepted procedure for the military. The plot of Castlemilk in southwest Scotland, 1547, is an example of a military sketch map drawn for the Duke of Somerset, the Protector of England (Merriman, p.60). The corps of "ingénieurs du Roi" set up by Henri IV in 1604 were assigned reconnaissance duties in France, and other European powers created similar groups along with the formation of standing armies, elaborate fortifications, and permanent navies.

As the formality of warfare increased in the 18th century, discernable map types became part of the reconnaissance category, among them road maps, town plans, harbour studies, communication routes, and spy maps. Normally, these were finished in the same drafting techniques applied to other military maps. In particular, the "coup d'oeil" became a recognized military device. Frederick the Great, King of Prussia, included it in his military instructions (1747), which were widely translated and published, and various handbooks such as that by Gottlob Friedrich von Brück (1777) were devoted to its exposition. In the 19th century, reconnaissance maps were designed to fill the military need for speed and were advocated as simple sketches which any officer could perform.

C BERTHAUT, Henry Marie Auguste (1902): *Les Ingenieurs géographes militaires, 1624-1831: Étude historique,* 2 vols. Paris, Service géographique de l'armée.

BRÜCK, Gottlob Friedrich von (1777): *Coup d'Oeil Militaire oder das Augenmerk im Kriege.* Dresden and Leipzig.

CAVE, Edward (1739): *Geography Reform'd: or, a New System of General Geography.* London.

FREDERICK II, called the Great (1747): *Die Instruction Friedrichs des Grosen für seine Generale von 1747.* Facsimile, Berlin, Reichsinstitut für Geschichte des neuen Deutschlands, 1936.

MERRIMAN, Marcus (1983): "Italian Military Engineers in Britain in the 1540s," in Tyacke, Sarah, ed., *English Map-Making, 1500-1650.* London, The British Library, 57-67.

2.132 Moral Statistics Map

A A map portraying the distribution of such data as school attendance, incidence of crimes, improvident marriages, illegitimate births, etc. Usually a shaded choropleth map showing rank values of the enumeration districts or departures above and below the overall average for a country.

The primary purpose of moral statistics maps was to show the possible geographical correlations among the various distributions, and a variety of hypotheses relating one array to another were put forward and debated with vigour.

B The first moral statistics map was made by Baron Charles Dupin in 1826 and published in 1827 in *Forces productives et commerciales de la France* (not in 1819 as reported by Funkhouser). It portrayed the number of persons per male elementary school child in the départements. A separate map of the Royaume des Pays-Bas, of almost identical design and using data of 1827, was made by H Somerhausen (1829?). In 1829 Adrien Balbi and André Michel Guerry made three small maps of France (one sheet, separate) showing elementary education, crimes against the person, and crimes against property. In 1831 Adolphe Quetelet made similar maps of France and the Low Countries, shown together, in his *Recherches sur le penchant au crime aux différens âges.* Quetelet's maps were not choropleth maps, but employed a system of variable shading, the darker the greater (see Shading,

5.191). These same maps were then included in Quetelet's better known *Sur l'homme et le développement de ses facultés, ou essai de physique sociale* (1835), and they were later redrawn in Germany and Britain for the translations of that book. In 1833 A M Guerry produced a series of much more elaborate maps in his *Essai sur la statistique morale de la France* (the first use of the term "moral statistic"), and from then on these kinds of maps became quite common. In 1849 Joseph Fletcher illustrated his study of the moral and educational statistics of England and Wales with a set of twelve maps, supplied at the request of Prince Albert (Fig.15).

C FUNKHOUSER, H Gray (1938): "Historical Development of the Graphical Representation of Statistical Data," *Osiris, 3,* 269-404.

ROBINSON, Arthur H (1982): *Early Thematic Mapping in the History of Cartography.* Chicago, University of Chicago Press.

2.161 Population Map

A A map portraying the distribution of numbers of people, either in absolute terms or relative to area occupied. The latter, formerly called specific population and now called density of population, shows the number of persons per square unit of area. Population statistics are obtained by enumeration districts.

B Innovations in methods are featured in the following. The earliest population maps simply had the numbers written on the maps. The first instance of so doing is not known, but two examples in the early 19th century are Carl Ritter's Map 6 in *Sechs Karten von Europa* (1806) and James Wyld's (the elder) separate 1815 "Chart of the World Showing the Religion, Population and Civilization of Each Country." The earliest map on which population was symbolized appears to be an 1830 dot map by A Frère de Montizon, "Carte Philosophique figurant la Population de la France," on which each dot represents 10,000 persons (têtes). Since then several other point symbols (cubes, spheres, etc.) have been employed to show numbers.

A small, crude choropleth map of the world showing three density categories was made in 1833 by George Poulett Scrope to accompany his *Principles of Political Economy ...* In 1836 Adolphe d'Angeville prepared a series of choropleth maps to accompany his *Essai sur la statistique de la population française ...* of which Map 1 showed population density.

In a map dated 1837 Henry Drury Harness portrayed separately the rural and urban populations of Ireland *(Atlas to Accompany the Second Report of the Railway Commissioners . . .).* Rural population was shown by a shaded dasymetric system (1.041), while urban populations were shown by proportional circles. The first detailed official choropleth (1.034) population maps appear to be those issued with the reports of the *Census of Ireland for the Year 1841* (Reports from Commissioners: 1843, Vol. 13, Dublin), made under the direction of Thomas A Larcom. In 1849 August Petermann prepared a separate population map of the British Isles, based on the census of 1841, in which he showed population density by smooth shading, the darker the greater, and also used dots and other symbols for settlements. He made similar maps for the census of Great Britain in 1851 which was published in 1852-53.

In 1857 Nils Frederik Ravn of Denmark made the first isopleth maps of population density, "Populations Kaart over Det Danske Monarki 1845, 1855" (in *Einleitung zu dem Statistischen Tabellenwerk,* New Series, *12,* Statistical Bureau, Copenhagen). Ravn's isopleth maps essentially complete the gamut of ways of presenting population numbers. Except for Petermann's smooth shading, which has rarely been used since, all methods have prospered. For some reason the dot method was slow to be adopted.

DU BUS, Charles (1931): *Démocartographie de la France, des origines à nos jours.* Paris, Librairie Félix Alcan.

ECKERT, Max (1925): *Die Kartenwissenschaft,* Vol. 2. Berlin, Walter de Gruyter, 152-211.

JARCHO, Saul (1973): "Some Early Demographic Maps," *Bulletin of the New York Academy of Medicine, 49,* 837-44.

KANT, Edgar (1970): "Über die ersten absoluten Punktkarten der Bevölkerungs-verteilung," *Lund Studies in Geography, Ser. B. Human Geography, No. 36.*

ROBINSON, Arthur H (1955): "The 1837 Maps of Henry Drury Harness," *Geographical Journal, 121,* 440-50.

ROBINSON, Arthur H (1971): "The Genealogy of the Isopleth," *Cartographic Journal, 8,* 49-53. (Also in *Surveying and Mapping, 32* (1972), 331-38. Translated into Dutch, "De Afstammung van de isopleth," *Kaartbulletin, 34* (1973), 5-14.)

ROBINSON, Arthur H (1982): *Early Thematic Mapping in the History of Cartography.* Chicago, University of Chicago Press.

2.162 Poverty Map

A A map of an urban area delimiting various regions or streets rated according to the quality of the housing or the economic and social status of the inhabitants. Such maps were sometimes referred to as sanitary maps (see 2.191), although the sanitary map focussed more directly on conditions associated with epidemic disease.

B The earliest poverty map, entitled a sanitary map, dealt with Dublin and was included in a section of a general report on health in Ireland ("Report upon the Tables of Deaths, Section V, Report upon . . . the Deaths, Occupations and Diseases of the various Districts . . . of the City of Dublin . . . their apparently Sanitary position . . . ," *Census of Ireland for the Year 1841* [Reports from Commissioners: 1843, Vol. 13, Dublin], map follows p. lxxiv). The streets were classified as 1st, 2nd, and 3rd class, private, shop, and mixed, and were identified with six colours. A map of Leeds by Robert Baker was included in Sir Edwin Chadwick's 1842 *Report . . . from the Poor Law Commissioners, on . . . the sanitary conditions of the labouring population of Great Britain* (House of Lords Sessional Papers, Session 1842, Vol.

123

26-28, map follows p. 160). The map shows three categories of housing: working class, trades people, and first class.

The most elaborate poverty map appears to be that accompanying Charles Booth's 1889 *Life and Labour, Vol. 1, East London.* The map was prepared from detailed notes of individual families in each house in each street made by School Board Visitors. Published at a scale of 1:10,280, the map employed six bright colours for categories ranging from "Well to do" to "very poor." This volume was followed in 1891 by a second volume and appendix covering the rest of London. The same survey method was used for the map, and a seventh category (nonexistent in the East End) was added, "upper-middle and upper classes, wealthy."

C BOOTH, Charles (1889): *Life and Labour, Vol. 1, East London.* London, William and Norgate.

BOOTH, Charles (1891): *Labour and Life of the People, Vol. 2, London.* London, William and Norgate. (Maps reprinted: London Topographical Society, *Publication No. 130,* 1984.)

ROBINSON, Arthur H (1982): *Early Thematic Mapping in the History of Cartography.* Chicago, University of Chicago Press.

2.163 Product Map

A A map showing the places where various kinds of production occur.

B Product maps had their beginnings in the general topographical maps (1.203) of the 16th century, in which a great assortment of land use, mining, quarrying, manufacturing, and other activities was portrayed. An early example is a map of Europe entitled "Nieuwe Paschaerte . . ." by Cornelis Doetszoon, published by Cornelis Claeszoon in 1602 (Karlsruhe, Badische Landesbibliothek; printed on vellum). It was evidently intended as a decorative wall map to hang in merchants' offices as a source of information on the products of European countries. These are described in legends (Schilder). When the Polish Jesuit and sinologue Michael Boym undertook the mapping of China, he included on his map, *c.* 1652, minerals denoted by symbols, together with rhubarb (in Tibet) and ginger (Radix Sinica), both shown as pictures (see Mineral map, 3.131).

By the 18th century more specialized maps were being prepared, and the first that was clearly a product map was published in 1782 by August Friedrich Wilhelm Crome, professor of statistics and political economy at Giessen. The map shows the occurrence of fifty-six commodities in Europe. It was well received and several editions were issued in the two decades after its first publication.

A more elaborate work was a product map of Austria published in 1796 by H Blum. It had twenty-eight signs for cities and associated features, sixty symbols for the products of mines, forests, and lakes, and thirty-eight signs showing the locations of economic activities. Another was the "New Map of Hungary, particularly of its

Rivers & Natural productions" by Johann Matthias Korabinsky, London, 1797. Such maps seem to have gone out of fashion until near mid-19th century when they again burgeoned, probably as a result of the industrial and international exhibitions that began in the 1840s. Product maps were prepared in connection with the Berlin Industrial Exhibition in 1844 by Johann Valerius Kutscheit and August Petermann, and by the latter in connection with the Great Exhibition held in London at the Crystal Palace in 1851.

C BLUM, H (1796): "Natur und Producten Karte von Österreich ob der Ems," in *Natur und Kunst Producten Atlas der Oesterreichischen Deutschen Staaten.* Wien.

CROME, A F W (1782): "Neue Carte von Europa welche die merkwürdigsten Producte . . ." Dessau.

ROBINSON, Arthur H (1982): *Early Thematic Mapping in the History of Cartography.* Chicago, University of Chicago Press, 54-56, 140-44.

SCHILDER, Günter (1976): "Willem Janszoon Blaeu's Map of Europe (1606), A Recent Discovery in England," *Imago Mundi, 28,* 9-20.

2.164 Promotional Map

A A map prepared to illustrate a proposal for raising money in support of a political or economic venture in the region depicted. Such maps do not necessarily give a favourable impression of the areas portrayed, and in general they must be identified from their archival or bibliographical locations rather than by any distinctive characteristics of style, content or format.

B Among the earliest promotional maps to be recognizable from their contexts are those associated with projects for the European exploration and colonization of the New World in the 16th century. A possible early example is Robert Thorne's discussion in 1527 of new routes to China as illustrated with a world map that was first published by Richard Hakluyt in 1582. Others include the maps produced in the 1570s and 1580s to support the theory of a navigable northern passage from Europe to the Far East. The most notable was a map illustrating Sir Humphrey Gilbert's *Discourse of a discoverie for a new passage to Cataia* (1576), which carried the title "A general map made onelye for the particular declaration of this discovery."

By this time a more specifically colonial map had been published to show the site of Sir Thomas Smith's proposed plantation in the Ards peninsula of northern Ireland: it appeared, presumably in 1572, in a tract, the title of which is "A Letter sent by I.B. Gentleman unto his very frende Mayster R.C. Esquire, wherein is conteined a large discourse of the peopling and inhabiting of the Cuntrie called the Ardes," where it was cited in the text as proof that the peninsula would be easy to defend against the Irish. In the 17th century the propagandist function of cartography was more generally recognized. In 1667 it was said of a new map of Jamaica that "if this map were printed . . ., and copies disposed to the several great cities of his Majesty's dominions, it might give great encouragement to become planters."

C BLACK, Jeanette D (1975): *The Blathwayt Atlas.* Providence, 188.

HAYES-McCOY, G A, ed. (1964): *Ulster and Other Irish Maps, c. 1600.* Dublin, 31.

SKELTON, R A (1958): *Explorers' Maps.* London, 99-135.

2.181 Religious Map

A (1) A map showing the religious allegiances and distributions, or the location of religious houses; (2) a map depicting religious beliefs about the world or the universe.

B The Reformation in Europe brought about the need to differentiate the religious allegiances of towns and cities. Thus Nikolaus Claudianus's map of Bohemia, 1518, denotes Catholic cities by crossed keys, Hussite by chalices. Pieter van der Beke on his map of Flanders, 1538, classifies religious establishments by initials (see Letter symbol, 5.122). Sebastian Münster's Bohemia in Ptolemy's *Geography,* Basle, 1545, uses symbols for the religious and civil status of towns (see Legend, 6.121). George Lily (GLA), an English Catholic exile at the Papal Court, indicated by conventional signs on his map of the British Isles, published at Rome in 1546, the status of archepiscopal and episcopal sees, and includes relevant information in the long legends. Gerard Mercator's map of Europe, 1554, which encompassed adjoining regions, has legends of religious significance on such topics as the travels of Jesus Christ, St Peter and St Paul. Abraham Ortelius on five maps in his *Theatrum Orbis Terrarum* (1570) uses symbols for religious sites (see Legend, 6.121). Maurice Bouguereau in his *Theatre françoys* (Paris, 1594), the first atlas of France, included a map of the diocese of Mans, entitled "Nova et integra Coe[n]oma[n]iae descriptio vulgo le Mans," with detailed ecclesiastical information (Dainville).

An early atlas of religion is "Chorographia Descriptio provinciarum, et conventuum Fratrum Minorum S. Francisci Capucinorum," based on the "Chorographia Descriptio" of Joannes a Montecalerio, first published in 1643, with new editions at least until 1712 (reissued 1721). The maps display the locations of Franciscan houses throughout Europe. James Wyld's "Chart of the World showing Religion . . . 1815" is an example of the enlarging field of thematic mapping in the 19th century (see Population map, 2.161). With the increase in statistical information from periodic censuses, more detailed mapping was possible; hence "Dr. Hume's Religious Map of England," in *Remarks on the Census of Religious Worship in England and Wales, with suggestions for an improved Census in 1861, and a map, illustrating the religious condition of the country,* by the Rev. A Hume (London, Liverpool, 1860).

When the Jesuit missionaries established themselves in Japan, it became important to indicate on maps the strength of the missions for reports to Europe. The chart of Japan – the first special map made by a European – in Fernão Vaz Dourado's atlas of 1568, folio 9 (Biblioteca Duques de Alba, Madrid), has flags of the cross to mark Christian centres. A manuscript map in Florence entitled "Japam," brought back by members of the Jesuit embassy of 1585, also shows by flags of the cross Christian missions (Florence, Medici papers). The earliest European printed map of Japan, published in Renwart Cysat's *Warhafftiger Bericht* at Freyburg, Switzerland, in 1586, marks the seminaries and novitiates of the Jesuit order (Kish). The Jesuits were also active in recording the strength of Christianity in China. A manuscript map of Sunkiang prefecture, drawn between 1661 and 1725, was evidently made for this purpose. More than 100 churches are recorded, most of them designated in Chinese "t'ien chu t'ang" (hall of the Lord of Heaven), and in Latin "templum Domini" (Church of the Lord) (Wallis *et al.*, C.20).

Maps depicting man's religious concept of the world date back to ancient times. The earliest, showing the Babylonian cosmos, 7th century BC, was found at Sippar in central Iraq (British Museum, Department of Western Asiatic Antiquities, No. 92687). It shows the earth as a circular disc surrounded by the "Earthly Ocean," entitled the "Bitter River," and has Babylon at the hub of the universe. An ancient Germanic map of the cosmos was found at Moordorf and comprises a gold disc believed to date from the middle of the second millennium (Hanover, Provinzial Museum; Unger).

One of the most celebrated Christian world maps of medieval times is the "Psalter map" in an illuminated Latin Psalter written in England in the 13th century. It comprises a "T-O" map with Jerusalem at the centre. The figure of Christ censed by two angels surmounts the circle of the world (British Library, Add. MS.28,681, f.9.); a second T-O map, more diagrammatic and filled with text, is on the verso (Fig. 7). Both date from about 1275. Another T-O world map represents the ideas of the Christian Moors in Spain. The map was completed in 1109, and is found in the second book of a commentary on the Apocalypse (written about 787), which was preserved in the monastery of Silos, in the diocese of Burgos, Old Castile (British Library, Add MS 11,695.f). Adam and Eve are portrayed in the Garden of Eden as on many medieval world maps, for which Jerusalem at the centre and the earthly paradise at the top of the map were major features. Both the Hereford map by Richard of Haldingham, *c.* 1285 (Hereford Cathedral) and the Ebstorf world map, *c.* 1235 (destroyed) displayed encyclopaedic information on Christian beliefs and traditions.

In and after the renaissance, cartographers concentrated their interest on historical mapping of Biblical topics. The site of Paradise remained a matter of continuing speculation. General Charles Gordon's manuscript maps (1882), which show Paradise sited on Praslin in the Seychelles, provide one of the more recent and more ingenious solutions (Plaut).

A long tradition of Buddhist maps exists in China, Japan, and Korea. The Buddhist prototype was the Go-tenjiku Zu (Map of the Five Indies), whose characteristic feature was the shield-shaped continent of Jambu-dvīpa, comprising the habitable world according to Buddhist cosmography. The sacred lake of Anavatapta lies at the centre, and the four rivers flowing in a whirlpool pattern represent the Ganges, Oxus, Indus, and Tarim. The map entitled "Go-tenjiku-Zu" (Map of the Five Indies), preserved in the Hōryuzi temple at Nara, Japan, is one of the most detailed and has been dated 1365 (Nakamura). The work of the scholar priest Zuda Rōkashi (Priest Hōtan) is also notable in illustrating the attempt to fuse Buddhist dogma and Western geographical knowledge, seen, for example, in his woodcut "Nansen Bushu Bankoku Shoku no Zu" (Outline map of all the countries in Jambu-dvīpa), Kyoto, published by Obei Bundaiken, 1710 (Wallis *et al.*, J.3). Korean world maps showing the Cosmic

Trees and derived from the Indian system of four continents were usually included in Korean world atlases of the 17th and 18th centuries. Tibetan mandela diagrams also represent the iconography of the cosmos (Mackay). Cosmic maps of the Hindus (18th century?) are preserved in Europe and India (e.g. Warsaw, National Library).

C DAINVILLE, François de (1956): *Cartes anciennes de l'église de France.* Paris, Librairie philosophique J. Vrin.

KISH, George (1949): "Some Aspects of the Missionary Cartography of Japan during the Sixteenth Century," *Imago Mundi, 6,* 39-47.

MACKAY, A L (1975): "Kim Su-Hong and the Korean Cartographic Tradition," *Imago Mundi, 27,* 27-38.

NAKAMURA, Hirosi (1947): "Old Chinese World Maps Preserved by the Koreans," *Imago Mundi, 4,* 3-22.

PLAUT, Fred (1984): "Where Is Paradise? The Mapping of a Myth," *The Map Collector, 29,* 2-7.

UNGER, Eckhard (1937): "From the Cosmos Picture to the World Map," *Imago Mundi, 2,* 1-7.

WALLIS, Helen (1975): "Missionary Cartographers to China," *Geographical Magazine, 47,* 751-59.

WALLIS, Helen, and others (1974): *Chinese and Japanese Maps.* London, British Library Board.

2.191 Sanitary Map

A A map showing the environmental conditions believed to be related to variations in the incidence of health and sickness, usually of urban areas. The name was used on occasion for maps that marked the locations of "sanitary cordons," quarantine lines intended to prevent the spread of disease. The term was applied loosely and even included poverty maps (2.162).

B A few maps relating sickness, environment, and quarantine locations were made in the late 17th century. An example is the map of the province of Bari in Italy by Filippo de Arrieta, Royal Auditor of the province, in his *Raguaglio historico del contaggio occorso nella Provincia di Bari negli anni 1690, 1691, e 1692* (Naples, 1694). The segregated area is delimited and centres of infection are indicated. The real development of sanitary maps, however, seems to have begun early in the 19th century with the mapping of cases of yellow fever in cities of the eastern United States. These maps were crude and amounted to little more than the plotting of the locations of cases on existing base maps (see Disease map, 2.041).

Sanitary maps worthy of the name began to appear soon after the cholera reached western Europe about 1830. The first such map seems to have been that by Robert Baker, *Report of the Leeds Board of Health, 1833,* showing the relation between sewerage, drainage, paving, and the incidence of cholera. Soon thereafter numerous such maps were made in most of the countries of western Europe, and by the 1840s it had become a common map type. Some were elaborate, such as Henry Wentworth Acland's map in his *Memoir on the Cholera at Oxford in the Year 1854 with Considerations Suggested by the Epidemic* (London, 1856). Sanitary maps were abundant until the late 19th century when the cause of cholera was learned.

Mapping the incidence of sickness along with environmental factors is still a common form of medical cartography, but the term "sanitary map" appears to have gone out of use by the end of the 19th century.

C GILBERT, E W (1958): "Pioneer Maps of Health and Disease in England," *Geographical Journal, 124*, 172-83.

JARCHO, Saul (1970): "Yellow Fever, Cholera, and the Beginnings of Medical Cartography," *Journal of the History of Medicine and Allied Sciences, 25*, 131-41.

JARCHO, Saul (1983): "Some Early Italian Epidemiological Maps," *Imago Mundi, 35*, 9-19.

JUSATZ, Helmut J (1939): "Zur Entwicklungsgeschichte der medizinisch-geographischen Karten in Deutschland," *Mitteilungen des Reichsamts für Landesaufnahme, 15* (1), 11-22.

JUSATZ, Helmut J (1969): "Medical Mapping as a Contribution to Human Ecology," *Bulletin, Geography & Map Division, Special Libraries Association, 78*, 19-23.

LEARMONTH, A T A (1969): "Viewpoints on Medical Cartography: A Selective Review," *Bulletin, Geography & Map Division, Special Libraries Association, 78*, 32-38.

SPENCER, F J (1969): " 'Woodworth's Tome,' a Biblio-geographical Contribution to Medical History," *Bulletin, Geography & Map Division, Special Libraries Association, 78*, 2-8.

STEVENSON, Lloyd G (1965): "Putting Disease on the Map: The Early Use of Spot Maps in the Study of Yellow Fever," *Journal of the History of Medicine and Allied Sciences, 20*, 226-61.

2.192 Satirical Map

A A map of a real or imaginary place used as a medium for ridicule or satire.

B Medieval examples of the satirical map are the work of Opicinus de Canistris, a clerk from Pavia, who lived at Avignon and saw himself as the Antichrist. His maps of the Mediterranean basin and Western Europe (based on portolan charts) and his diagrams of the world incorporated figures satirizing the Church and the Papacy. In the most famous, one of the Mediterranean maps, 1335-6 (Vatican Library, Lat.

6435, fol 77ʳ), he represents Europe as a female and Africa as a male figure, who like Adam and Eve face each other across the Strait of Gibraltar. A well-known religious satirical map of the renaissance, probably by Jean-Baptiste Trento, is the "Mappe-monde nouvelle papistique," Geneva, 1566. This anti-papal map depicts in carto-graphic form the Pope's tyrannical control over men's souls (Hill, p.51). The "fool's cap world," which shows a map based on Ortelius within the face of a jester's head, first appeared as a woodcut by Jean de Gourmont, Paris, c. 1575, under the motto "Congnois toy toy-mesme." The more famous and larger copperplate version appeared probably at Antwerp, c. 1590, and its author is named by Robert Burton in *The Anatomy of Melancholy* (1621) as Epichthonius Cosmopolites, as yet uniden-tified (Shirley, pp. 157-8, 189-90).

Under a pseudonym, Bishop Joseph Hall in his *Mundus alter et idem* (Frankfurt [ie London?], 1605; London, 1609), uses the hypothetical "Terra Australis Incognita" to portray an imaginary southern continent of vice. Matthias Seutter in his "Accurata Utopiæ Tabula. Das ist . . . des . . . Schlaraffenlandes Neu-erfundene lächerliche Land-Tabell," Augsburg, c. 1730, satirized the life of idleness in the German equivalent of the Land of Cockaigne. His satire on matrimony and love, "Representa-tion Symbolique et ingenieuse projettée en Siege et en Bombardement, comme il faut empecher prudemment les attaques de l'Amour," Augsburg, c. 1730, represented men in a central fortress under siege from the fair sex.

In the 18th and early 19th centuries artists used maps in political cartoons as part of satire rather than creating actual satirical maps (British Museum, Department of Prints and Drawings, *Political and Personal Satires*). In the middle of the 19th century a new genre came into vogue, in which cartoons appeared in the form of maps; for example, "The evil genius of Europe" (anon, 1859), showing Napoleon III trying to pull on the "boot" of Italy. One particularly popular map was Joseph Goggins's "Novel Carte of Europe designed for 1870," which was published not only in English but also in French, German, and Danish.

One of the most prolific "map cartoonists" in England was Frederick Rose, whose so-called "Octopus map," showing Russia as an octopus, was first published in 1877. It was redrawn at least twice to suit the changing political situation, and in 1904 yet another version appeared in Japanese.

C HILL, Gillian (1978): *Cartographical Curiosities.* British Museum Publications for the British Library.

POST, Jeremiah B (1973, 1979): *An Atlas of Fantasy.* Baltimore, Mirage Press; London, Souvenir Press.

SALOMON, Richard (1936): *Opicinus de Canistris.* London, Warburg Institute.

SHIRLEY, Rodney W (1983): *The Mapping of the World.* London, Holland Press.

TOOLEY, Ronald Vere (1963): *Geographical Oddities.* Map Collectors' Series No 1. London, Map Collectors' Circle.

2.201 Tithe Map

A Tithe, the long-established practice of contributing one-tenth of the income from lands to the local parish, was commuted in England for a rent charge by the Tithe Commutation Act of 1836. Maps made for 11,764 tithe districts between 1836 and 1855 depict exactly the rural landscape of England and Wales in these field-by-field surveys. In conjunction with their associated apportionment rolls and files, they record the ownership, occupation, name, state of cultivation and acreage in statute measure of every parcel of titheable land in England and Wales. They show the boundaries of fields, usually distinguishing between open strips and enclosures, woods, roads, streams and the position of inhabited and uninhabited buildings. Some use a well-established system of colour conventions (e.g., brown for arable) to record land use.

B Tithe surveys were compiled between 1836 and 1855 by local land surveyors employed by parish land owners. An original and two certified copies were produced. All the original maps are now in the custody of the Public Record Office, Kew Gardens, Surrey, and about 10,000 diocesan and parish copies are preserved in County Record Office collections.

The survey was organized by Lieutenant Robert Kearsley Dawson, who was seconded from the Royal Engineers to the Tithe Commission in 1836. He recommended a scale of three chains to an inch for the maps (1:2,376) to enable the size of parcels easily to be computed directly from the plans. Drawn at this scale, maps of large parishes measure a hundred square feet or more. Dawson hoped that the tithe maps might be assembled and published as a General Survey of the whole country, but for financial reasons the government decided against this. It would have been expensive to survey areas where maps of sufficient accuracy for tithe commutation purposes already existed. Furthermore, to compel land owners to resurvey such estates would have run counter to the spirit of the 1836 act which tried to encourage voluntary agreements for commutation of tithe.

Although the tithe maps never formed an official cadaster, it was important that they were sufficiently accurate to serve as evidence in tithe disputes. Dawson produced elaborate specifications for surveyors to follow which included a number of diagrams suggesting how systems of internal triangulation of parishes might be laid out, how a parish boundary might be accurately plotted, and how existing surveys might be tested for accuracy.

Most tithe maps contain some biographical information to help unravel their ancestry. Quite often they bear the name of their surveyor, usually a local man whose tithe survey work extended over only a limited area. Further information on the organization of the tithe survey at a particular place will be found in the parish tithe file. All the maps were tested for accuracy in Lieutenant Dawson's office in London, where a series of set procedures was followed. This focussed essentially on comparisons between measurements recorded in surveyors' field books and construction lines required by law to be left on the original map. Maps which passed the rigorous checks were known as "first class maps." Others, which failed on some count or were simply drawn at a scale smaller than the recommended three chains to the inch but which were sufficiently accurate for the immediate purposes of tithe commutation, were known as "second class maps." The 2,333 first class maps can be identified by the presence of the Tithe Commissioners' seal and a statement of accuracy signed by two of them.

The plans accepted by the Tithe Commissioners (that is, both first and second class maps) are on a variety of scales and from a variety of dates. Some are accurate plans, others are little more than crude sketches in comparison with later Ordnance Survey large-scale plans. Some are beautiful examples of the cartographer's art employing the system of conventional signs recommended by Dawson, but many show little more than boundaries and buildings. Although much 19th century informed opinion was critical of the body of tithe maps, recent work has shown that even many second class maps possess levels of planimetric accuracy adequate for most uses as sources in historical studies. In fact their accuracy is sufficient to warrant the continued use of tithe maps as evidence in courts of law, while their uniformity and comprehensiveness is surpassed only by the Land Utilisation Survey of the 1930s. Indeed, they rank as the most complete record of the agrarian landscape of England and Wales at any prior period of history.

C BEECH, Geraldine (1985): "Tithe Maps," *Map Collector,* No. 33, 22-25.

EVANS, Eric John (1976): *The Contentious Tithe: The Tithe Problem and English Agriculture 1750-1850.* London, Routledge & Kegan Paul.

HARLEY, J B (1967): "Enclosure and Tithe Award Maps," *Amateur Historian, 7,* 265-74.

HOOKE, Janet, and PERRY, R A (1976): "The Planimetric Accuracy of Tithe Maps," *Cartographic Journal, 13,* 177-83.

KAIN, Roger J P (1974): "The Tithe Commutation Surveys," *Archaeologia Cantiana, 89,* 101-18.

KAIN, Roger J P (1975): "R. K. Dawson's Proposal in 1836 for a Cadastral Survey of England and Wales," *Cartographical Journal, 12,* 81-8.

KAIN, Roger J P, and PRINCE, Hugh C (1985): *The Tithe Surveys of England and Wales.* Cambridge, Cambridge University Press.

PRINCE, Hugh C (1959): "The Tithe Surveys of the Mid-Nineteenth Century," *Agricultural History Review* (Reading), *7,* 14-26.

2.202 Tourist Map

A (MDTT 822.13) A collective term for a variety of maps designed for uses related to tourism and holiday making on which features of interest such as routes, distances, accommodations, viewpoints, antiquities, and recreational facilities are emphasized. Tourist maps have a long history, but the word is relatively modern. The Oxford English Dictionary shows the first recorded use of the term "tourist" in English to be in 1800.

B The first maps for "tourists" (although not so-called) were itineraries or route maps (1.1810). There are medieval copies of Roman itinerary-maps of which the outstanding example is the so-called Peutinger map, now known in a copy dating from the 11th or 12th century made from a map of the Roman empire of the 4th or 5th century, which itself was probably a revised version of an older original. Hostelries and staging posts are marked, and Levi and Trell have identified three categories of recognized stopping places.

Medieval examples survive of route maps designed for pilgrims, such as the itinerary-map from London to Apulia by Matthew Paris of St Albans, dating from the mid-13th century. Another type of medieval tourist map is Matthew Paris's maps of the Holy Land, *c.* 1250. One of the earliest printed travel books with maps is the *Peregrinatio in Terram Sanctam* of Bernhard von Breydenbach (Mainz, 1486), with a map of Palestine and views of cities by Erhard Reuwick. The "Isolarios" or island atlases (8.0110c), first produced in manuscript in 1420 and in printed form in 1485, may also be regarded as a type of tourist guide. Erhard Etzlaub's "Romweg" map, which shows routes from northern Europe to Rome in the Holy Year 1500, with each dot on the roads equal to a German mile, is notable as one of the first printed tourist maps. The first printed atlas for travellers in the Christian world is a small atlas in four parts, *Itinerarium Orbis Christiani,* 1579-80. A new edition of the part relating to Germany was printed at Cologne in 1598 (see Road map, 1.1810d).

Father Vincenzo Coronelli, cosmographer of Venice, published in his *Viaggi,* 1697, accounts of his travels in Italy, to London, and on the continent of Europe, which were illustrated with maps and plans. In his introduction he discussed the utility of travel and gave practical advice, recommending the use of modern topographical maps. This work may be regarded as the forerunner of the Baedeker guides of the 19th century.

In China it was traditional for the emperor to have maps and plans made for all tours as well as military campaigns, as illustrated by "The Album by Ch'ien Wei-ch'eng of the Fifth Journey of the Emperor Ch'ien-lung," 18th century, MS. (British Library, Department of Oriental Manuscripts and Printed Books, Or. 12895). The resting places of the Emperor Ch'ien-lung on his provincial tours had to be recorded with accuracy, as all the actions of the emperor had the utmost sanctity. An early example of the Japanese guide map is by Hishikawa Moronobu of the road known as the Tokaidō between Kyoto and Edo (Tokyo), dating from 1690 (Fig. 5).

C CORONELLI, Vincenzo (1697): *Viaggio d'Italia in Inghilterra,* 2 vols. Venice.

LEVI, Annalina C, and TRELL, Bluma (1964): "An Ancient Tourist Map," *Archaeology, 17,* 227–36.

2.203 Traffic Map

A A map showing the volume of passenger or freight movement along routes of transportation. Commonly the magnitudes were symbolized by bands or lines, the widths of which were made proportional to the amounts. Traffic maps are also called flow maps.

B The first published traffic maps were by Henry Drury Harness portraying passenger and goods traffic in Ireland, dated 1837. Harness was followed in 1845 by Alphonse Belpaire (Belgium) and Charles Joseph Minard (France) (see Flow Maps, 1.062).

Minard was by far the most prolific early producer of traffic maps, making many maps of France, Europe and the world, and he is certainly the individual who popularized this type of map. His subject matter ranged widely, from the export of British coal and French wines to passenger traffic on European railways. His series of extremely detailed maps showing the tonnage of merchandise which circulated in France on all waterways and railways began in 1852 and was followed by more than a dozen similar maps for later years. Minard's traffic maps were recognized by the International Statistical Association in 1857, and they were widely used in French official quarters.

C ROBINSON, Arthur H (1955): "The 1837 Maps of Henry Drury Harness," *Geographical Journal, 121,* 440-50.

ROBINSON, Arthur H (1967): "The Thematic Maps of Charles Joseph Minard," *Imago Mundi, 21,* 95-108.

GROUP 3

Maps of Natural Phenomena

3.011 Astronomical Map

A A flat map of the celestial sphere, or of some part thereof. Information displayed may include stars, constellations, and co-ordinates. Compare with celestial globe (1.0710a).

B In view of the importance of the stars in determining the agricultural calendar and navigation and, to a lesser degree, for time-reckoning in primitive society and in early times, it is not surprising that representations of individual constellations or parts of the night sky have been discerned in prehistoric cup-marks on monoliths and standing stones, though few so far known could be called maps.

The astronomical sciences of ancient Babylon are believed to have been a common source of early Chinese and Hellenistic astronomical systems. Joseph Needham shows that diagrams on cuneiform tablets dating from the late 2nd millennium BC, preserved in the library of King Assurbanipal (668-626 BC) at Ninevah, may represent primitive planispheres showing circumpolar stars and corresponding equatorial "moon-stations." Planispheres have also been identified from the Shang period (1400 to 1000 BC) (Needham, pp. 254-6). Babylonian cosmological ideas were transmitted to China and also to India by way of Persia.

China claims the longest and most continuous tradition of celestial mapping, producing astronomical maps based on cartographic projection from which positions and angular distances could be read. Astronomers determined star positions as early as the 4th century BC, and had mapped them by the early 4th century AD (Needham,

135

p. 263). The oldest extant Chinese astronomical map, entitled "Ch'i chich mieh chi tien ching," is dated 940. It was found in the early years of the 20th century by the archaeologist Sir Aurel Stein in a great library of early Chinese and Central Asian manuscripts at Tunhuang on the edge of the Gobi desert (British Library, OMPB S.3326). The map illustrates the traditional Chinese method of indicating constellations by means of circles linked by lines and seems to be the earliest extant example of a colour notation being used to distinguish astronomical systems. The oldest extant printed Chinese astronomical map is the Suchow planisphere prepared in 1193 and engraved on stone in 1247. It is now preserved in the Confucian temple in Suchow, Chiangsu.

Although ancient and medieval astronomers in the West seem to have used celestial globes rather than maps, it is not always possible to determine whether early references are to spheres or to charts. Anaximander of Miletus (early 6th century BC) is said by Diogenes Laertius (writing in the earlier decades of the 3rd century AD, Book II. chapter 1) and Pliny *(Natural History,* vii, 20, published AD 77), to have made a "sphere" which may, however, have been a chart of the heavens (Kahn, pp. 60, 89; see also Celestial globe, 1.0710a). Flat projections of the skies were made as the retes of astrolabes. Early planispheres which have survived are stylized representations from which positions cannot be read. Thus the planisphere of Bianchini, a marble tablet of the 2nd or 3rd century AD, now in the Louvre, has a general zodiac scheme of concentric circles and intersecting radii presumably characteristic of Greek planispheres and globes such as Anaximander's (Kahn, p.89). Illustrations of individual constellations appear in medieval copies of two popular classical works, the poetic description of the heavens, known as the *Phaenomena*, by the Hellenistic poet Aratus of Soli (3rd century BC), and the Latin *Poeticon astronomicon* of Caius Julius Hyginus (1st century BC). The stars are often correctly placed, and the forms clearly are derived from classical prototypes. Notable examples are the 10th century manuscript al-Sufi's "Book of the Fixed Stars" (Bodleian Library, Oxford, Marsh 144), and the early medieval manuscript known as the "Harley Aratus" (British Library, Harley MS 647). The main part of the Harleian manuscript, which includes depictions of the individual constellations, is in a French mid-9th century hand and was probably written under the direction of Lupus of Ferrières, c. 805-62 (Saxl and Wittkower, p.30). The manuscript is known to have reached Canterbury Cathedral by the year 1000, and the celestial planisphere which completes the volume (fol 21b) is distinctively English and evidently dates from the early 11th century (Saxl and Meier, 1953, part 2, pp. 149-153). The work became available to other monasteries, and through copying by English monks up to the middle of the 12th century was one of the main sources of English knowledge of the stars.

Advances in celestial cartography in the European renaissance date from the early 15th century with such works as Conrad of Dyffenbach's maps of 1426 (Vatican, codex Palat. Lat. 1368), Paolo Toscanelli's maps of the paths of comets, 1433-1457, shown against the fixed stars (Firenze, Biblioteca Nazionale Centrale, Codice Magliabechiano classe xi; Jervis), and planispheres of the Ptolemaic stars, dated about 1440 (Vienna, Nationalbibliothek, codex 5415).

Albrecht Dürer's woodcut planispheres published at Nuremberg in 1515 were the first printed astronomical maps in the West. Their co-ordinates were drawn by the Imperial mathematician Johann Stabius (d. 1522) and the stars fixed by Conrad Heinfogel (1470-1530). These represent the stellar universe as seen from outer space.

Johannes Honter of Kronstadt (1498-1549) achieved a further advance in showing for the first time on printed (as distinct from manuscript) planispheres the celestial sphere as seen from the earth. His pair of woodcut planispheres bearing the initials "I.H.C.," were published at Basel in 1532. The works of both Dürer and Honter were used in turn by Peter Apian, professor of mathematics at Ingolstadt, for his own map of 1536, which he included in his *Astronomicum Caesarum,* published at Ingolstadt in 1540. This remarkable but somewhat obsolete work comprised a graphical representation of the universe according to the Ptolemaic system.

The Italian astronomer Giovanni Paolo Gallucci produced the first printed star atlas from which co-ordinates could be read: the *Theatrum Mundi, et Temporis,* Venice, 1588. This volume thus predated the work popularly known as the first star atlas, the *Uranometria omnium asterismorum* of Johann Bayer, a lawyer of Augsburg, which was published at Augsburg in 1603.

An early map of the moon was made from observations by the naked eye by William Gilbert in 1603, and published posthumously in his *De Mundo,* Amsterdam, 1651. With the invention of the telescope in about 1608 the more detailed mapping of the moon became possible. The earliest astronomical map made by observation with a telescope was Thomas Harriot's map of the moon, drawn on 26 July 1609. He made a further series of observations and maps in 1610, all of which remained in manuscript (Petworth House, Sussex, HMC, 241, ix,ff.1-46). Galileo Galilei also observed the moon by telescope and included a sketchy map in his *Sidereus nuncius,* Venice, 1610. The first detailed map of the moon, published in Matthias Hirzgarter's *Detectio dioptrica corporum planetarum verorum,* Franckfurt am Mayn, 1643, was otherwise of little note (see John North in *Dictionary of Scientific Biography,* vol 6, 1972, 368-69). Johannes Hevelius, the great astronomer of Danzig (1611-87), published the first atlas of the moon in the *Selenographia* (Danzig, 1647), recording his lunar observations and providing a new nomenclature. (See also Lunar globe, 1.0710b.)

In the 17th and 18th centuries the traditional styles of celestial cartography were elaborated with new iconographies, and the rapidly expanding knowledge of the stars led to the incorporation of many new constellations. A move to reform the system led to the production of functional maps appropriate to 19th century scientific education and research.

C BROWN, Basil (1932): *Astronomical Atlases, Maps & Charts.* London. (Facsimile, 1968.)

DELANO SMITH, Catherine (1982): "The Emergence of 'Maps' in European Rock Art," *Imago Mundi, 34,* 9-25.

JERVIS, Jane L (1985): *Cometary Theory in Fifteenth-Century Europe.* Studia Copernicana, 26. Wroclaw, Polish Academy of Sciences Press.

KAHN, Charles H (1960): *Anaximander and the Origins of Greek Cosmology.* New York, Columbia University Press.

LAERTIUS, Diogenes (1925): *Lives of Eminent Philosophers.* With an English translation by R D Hicks. London.

NEEDHAM, Joseph (1959): *Science and Civilisation in China,* vol 3. Cambridge, At the University Press.

OTTLEY, William Young (1836): "On a MS of Cicero's translation of Aratus," *Archaeologia, 26,* 47-214.

SAXL, Fritz (1915): "Verzeichnis astrologischer und mythologischer illustrierter Handschriften des lateinischen Mittelalters," parts 1 and 2, in *Sitzungsberichte der Heidelberger Akademie der Wissenschaften,* Philosophisch-historische Klasse, Abh 6-7 (1915); Abh 2 (1925-26); part 3 (with Hans MEIER), London, The Warburg Institute, 1953.

SAXL, Fritz, and WITTKOWER, K (1948): *British Art and the Mediterranean.* London, New York.

SCHÜTTE, Gudmund (1920): "Primaeval Astronomy in Scandinavia," *Scottish Geographical Magazine, 36,* 244-54.

SHIRLEY, John W (1978): "Thomas Harriot's Lunar Observations," in *Science and History: Studies in Honor of Edward Rosen.* Studia Copernicana, 16. Wroclaw, Polish Academy of Sciences Press, 283-308.

WALLIS, Helen, and others (1974): *Chinese and Japanese Maps.* London, British Library Board.

WARNER, Deborah Jean (1979): *The Sky Explored.* New York.

3.021 Block Diagram

A A perspective map of a segment ("block") of the land portrayed as if removed from its adjacent areas and showing the underlying rock formations on the front and sometimes also the side of the block. The primary purpose of the block diagram is to show the relation between the subsurface structure and the land surface, although a series of block diagrams can illustrate geomorphic evolution.

B Perspective views of the land surface or features of it, such as mountains, are as old as cartography, and profiles and sections showing lodes illustrate books as early as 1500. Ulrich Rülein von Calw's *Ein nutzlich bergbuchleyn,* printed *c.* 1500 at Leipzig, includes woodcut engravings of the trends of lodes in perspective view. Georgius Agricola improved on this in his . . . *de re metallica libri XII* (book 3), Basle, 1556, distinguishing the different directions of the lodes, and thus employing the techniques of descriptive geometry (H Prescher and P Schmidt, "The Importance of the Illustration of Ore Lodes in Books by Ulrich Rülein von Calw around 1500 and Georgius Agricola 1556 . . . ," in Dudich, pp. 411-22). Nicolaus Steno in his . . . *Prodromus* (1669) includes a sectional diagram showing six successive stages in the geological history of Tuscany. Lazzaro Moro's *De Crostacei . . .* (1740) shows sections with structure and the land surface.

James Hutton's *Theory of the Earth* (Edinburgh, 1795, plate 3) depicts an unconformity with a perspective view of the landscape above it. Hutton planned to produce a third volume with many plates, but he died in 1797 before the engravings were completed. Nothing more was known of the drawings until 1968 when they were found among the papers of John Clerk of Eldin, the artist who had prepared them from his own original sketches. These include a number of "extraordinarily accurate" block diagrams, *c.* 1785 (Craig, et al, pp. 6, 29).

Probably the first to show rock structure with a perspective view of the landform in relation to it, not just landscape, was William Smith in 1815. J Peter Lesley's *Manual of Coal and Its Topography* (Philadelphia, 1856) shows a perspective diagram of the outcropping of plunging folds. John Wesley Powell's *Exploration of the Colorado River of the West* (1875) includes several perspective sections and drawings, and this practice was followed by G K Gilbert and C E Dutton. In Europe at this time G de la Noë and E de Margerie's *Les Formes du terrain* (1888) includes a variety of block diagrams.

Although the block diagram clearly evolved over a considerable period, William Morris Davis popularized it as a graphic device to show geomorphologic interrelationships among structure, time, and surface. He began his lengthy series of publications in the 1880s, but it was probably his *Physical Geography* (1898) that was most influential in elevating the block diagram from the status of a rarely used technical illustration in scientific writings to a common graphic device.

C ADAMS, Frank D (1938): *The Birth and Development of the Geological Sciences.* Baltimore, Williams and Wilkins.

BROWN, Charges Barrington, and DEBENHAM, Frank (1929): *Structure and Surface.* London, Edward Arnold.

CASTELNAU, Paul (1912): "La theorie du bloc-diagramme," *Bulletin de la Société de Topographie de France,* July-August, 121-36.

CHORLEY, Richard J, DUNN, Antony J, and BECKINSALE, Robert P (1964, 1973): *The History of the Study of Landforms, or the Development of Geomorphology,* 2 vols. London, Methuen; New York, John Wiley.

CRAIG, G Y, McINTYRE, D B, and WATERSTON, C D (1978): *James Hutton's Theory of the Earth: The Lost Drawings.* Edinburgh, Scottish Academic Press.

DUDICH, E, ed. (1984): *Contributions to the History of Geological Mapping: Proceedings of the Xth INHIGEO Symposium . . . 1982.* Budapest, Akadémiai Kiadó.

ECKERT, Max (1921): *Die Kartenwissenschaft.* Berlin and Leipzig, Walter de Gruyter, vol. 1, 450-53.

3.022 Botanical Map

A A map that indicates the typical plants of an area or a distribution map that portrays the variations in vegetation from place to place or the common or dominant elements.

B Plants have appeared on maps from classical times. The Peutinger map, the Madaba mosaic map, the Beatus maps, the Ptolemaic maps, and others up to the 18th century show kinds of trees, such as palms, and often forests. Detailed delineation of forests and vegetation categories began to appear on the 16th century regional maps, such as Philipp Apian's *Bairische Landtaflen* (1568), and larger scale topographical maps, such as the estate and county maps of England and the *plans terriers* of France. The more detailed, large-scale topographical map series, which began in the 18th century, included vegetation categories as an important element. In all such maps vegetation was recorded as only one of many items of interest.

Local forestry maps of areas of Germany, France, and Alsace date from the 16th century. The Nuremberg cartographer Erhard Etzlaub is believed to be the author of a map, "Nürnberg im Reichswald," 1516, showing the administrative divisions of the Nuremberg imperial forests (Schnelbögl, p.22). The artist and mapmaker David Kandel (1524-96) made two coloured plans of hardwood areas between Molsheim-Altdorf-Dorlisheim. An anonymous map, *c.* 1573, of the area between Schirmeck and Grendelbruch is identified as the first of the local forestry maps of Alsace (preserved in the Archives du Bas-Rhin, Strasbourg). The "Atlas des Plans des Forêts," MS, by the Paris mapmaker Nicolas Lallement, late 17th century, was prepared in connection with an inventory of all the forests in royal ownership, as one of the economic measures of the "Reformation Colbertienne" (1661-90) (Grenacher, pp. 61, 63).

The first thematic botanical maps, in which the distribution of vegetation types is the primary focus of interest, seem to be two small maps by Carl Ritter for his atlas *Sechs Karten von Europa* (1804-06). These are simple outline maps on which the various names of the plants are lettered here and there and on which the approximate limits of some species are noted. One map shows cultivated plants, the other wild trees and shrubs. Ritter's maps are relatively crude and are similar to Zimmermann's earlier zoological maps (3.262).

Far more sophisticated and innovative are the maps of J F Schouw, a Danish botanist who published in 1823 a separate atlas to accompany his large treatise *Grundzüge einer allgemeinen Pflanzengeographie* (1823), which contained 12 sets of paired eastern and western hemispheres showing by colours and patterns the distributions of various classes of plants, such as cereals and palms. In 1833 Schouw published a physical-geographical description of Europe with an accompanying atlas. The atlas included two simple maps showing the distribution of wild trees and shrubs and important cultivated plants. Another early botanical map is Charles Pickering's map of the distribution of plants in North America published in 1830.

Botanical maps rapidly became sophisticated in the 1830s. The *Physikalischer Atlas* by Berghaus (Part I, 1845), contained six maps on botanical geography made between 1838 and 1841. After the mid-1840s botanical maps were relatively common.

C BERGHAUS, Henrich (1845): *Physikalischer Atlas oder Sammlung von Karten* . . . (Part I). Gotha, Justus Perthes.

DAINVILLE, François de (1964): *Le langage des géographes.* Paris, Éditions A et J Picard, 190-92.

ECKERT, Max (1925): *Die Kartenwissenschaft,* vol. 2. Berlin and Leipzig, Walter de Gruyter, 385-406.

FONCIN, Myriem (1961): "La représentation de la végétation sur les cartes anciennes," in *Colloques du C.N.R.S. Méthode de la cartographie de la végétation,* Toulouse, 1960.

GRENACHER, Franz (1964): "Current Knowledge of Alsatian Cartography," *Imago Mundi, 18,* 60-77.

PICKERING, Charles (1830): "Map of North America," in "On the Geographical Distribution of Plants," *Transactions of the American Philosophical Society, 3,* 113-17.

RITTER, Carl (1806): *Sechs Karten von Europa mit erklärendem Texte . . .* Schnepfenthal.

ROBINSON, Arthur H (1982): *Early Thematic Mapping in the History of Cartography.* Chicago, University of Chicago Press, 100-105.

SCHNELBÖGL, Fritz (1966): "Life and Work of the Nuremberg Cartographer Erhard Etzlaub," *Imago Mundi, 20,* 11-26.

SCHOUW, Joakim F (1823): *Pflanzengeographischer Atlas zur Erläuterung von Schouws Grundzügen einer allgemeinen Pflanzengeographie.* Berlin, Reimer.

SCHOUW, Joakim F (1833): *Atlas zu Schouws Europa.* Copenhagen.

WALLIS, Helen (1981): "The History of Land Use Mapping," *Cartographic Journal, 18,* 45-48.

3.031 Compass Declination Map

A A map showing the compass declination for numerous places, i.e., the angular value of the difference between true north and magnetic north. Variation is usually indicated by means of lines of equal compass declination called isogones (5.0910e), but it can be shown by arrows or compass roses which lie at an angle with the meridians over the map.

B The phenomenon of compass variation was observed by Europeans in the thirteenth century. Roger Bacon commented on it in his "Opus Minus," 1266, and Petrus Peregrinus in his "Epistola de magnete," 1269. The claim that Christopher Columbus discovered on his first voyage to the West Indies, 1492, the general "space-variation" of magnetic declination is now rejected (Mitchell, 1937). The fact that there was an easterly variation in northwest and central Europe was in fact well known before Columbus sailed. It was indicated on the map entitled "Das ist der Romweg," made by the Nuremberg mapmaker Erhard Etzlaub, *c.* 1500 (see Magnetic north, 4.132).

The earliest attempt to map the variations in declination apparently was in 1536 by the Spaniard Alonso de Santa Cruz, who had the idea that magnitudes of equal declination paralleled the meridians and who noted the supposed amounts of declination on the meridians of the world map. According to Martin Fernandez de

Navarrete, 1846, the pilots could agree on only three points at which the declination had been determined – Santo Domingo, Havana and New Spain – but they may have been concerned only with the West Indian region (Mitchell, 1937, pp. 270, 280). The earliest surviving manuscript chart to show magnetic variation is a map of the Atlantic, dated the first of June 1576, prepared by William Borough, the English navigator, for the use of Martin Frobisher on his first voyage in search of the Northwest Passage, and marked with his discoveries, 1576, and additions to 1578. Five arrows indicate the increasing westerly deviation observed by Frobisher on his course to Greenland. Sir Robert Dudley, the English hydrographer and voyager, made extensive records of variation by written words, not symbols, on his manuscript charts, c. 1636 (preserved in the Bayerische Staatsbibliothek, Munich), which were later engraved and published in his sea atlas of the world *Dell' Arcano del Mare* (Florence, 1646-47). At the same time that Dudley was compiling records from his own and other voyages, the English mathematician Edward Wright on his manuscript chart (c. 1595, at Hatfield House) and printed chart of the Northeast Atlantic, 1599, and on his world chart, 1610, included extensive records of variation with the purpose of warning seamen of errors in navigation.

The Portuguese chartmaker Luís Teixeira seems to have been the first to draw curved lines of equal declination on a map. His lines drawn c. 1585 appear on a manuscript chart of the Pacific which was probably part of a world map and is preserved in the Museu de Marinha, Lisbon (Cortesão and Teixeira da Mota, pp. 71-72, Plate 363). The lines are labelled for variations in bearings and thus seem to be an early form of isogone. The chart may represent a late version of the magnetic theory expounded by João de Lisboa in his *Tratado da Agulha de Marear* (1514).

The Jesuit mathematician and geographer Athanasius Kircher reported *(Magnes; sive de arte magnetica opus tripartitum . . .* , 1641) that the Milanese Jesuit scientist Father Cristoforo Borri sometime in the first half of the 17th century and probably while in Portugal had constructed a map, "mappa geographico-magnetica," with lines of equal declination which were parallel with one another. Kircher, on the basis of the information obtained from the archives of earlier correspondence in Rome and from missionaries all over the world, deduced that the isogonic lines would not be parallel. In the 1654 edition (3rd) of that work Kircher reports that his student, Jesuit Father Martinus Martini, had drawn lines of equal declination on a world map. Kircher also commented that a complete map was by then possible, and he would have added it to his work if time and cost had allowed. The English mathematician William Whiston in *The Longitude and Latitude Found by the Inclinatory or Dipping Needle* (London, 1719) alludes to a manuscript map of declination of c. 1640.

The first published map of compass declination is by Edmond Halley (1701) based on his own observations in the Atlantic (A New and Correct Chart shewing the Variations of the Compass in the Western and Southern Oceans") (Fig. 16). This was followed a year later by Halley's declination chart of the world (lacking isogones in the Pacific area). The world chart became widely known and in revised form reappeared regularly for some half century. Halley's maps of compass declination rank among the most influential maps made; for his use of the "curve line," a line along which some mathematical relationship remains constant, was the model for many such isolines in the following centuries.

C CORTESÃO, Armando, and TEIXEIRA DA MOTA, Avelino (1960): *Portugaliae Monumenta Cartographica*, Vol. 3. Lisbon.

HELLMANN, Gustav (1895): *Neudrucke von Schriften und Karten über Meteorologie und Erdmagnetismus*, No. 4. Berlin, A. Asher.

HELLMANN, Gustav (1909): "Magnetische Kartographie in historisch-kritischer Darstellung," *Veröffentlichungen des Königlich Preussisches Meteorologisches Institut*, No. 215, Bd. III (3), 5-61.

MITCHELL, A Crichton (1932, 1937, 1939): "Chapters in the History of Terrestrial Magnetism," *Terrestrial Magnetism and Atmospheric Electricity, 37*, 105-46; *42*, 241-80; *44*, 77-80.

ROBINSON, Arthur H (1971): "The Genealogy of the Isopleth," *Cartographic Journal, 8*, 49-53. (Also in *Surveying and Mapping, 32* (1972), 331-38. Translated into Dutch, "De Afstammung van de Isopleth," *Kaartbulletin, 34* (1973), 5-14.)

THROWER, Norman J W (1969): "Edmond Halley as a Thematic Geo-Cartographer," *Annals of the Association of American Geographers, 59*, 652-76.

WALLIS, Helen (1973): "Maps as a Medium of Scientific Communication," *Études d'Histoire de la Géographie et de la Cartographie*, Wroclaw, Warszawa, . . . Poliskiej Akademii Nauk, 251-62.

3.051 Eclipse Map

A A cartographic representation of the predicted areal extent (shape), track or path, and time of passage of the shadow of the moon on the Earth's surface. This lunar shadow results from a total eclipse of the sun.

B Eclipse maps became popular in the 18th century after the publication of Edmond Halley's map "A Description of the Passage of the Shadow of the Moon over England in the Total Eclipse of the Sun on the 22nd Day of April, 1715 in the Morning." This is apparently the earliest such map which was made before the event it depicts, and it ranks as the first thematic map to have wide popular appeal (A H Robinson).

C HALLEY, Edmond (1715): "Observations of the late Total Eclipse of the Sun on the 22nd of April . . . ," *Philosophical Transactions* (Royal Society of London), *29*, no. 343, 245-62.

THROWER, Norman J W (1969): "Edmond Halley as a Thematic Geo-Cartographer," *Annals of the Association of American Geographers, 59*, 652-76.

3.071 Geological Map

A A broad category encompassing maps portraying the age and distribution of kinds of surface and sub-surface rock materials. Not to be confused with mineral (3.131) and morphological (3.132) maps.

B The first geological map is generally alleged to be that of Christopher Packe who in 1736 exhibited his ". . . Philosophico-Chorographical Chart of East Kent . . ." to the Royal Society. The map was completed and published in 1743. Among a variety of other features, it showed the character of some surface materials. A more valid claim can be made for two maps made by Philippe Buache, dated 1746, to illustrate a paper by the geologist Jean-Étienne Guettard which appeared in 1751 ("Mémoire et carte minéralogique sur la nature et la situation des terreins qui traversent la France et l'Angleterre," *Mémoires de l'Academie Royale des Sciences, Année 1746*, Paris, 1751, pp. 363-92, plates 31 and 32). In addition to showing the occurrence of various minerals and rocks by point symbols, shaded areas show the distribution in western Europe, mainly France, of three zones or *bandes* of rock types: *Schiteuse ou metallique, marneuse,* and *sabloneuse.* Several other similar maps made by Buache for Guettard appeared in subsequent years. In 1766 Antoine Laurent Lavoisier, in collaboration with Guettard and A-G Monnet, introduced the concept of the geological cross-section as an aid to understanding geological history on a map in the *Atlas et description minéralogiques de la France* (1780).

In 1762 G C Füchsel made a geological map of a considerable part of Germany, "Historia terrae et maris, ex historia Thuringiae per montium descriptionem erecta" *(Transactions, Electoral Society of Mayence, 2,* 44-209). The first geological map to use colour to distinguish among various formations appears to have been one by Friedrich G Gläser in 1774 to accompany his *Versuch einer mineralogischen Beschreibung der gefürsteten Grafschaft Henneberg chursächsischen Antheils . . . ,* Leipzig, 1775. This was soon followed by Johann F W Charpentier's 1778 map to accompany his *Mineralogische Geographie der chursächsischen Lände,* Leipzig, which used eight colours to distinguish among rock types.

After the last quarter of the 18th century, geological maps became increasingly common and more sophisticated. They may be said to have reached maturity with the publication of the map of the Paris basin, "Carte géognostique des environs de Paris," by Georges Cuvier and Alexandre Brongniart in 1810 and of William Smith's well-known 1815 map "A Delineation of the Strata of England and Wales . . ."

C CAMPBELL, E M J (1949): "An English Philosophico-Chorographical Chart," *Imago Mundi, 6,* 79-84.

GEIKIE, Archibald (1905): *The Founders of Geology,* 2nd edition. London, Macmillan and Company.

IRELAND, H Andrew (1943): "History of the Development of Geologic Maps," *Bulletin of the Geological Society of America, 54,* 1227-80.

KISH, George (1976): "Early Thematic Mapping: The Work of Philippe Buache," *Imago Mundi, 28,* 129-36.

RAPPAPORT, Rhoda (1969): "The Geological Atlas of Guettard, Lavoisier, and Monnet: Conflicting Views of the Nature of Geology," in Schneer, Cecil J, ed., *Towards a History of Geology.* Cambridge, Mass., MIT Press, 272-87.

3.081 Hypsometric Map

A A map showing the measured or inferred elevations of all the land surface above a reference datum. A hypsometric map should not be confused with a map showing only spot heights (5.193) or a map implying relative elevation as by plastic shading or hachuring (5.081).

B A hypsometric map could not be made until elevations could be measured and a symbolic system had been devised to portray them. By the mid-18th century techniques of measuring heights by barometer and theodolite were in use, but the symbolic system of contouring (5.0910a) was required before a hypsometric map could be made. Occasional suggestions for using "level lines" were made from the early 1740s on, but the formal system of contouring was proposed by Marcellin Du Carla in 1771 and published by Jean Louis Dupain-Triel in 1782: *Expression des nivellemens, ou méthode nouvelle pour marquer rigoureusement sur les cartes terrestres & marines les hauteurs & les configurations du terrein,* Paris.

The first hypsometric map appears to be the 1791 map by Dupain-Triel, "La France considérée dans les différentes hauteurs . . ." The data available were insufficient to produce more than a general portrayal of the elevations of the land surface. Dupain-Triel published another map at the same scale and with the same contours in 1798-1799 (Year VII), but with several elevational tints of the same hue between the contours (see Layer tints, 5.121).

In 1813 Göran Wahlenberg, a Swedish botanist, made a map of the Tatra Mountains which used different hues to portray altitudinal-vegetative zones: Mappa Physicogeographica Carpatorum Principalium e quibus Wagus et Dunajetz origines trahunt . . ." to accompany his *Flora Carpatorum Principalium* (Göttingen, 1814). In 1830 the Swedish cartographer Carl Forsell prepared a map of the southern part of the Scandinavian peninsula in which he used four hues, including white, as altitude tints to show the hypsometry.

In 1824 two Danes, Oluf Nikolay Olsen and J H Bredsdorff prepared a manuscript quasi-hypsometric map of the distribution of elevations in Europe to accompany their entry in a competition announced by the Geographical Society of Paris. This map was revised by Olsen in 1830 and published by him in 1833; "Esquisse Orographique de l'Europe, par J H Bredsdorff et O N Olsen en 1824; corrigée . . . par O N Olsen, 1830." In 1842 it was copied in its essentials, with some additions, by Heinrich Berghaus for his *Physikalischer Atlas.*

Leadership in the application of contours to large-scale military and topographic mapping was assumed by France in the second decade of the 19th century. Contours began to be adopted in the 1830s and 1840s by many surveys, and by 1850 hypsometric maps had become relatively common at all scales.

With the perfection of lithography and colour printing, General Ritter von Hauslab (1864) of Vienna experimented with altitude tints, colouring each zone within adjacent contour lines with a slightly different colour. In the first experiments the darker colours were used for the lower elevations, but the system was reversed because the lower altitudes had more detail about man's works and buildings. The system is said to have been first employed by him on a map of Turkey published in 1830 (Lyons, 1914).

C DAINVILLE, François de (1958): "De la profondeur à l'altitude," in *Le Navire et l'économie maritime du moyen âge au XVIIIe siècle principalement en Méditerranée,* Travaux du Deuxième Colloque international d'histoire maritime, Paris, 195-213. (Reprinted in *International Yearbook of Cartography, 2* (1962), 151-62. Translated into English in *Surveying and Mapping, 30* (1970), 389-403.)

FONCIN, M (1961): "Dupin Triel and the First Use of Contours," *Geographical Journal, 127,* 553-4.

HAUSLAB, Josef Ritter von (1864): "Ueber die graphischen Ausführungsmethoden von Höhenschichtenkarten," *Mittheilungen der k.–k. geographischen Gesellschaft* (Wien), *8,* 30-37.

LYONS, H G (1914): "Relief in Cartography," *Geographical Journal, 43,* 233-48; 395-407.

STEINHAUSER, Anton (1858): "Beiträge zur Geschichte der Entstehung und Ausbildung der Niveaukarten, sowohl See- als Landkarten," *Mittheilungen der k.–k. geographischen Gesellschaft* (Wien), *2,* 58-74.

SZAFLARSKI, Józef (1959): "A Map of the Tatra Mountains Drawn by George [sic] Wahlenberg in 1813 as a Prototype of the Contour-Line Map," *Geografiska Annaler, 41,* 74-82.

3.131 Mineral Map

A Mineral maps display the location and character of mineral and ore-bearing rocks, coal seams, and mineral workings such as mines and quarries. They preceded geological maps (3.071) but are closely related to them.

B The earliest surviving map, *c.* 1200 BC, containing details of a mineralogical nature is the Turin Papyrus which shows the ancient Egyptian gold mining centre at Hammamat-Fawakhir during the period 1350-1205 BC. On this highly stylized manuscript map, colour washes are used to indicate the gold-bearing lode and veins.

Central European manuscript maps from the 15th century documemt mining works, and "contain the first evidence of some geological knowledge" (J Urban in Dudich, p. 103). Maps from the 16th century are more frequent and include Kohler's map of mining in Freiberg, Saxony, 1529; Sigismund (Zikmund) Prášek's map of the silver mines of Poličany Kutná Hora, Bohemia, *c.* 1534; and the map of the distribution of ore veins in the upper Nahe River area, Germany, 1574 (W Langer in Dudich, p. 83). One of the earliest English mineral maps is a manuscript sketch map of southwest England, which was apparently drawn by William Cecil, treasurer to Elizabeth I, *c.* 1570. The location of lead and copper is indicated by the word *plumbum* or *cuperum* and marked with a cross symbol (Hatfield House, C.P.M. supp 1; reproduced in R A Skelton and John Summerson, *A Description of Maps and Architectural Drawings in the Collection Made by William Cecil, First Baron Burghley, Now at Hatfield House,* Oxford, 1971).

Probably the first printed map to show the location of precious minerals and ores was Olaus Magnus's map of Scandinavia, "Carta Marina," published in Venice (1539) and reissued by Antonio Lafreri in Rome (1572). Olaus used four different point symbols to denote the general locations of gold, copper, iron and silver deposits. During the 17th and 18th centuries, Flemish, Dutch and German cartographers adopted a universal system of abstract signs for depicting mineral resources of economic interest. These signs included the miner's mallet, introduced by Kaspar Henneberger on his map of Prussia (1576), and planetary signs devised by alchemists and first used by Paul Aretin on his "Regni Bohemiae Nova et Exacta Descriptio" (1619). Perhaps the earliest printed maps to include cartouches and pictorial details of mining activities were prepared by Marcin German, engraved by Willem Hondius, to show the salt mines at Wieliczka, Poland, in 1645, "Miasto Wieliczka/Hae Admirandae Salinarum Fodinae" (Figs. 10 and 11).

European interest in mineral resources overseas resulted in one of the earliest maps of China to show minerals. The Polish Jesuit and sinologue Michael Boym with an assistant, the Chinese scholar Andreas Chên, produced a detailed manuscript map *c.* 1652 (Phillipps Collection, London) indicating minerals (by symbols), as well as other economic data.

Among the first large scale mineralogical maps to be based on direct field surveys were those prepared by Count Luigi Ferdinando Marsigli in 1726 to accompany his study of the Danube basin, *Danubius Panonico-Mysicus* (Vol. 3, "De Mineralibus"), The Hague and Amsterdam (1726). On these maps, planetary and alchemy signs were used as point symbols to depict the locations of fourteen different minerals and ore-bearing rocks.

In 1746 Jean-Étienne Guettard presented a memoir to the Académie Royale des Sciences (Paris) accompanied by two mineralogical maps, prepared by Philippe Buache, which employed nearly fifty symbols to show the occurrence of rocks and minerals (see Geological map, 3.071).

By the second half of the 18th century, large scale mineralogical maps were being prepared in most European countries, and in 1780 Guettard and A-G Monnet published the first mineralogical atlas.

C DUDICH, E, ed. (1984): *Contributions to the History of Geological Mapping: Proceedings of the Xth INHIGEO Symposium . . . 1982.* Budapest, Akadémiai Kiadó.

EHRENSVÄRD, Ulla (1977): "Gruvor På Kartor," in *Vilja och kunnande: teknik historiska uppatser tillägnade Torsten Althin På Hans Åttioårsdag Den 11 Juli 1977 AV Vänner.* 172-188.

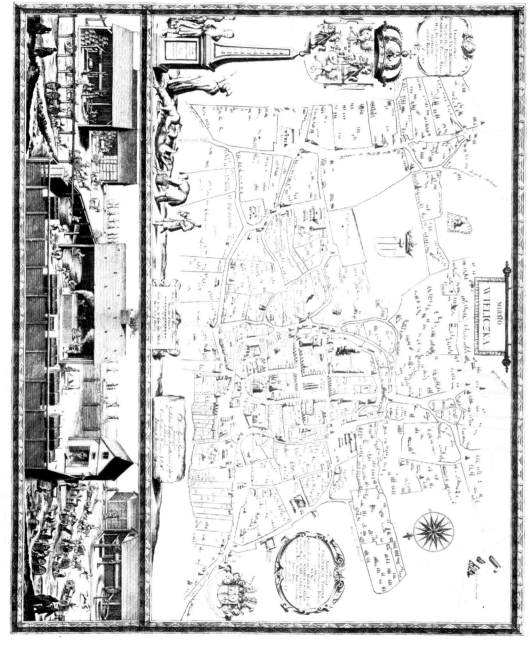

Fig. 10 – Plan of the famous salt mining town of Wieliczka in Poland, with a surface view of the mine works. Surveyed by Marcin German, drawn and engraved by Willem Hondius. Gedani (Danzig), 1645. The first plate in an atlas of plans of the Wieliczka mines. By permission of the British Library. (Mineral map. 3.131)

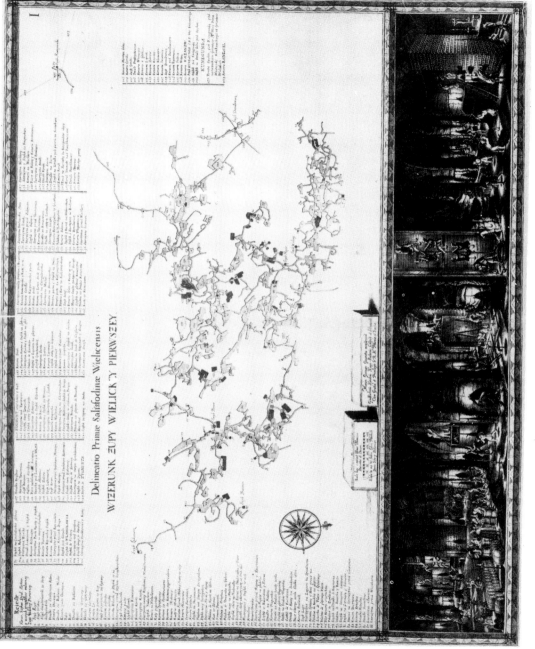

Fig. 11 – The first underground level of the Wieliczka salt mines, with interior view. Plan I in the atlas by Marcin German (see Fig. 10).

EYLES, V A (1972): Mineralogical Maps as Forerunners of Modern Geological Maps,"
  Cartographic Journal, 9, 133-5.

GRANLUND, John (1951): "The *Carta Marina* of Olaus Magnus," *Imago Mundi, 8,*
  35-43.

GUETTARD, J-É, and MONNET, A-G (1780): *Atlas et description minéralogiques de
  la France.* Paris.

KUCHAŘ, Karel (1961): *Early Maps of Bohemia, Moravia and Silesia.* Prague.

3.132 Morphological Map

A (MDTT 821.8) "A map showing the distribution of the surface forms of the Earth as distinct from the quantitative expression of relief. Such maps may either depict areas occupied by a characteristic landscape type or represent and classify individual features or facts. Not to be confused with Relief Map (811.2)." A map depicting the form of the land surface. The representation of landform shapes on maps has been variously called the "physiographic," "morphographic," "geomorphic," "morphologic," and more rarely the "landscape" method. Because shape is a quality depicted on all maps that include the relief of the earth's surface, whether a map is morphological or not depends usually on the extent to which morphological features dominate it.

B The traditional pictorial style of depicting mountains on medieval maps as conical molehills, sugar loaves, and caterpillars was adapted to printed maps in the 15th century and persisted until as late as *c.* 1850. The maps of the world and the Holy Land in the world history, *Rudimentum novitiorum,* printed by Lucas Brandis at Lübeck in 1475, show the land regions as a myriad of hummocks. Ptolemy's *Geography* (Bologna, 1477) has caterpillar mountain ranges with shaded slopes. The caterpillar style was more pronounced in the Ulm Ptolemy of 1482, edited by Donnus Nicolaus Germanus. The "Tabulae novae" or "Tabulae modernae" in the Strassburg edition of Ptolemy (1513) deploy both the caterpillar and hummock symbols (by then the commoner style) with shading on the eastern sides.

These naturalistic styles reflect a universal iconographic tradition. They are also found, for example, on the early maps of China. Mountains appear as cones on the earliest "printed" map of China extant, the "Ti Li Thu" (General Map of China), made by Huang Shang about 1193 and engraved on a stone stele at Suchow by Wang Chih-Yuan in 1247 (Needham, fig. 229).

The hill-shading method to enhance relief forms at the end of the 18th century in Europe developed into hachuring, in which short lines follow the direction of maximum slope and so portray relative relief. By the close of the 18th century the pictorial or qualitative insertion of hachures had begun to be replaced by a more rigorous method in which the thickness and spacing of the hachures were directly proportional to the degree of slope (see Hachures, 5.081). The first printed map on which this improved technique was used was Johann Georg Lehmann's map of the Hemsdorf district (1798).

In the 1770s another technique, contouring, was introduced. Contours (5.0910a) are horizontal lines of equal altitude above some horizontal datum. On the two dimensional map, contours have necessarily to be portrayed in one plane, and they present to the reader a horizontal plan of the morphology as it would be seen when viewed from directly overhead. In 1798-99 the device of enhancing the spatial intervals between certain successive contours by "layer tinting" in a gradation of colours was invented, to enable the highlands and lowlands to be seen more clearly (see Layer tints, 5.121; Hypsometric map, 3.081).

With the exception of perspective maps all other developments in morphologic mapping, such as kantography, inclined contours, relief contours, physiographic diagrams, landform maps, and landform-type maps, are innovations of the 20th century.

C DAINVILLE, François de (1964): *Le langage des géographes*. Paris, Éditions A et J Picard.

ECKERT, Max (1921, 1925): *Die Kartenwissenschaft*, 2 vols. Berlin and Leipzig, Walter de Gruyter.

IMHOF, Eduard (1965): *Kartographische Geländedarstellung*. Berlin, Walter de Gruyter.

IMHOF, Eduard (1982): *Cartographic Relief Representation*. English language version translated and edited by H J Steward. Berlin, New York, Walter de Gruyter.

NEEDHAM, Joseph (1959): *Science and Civilisation in China*, vol. 3. Cambridge, At the University Press.

3.151 Ocean Current Map

A A map of sea or ocean areas on which currents are shown in some distinctive manner.

B Jodocus Hondius on his world map "Novissima ac exactissima totius orbis terrarum descriptio magna," *c.* 1611, published at Amsterdam in 1618, claims that he is the first to represent ocean currents. Each current is indicated by a brief description within a banner trending in the stated direction: thus Northoostten Oost is located off Florida; Noord-Oost Courrant lies east of Nova Scotia; West Courrant runs south of Greenland. Twenty-five years later, in *c.* 1636, Sir Robert Dudley marked currents on the charts of his world atlas *Dell' Arcano del Mare* (published at Florence, 1646-47), along with winds, soundings, and magnetic variation. The currents are indicated by brief verbal descriptions in the appropriate locations: for example, la corrente verso ponente, written between 19° and 20°S in the Pacific off South America.

The earliest thematic charts of currents in the open ocean were devised by the Jesuit geographer Athanasius Kircher to illustrate his physical geography of the earth, *Mundus subterraneus* (Amsterdam, 1665). They comprise a chart of the horizontal circulation of the ocean waters of the world, "Tabula geographico-hydrographica motus oceani currentes . . . ," and six charts of the Atlantic and the Mediterranean. Currents are represented by belts of linear shading, a type of flow line, but lack any indication of direction, so the maps can only be fully understood with the help of the accompanying text. Eberhard Werner Happel adapted Kircher's work in his chart "Die Ebbe und Fluth auff einer Flachen Land-Karten fürgestelt," first published in his . . . *Relationes curiosae* (Hamburg, 1685) and reprinted in his *Mundus mirabilis tripartitus* (Ulm, 1687). Neither his maps nor Kircher's became well known.

During the 1660s the English Captain Richard Bolland conducted detailed observations of current flow in the Strait of Gibraltar. His map delineated along both the northern and southern shores the currents whose directions are dependent on the tidal flow, and in the middle of the strait another current with constant direction unrelated to the moon. All Bolland's information is written on the map; no symbols are used.

In the 18th century small arrow symbols came into use to show the direction of currents. For example, the first printed map to show the Gulf Stream in the Florida Straits employed a train of arrows to inform navigators of the current's set or direction ("A Chart of the Bahama Islands with a description of the Gulf of Florida . . . ," by "J.C.," in *A Description of the Windward Passage, and Gulf of Florida . . . ,* London, 1739).

Later in the 18th century the importance of speedy navigation across the North Atlantic led to studies of the Gulf Stream. William Gerard de Brahm in *The Atlantic Pilot* (London, 1772), a volume of sailing directions for navigation from the Gulf of Mexico through the New Bahama Channel to the British colonies in North America and on to Europe, includes a "Hydrographical Map of the Atlantic Ocean" which shows the "Settings and Changes of the Currents in the Ocean." The American statesman and scientist, Benjamin Franklin, while Deputy Postmaster General for the American colonies, had been studying the current from 1768. The map entitled "A Chart of the Gulf Stream" was published with his memoir in *Transactions of the American Philosophical Society* (Philadelphia, 1786). On their published charts both De Brahm and Franklin employed patterns of closely spaced gently undulating lines to indicate the flow of the Gulf Stream. Franklin also placed small directional arrows and velocity figures within his flow patterns.

In a later unpublished chart, De Brahm abandoned lines to show flow and adopted patterns of tiny arrows to show the Atlantic currents. Closely following Franklin's lead, Thomas Pownall, the colonial governor and antiquary, made a chart of the Gulf Stream, published in London, 1787, which employed a pattern of graduated shading, darkest where the edges of the Gulf Stream were most distinct and faintest where they were least perceptible. In the 19th century Alexander Dallas Bache, Director of the U.S. Coast Survey, used shading to indicate temperature variations within the current.

By 1900 small arrow symbols showing current direction with associated figures to indicate velocity were being employed on the ocean current charts being produced by most charting agencies. In atlases flow patterns or arrows were employed.

C BOLLAND, Richard (1675): "A Draught of the Streights of Gibraltar with Some Observations upon the Currents thereunto belonging," in *A Collection of Voyages and Travels* (1704). London, A and J Churchill, Vol. 4, 846-48.

DE VORSEY, Louis (1976): "Pioneer Charting of the Gulf Stream: The Contributions of Benjamin Franklin and William Gerard De Brahm," *Imago Mundi, 28,* 105-20.

KRUG, Martha (1901): *Die Kartographie der Meeresströmungen in ihren Beziehungen zur Entwicklung der Meereskunde.* Heidelberg Dissertation, reprinted in *Acta Cartographica, 11* (1971), 227-99.

POWNALL, Thomas (1787): *Hydraulic and Nautical Observations on the Currents in the Atlantic Ocean.* London, Robert Sayer.

3.161 Precipitation Map

A A map showing the amount of precipitation as measured at observation points or inferred for an entire area on the basis of the point sample. Various phenomena, ranging from rain to snow (or expressed as liquid equivalent), and many sorts of averages (annual, seasonal, monthly, etc.) can be represented.

B The first "hyetographic" or precipitation maps seem to be two prepared by the Danish cartographer Oluf Nikolay Olsen dated 1839 for an atlas to accompany a study of the climate of Italy by the botanist J F Schouw. One map simply shows the numerical observations at their locations, while the other is a shaded map of parts of Europe and Africa showing seasonal concentrations and orographic effects. Heinrich Berghaus included a shaded "hyeto-graphic" map of the world, dated 1841, in his *Physikalischer Atlas* in which the darker the tone the greater the precipitation, but with no amounts given. A similar "Rain Map of the World" appeared in A K Johnston's *The Physical Atlas* in 1848. Berghaus also included in his atlas a precipitation map of Europe, dated 1841, showing by isarithms a variety of data, from annual amounts of precipitation to numbers of rainy days, days with snow, directions of "rain winds," etc.

After the early 1840s precipitation maps became quite common.

C BERGHAUS, Heinrich (1845; 1848): *Physikalischer Atlas oder Sammlung von Karten . . .* Gotha, Justus Perthes.

JOHNSTON, A K (1848): *The Physical Atlas . . .* Edinburgh, W and A K Johnston.

OLSEN, O N (1839): *Atlas pour le tableau du climat d'Italie.* Copenhagen, Chez Gyldendal.

ROBINSON, Arthur H (1982): *Early Thematic Maps in the History of Cartography.* Chicago, University of Chicago Press.

3.201 Tidal Map

A A map of sea or ocean areas, on which tidal phenomena are represented usually by means of arrows, numerals, and isarithms (5.0910). Such maps were more commonly made of relatively shallow seas and coastal areas, rather than of the open ocean.

B The earliest tidal diagram, drawn in the form of a windrose, is found in the Catalan Atlas, 1375. Fourteen places in Brittany, France, and England are named within concentric circles, and the letter P within the circles indicates the time of high water on the days of the full and new moon. Pierre Garcie's *Le Routier de la mer* (soon after 1500) gives the tidal information in words not diagrams or maps. The Luso-French atlas known as "The Hague" atlas, *c.* 1545-47, has an elaborate tidal diagram giving information for 81 ports and harbours from Brittany and Normandy to Scotland and Flanders (The Hague, Koninklijke Bibliotheek, Atlas MS.129.A.24).

The first charts of tides were published by Guillaume Brouscon of Le Conquet, Brittany, in or a little before 1546 (Fig. 12). They comprise simple diagrammatic woodcut maps in pocket almanacs made for the guidance of Breton seamen, many of whom were illiterate. The basic chart, which was cut into four parts, covered the coasts from Finisterre to Flanders, including the British Isles. The "establishment of the port," i.e. the time of high water at that port on days of full and new moon, is expressed as the compass bearing of the sun at the time in question. Lines run from each port to the appropriate compass point on adjacent windroses (Howse, 1980). Brouscon's tidal almanacs proved very popular and rank as a major innovation in navigational aids for northwest Europe. In England, where copies were made, the almanacs were used by leading navigators such as Sir Francis Drake.

Brouscon's charts established a tradition which entered the main stream of national French cartography. Thus Jean Le Clerc in his *Theatre géographique du Royaume de France* (Paris, 1619), marks tides on his map of Brittany, "Description du Pays armorique à près Bretaigne." Tidal dials inscribed with the days of the month, the months of the year, and the phases of the moon, etc., emit octopus arms leading to the coasts, to represent the "establishment of the port."

The first truly scientific tidal chart was devised by the English mathematician and natural philosopher Edmond Halley in 1701-2. Based on observations carried out on his voyage of scientific investigation in the English Channel made between June and October 1701, the chart shows the tidal direction and state by means of Roman figures giving hours of high water and arrows showing the direction of tidal currents. The chart embodying Halley's results was published by Richard Mount and Thomas Page in London in two versions in 1702, and achieved widespread fame and use. The tidal information as given on these charts has been verified in its main essentials by the much more complete data available today.

Halley's chart was much imitated, but no further innovations in tidal mapping appear to have been made until the 1830s and 1840s. William Whewell's first two charts of tides using "cotidal lines," one covering the world between 60°N and 55°S, the other the British Isles, accompanied his "Essay towards a First Approximation to a Map of Cotidal Lines" (Whewell, 1833). F W Beechey extended the range of phenomena represented, showing lines of direction of stream, lines of equal range

Fig. 12 – Tidal charts in the nautical guide and almanac of Guillaume Brouscon of Le Conquet, Brittany, 1546. (left) Bay of Biscay, with north to the right. (right) Northern France, with north to the left. By permission of the British Library, Add. MS. 22721. (Tidal map. 3.201)

and the direction of turn (rotary streams) on his maps of the tidal streams in the English Channel and the Irish Sea, and of the North Sea (Beechey, 1848 and 1851).

C BEECHEY, F W (1848): "Report of Observations made upon the Tides in the Irish Sea," *Philosophical Transactions of the Royal Society of London, 138,* pt 1, 105-60.

BEECHEY, F W (1851): "Report of Further Observations upon the Tidal Streams of the North Sea and English Channel," *Philosophical Transactions of the Royal Society of London, 141,* 703-18.

BULLARD, Sir Edward (1956): "Edmond Halley (1656-1741)," *Endeavour, 15,* No. 60, 189-99.

DUJARDIN-TROADEC, Louis (1966): *Les cartographes bretons du Conquet. La Navigation en images 1543-1650.* Brest.

HINKS, Arthur H (1941): "Edmond Halley as Physical Geographer," *Geographical Journal, 98,* 293-96.

HOWSE, Derek (1980): "Brouscon's Tidal Almanac, 1546." Explanatory sheet in *Sir Francis Drake's Nautical Almanack,* London.

HOWSE, Derek (1985): "Some Early Tidal Diagrams," *Revista da Universidade de Coimbra, 33,* 365-85.

PROUDMAN, J (1942): "Halley's Tidal Chart," *Geographical Journal, 100,* 174-76.

THROWER, N J W (1978): "Edmond Halley and Thematic Geo-Cartography," in *The Compleat Plattmaker, Essays on Chart, Map, and Globe Making in England in the Seventeenth and Eighteenth Centuries,* N J W Thrower, ed. Berkeley and Los Angeles, University of California Press, 195-228.

THROWER, Norman J W, ed. (1981): *The Three Voyages of Edmond Halley in the Paramour, 1698-1701.* London, Hakluyt Society, 2nd ser., no. 156, and 157, with a facsimile of Halley's Tidal Chart of the Channel.

WHEWELL, W (1833): "Essay towards a First Approximation to a Map of Cotidal Lines," *Philosophical Transactions of the Royal Society of London, 123,* pt 1, 147-236.

3.231 Weather Map

A A synoptic map (or chart) showing the principal meterological conditions, such as air pressure, temperature, winds, precipitation, cloudiness, fronts, and air masses.

B The preparation of a map combining meteorological elements was accomplished in 1820 by H W Brandes of Breslau. Working from the records of the Societas Meteorologica Palatina (Mannheim), he mapped the wind directions and equal deviations from normal atmospheric pressure that had been observed during a violent storm over Europe in 1783. The next mapping of synoptic meteorological elements was by Elias Loomis who studied two 1842 storms in the United States and made a series of five charts.

The first weather reports published on the basis of telegraphic reports appeared at almost the same time in 1849 in England and the United States. The first attempt at the publication of daily weather maps was in England from 8 August to 11 October, 1851, in connection with the Great Exhibition in London. In 1861 weather maps were published in England, and in September 1863 the meteorological service of the Paris observatory first published maps of "interesting" days and then daily maps after 16 September.

C DAVEY, Marié (1866): *Les mouvements de l'atmosphere*. Paris.

ECKERT, Max (1925): *Die Kartenwissenschaft,* vol. 2. Berlin and Leipzig, Walter de Gruyter, 341-46.

HARRINGTON, Mark W (1895): "History of the Weather Map," in *Report of the International Meteorological Congress, 1893.* Washington.

HELLMANN, Gustav (1897): "Meteorologische Karten," in his *Neudrucke von Schriften und Karten über Meteorologie und Erdmagnetismus,* No. 8. Berlin, A Asher.

LOOMIS, E (1846): "On two Storms which were experienced throughout the United States, in the month of February, 1842," *Transactions of the American Philosophical Society, 9,* 161-84.

3.232 Wind Map

A A cartographic representation of the circulation of air currents averaged for the year, a season, etc. Arrows are often used to indicate the direction of air movement. See also Windrose, 5.231.

B Greco-Roman windroses and sundial gnomons gave indications of winds. The Boscovich anemoscope, found in 1759 near the Porta Capena, Rome, and now in the Oliveriano Museum, Pesaro, comprises a flat stone cylinder on which is engraved a map combining diagrammatic climatic zones and winds, perhaps of the 3rd to 4th century. The twelve winds are asymmetrically placed at the points where the equator and the three equally spaced climatic zones on each side of it meet the circumference. It was mounted on a column, with a central hole, probably to insert a wooden pole with flag, and has south at the top. The world map, probably by Agathodaemon, in a Greek manuscript of Ptolemy's Geography, 13th century, formerly in the Vatopedi monastery (now in the British Library, Department of Manuscripts, Add. MS. 19391), may record knowledge of winds from the Greco-Roman world (Fig. 14). There are ten faces blowing winds, north and south poles, and a similar delineation of climate zones. It displays a combination of geographical and cosmological names and its archetype is believed to have been of similar date to the Boscovich anemoscope.

Jodocus Hondius on his map of the world, 1611 (Amsterdam, 1618), includes data on winds, claiming that he was the first to represent the winds and currents of the ocean. For example, in the Atlantic Ocean, "Subsolanus continuo flat (*sic,* i.e. fluit] inter utrumque Tropicum in mari aperto del Nort." A long legend explains the winds of the Pacific. Sir Robert Dudley also marked winds as well as currents on the manuscript charts for his maritime atlas *Dell' Arcano del Mare, c.* 1636, published in 1646-47.

The first wind map as such is that by Edmond Halley, untitled, which he issued in 1686 accompanying his article on the Trade Winds ("An Historical Account of the Trade Winds, and Monsoons, observable in the Seas between and near the Tropicks, with an attempt to assign the Phisical Cause of the said Winds," *Philosophical Transactions* (Royal Society, London), Vol. 16, No. 183, 1686 (1688), 153-68). This map shows average annual conditions with the symbols for the monsoons interdigitating. Halley used arrows only in the Cape Verde area, but elsewhere employed tapering "stroaks." Halley's map is confined to the low latitudes of the Atlantic and Indian Oceans. Inspired by Halley's example, the English navigator William Dampier illustrated "His Discourse of the Trade Winds, Breezes, Storms . . ." with two maps, "A View of the General & Coasting Trade-Winds in the Atlantick & Indian Oceans," and a similar map for the "great South Ocean" (the Pacific). The winds are indicated by arrows (W Dampier: *Voyages and Descriptions,* II, pt. III, London, 1699). Such was the widespread influence of Halley's paper that it was common henceforward for even English pocket globes to be marked with the trade winds.

In 1846 onwards the American naval captain Matthew Fontaine Maury introduced a new technique with his Monsoon and Trade Wind charts for the oceans, published at Washington, D.C., on which windroses indicate the annual regime of the winds.

Maps of winds blowing over land masses were slower to appear, as such knowledge was not urgently required, as at sea. The publication of "Carte des vents qui sont propres à la Province de Languedoc" in Jean Astruc's *Mémoires pour l'histoire de la Province de Languedoc,* Paris, 1737 (pp. 336-7) thus marks a notable innovation. Astruc explains in the text that he was concerned in particular to represent the irregularity of the winds. He distinguishes various named winds by different types of flow line.

C BÖKER, R (1958): "Windrosen," in Pauly, A F von, and Wissowa, G, *Real-Encyclopädie der classischen Altertumswissenschaft, 8,* A², cols. 2326-80, esp. 2358-60.

CHAPMAN, Sydney (1914): "Edmond Halley as Physical Geographer and the Story of His Charts," *Occasional Notes of the Royal Astronomical Society* (London), *9.*

NEUGEBAUER, O (1975): "A Greek World Map," in *Le monde grec: . . . Hommages à Claire Préaux,* ed. Jean Bingen, *et al.* Bruxelles, pp. 312-17 and Pl. III.2.

THROWER, Norman J W (1969): "Edmond Halley as a Thematic Geo-Cartographer," *Annals of the Association of American Geographers, 59,* 652-76.

ZICÀRI, I (1954): "L'anemoscopio Boscovich del Museo Oliveriano di Pesaro," *Studia Oliveriana, 2,* 69-75.

3.261 Zone Map

A A type of hemispherical mappa mundi (1.2320b) which divides the earth's surface into five zones in accordance with ancient literature. These zones and the circles themselves were called by Latin writers *septentrionalis* (north frigid zone or Arctic circle), *solstitialis* (north temperate zone or Tropic of Cancer), *equinoctialis* (torrid zone or Equator), *brumalis* or *hyemalis* (south temperate zone or Tropic of Capricorn), and *australis* (south frigid zone or Antarctic circle). Of these, only *solstitialis* was considered to be inhabited.

B Through astronomical observation the ancient Greeks Thales of Miletus (*c.* 580 BC) and Pythagoras (*c.* 525 BC) divided the sphere of the universe into five zones. Parmenides (*c.* 490 BC) is believed to have been the first to transpose these celestial zones onto the earth's surface. Eratosthenes (*c.* 275–*c.* 194 BC) was probably the first to put the terrestrial theory of zones on a firm scientific footing by determining exactly the positions of the fixed circles on the globe. The opinion of most Greek writers was that the two polar caps were too cold and the equatorial region too hot to support life. This led to a theory first advanced by the Pythagoreans, refined by Crates of Mallos (*c.* 150 BC), and introduced into Western knowledge through the works of Macrobius and Martianus Capella (both 5th century), which held that the oecumene, or inhabited portion of the world, was one of four similar bodies of land on the globe separated by two encircling oceans flowing at right angles to one another. Involving as it did the potentially heretical doctrine of the Antipodes, this theory provoked the indignation of the Church and, as a result, zone maps did not become popular until the 12th and 13th centuries, although earlier examples do exist. Maps of this type can be found in the works of Guillaume de Conches (*c.* 1080-*c.* 1154) and the medieval texts of Macrobius.

Related to the zone maps, but forming a distinctly different class, are the climatic mappae mundi, which divided the habitable world into the seven climata of the Greek and Alexandrian geographers. Neither numerous nor elaborate, this group of maps indicates in manner of orientation and in the predominance given to the mythical world centre of Arin (or Arim) the influence of Arab geography in the later Middle Ages. Examples of this type can be found in the *Dialogus contra Judeos* of Petrus Alphonsus (12th century) and in the various editions of *Ymago Mundi* by Petrus de Alliaco (1410; first printed in 1483).

C ANDREWS, Michael C (1926): "The Study and Classification of Medieval Mappae Mundi," *Archaeologia, 75,* 61-76.

BAGROW, Leo (1964): *History of Cartography.* Revised and enlarged by R A Skelton. London, C A Watts; Cambridge, Mass., Harvard University Press.

BROWN, Lloyd A (1949): *The Story of Maps.* Boston, Little, Brown and Company.

DESTOMBES, Marcel, ed. (1964): *Mappemondes: A.D. 1200-1500,* Vol. 1 of *Monumenta Cartographica Vestutoris Aevi.* Amsterdam, N. Israel.

MILLER, Konrad (1895-1898): *Mappaemundi, die ältesten Weltkarten,* 6 vols. Stuttgart, Jos. Roth'sche Verlagshandlung.

WRIGHT, John K (1925): *The Geographical Lore of the Time of the Crusades.* New York, American Geographical Society. (Reprinted, New York, Dover Publications, 1965).

3.262 Zoological Map

A A map that indicates the typical animals of an area or a distribution map that portrays the extent of an animal's range.

B Many ancient maps record animals. On the clay tablet bearing a Babylonian map of the world, 7th century BC, the circular disc of the earth is surrounded by the Heavenly Ocean with animal constellations such as a dragon, lion, bull, etc. (British Museum, Department of Western Asiatic Antiquities, no. 92687). Fish swim in the oceans of the Beatus maps of the 12th and 13th centuries. Typical animals roam across the continents on Gervaise of Tilbury's world map (the Ebstorf map, *c.* 1235) and Richard of Haldingham's world map (the Hereford map, *c.* 1285). Animals feature prominently on European maps of the world and its regions in the 16th and 17th centuries. Olaus Magnus's "Carta Marina" of 1539 with its land animals and abundant sea creatures ranks as a major contribution to the natural history of northern Europe and the northeast Atlantic Ocean. The fauna of the newly discovered countries of the world was of particular interest, and many maps, especially those of the Americas, were in advance of the bestiaries of their time. The atlases and charts (*c.* 1540-80) of the Dieppe hydrographers and their associates, such as Jean Rotz, Pierre Desceliers and Guillaume le Testu, were notable for their painted miniatures of exotic scenes in which animals appeared in their natural environment.

By the middle of the 16th century the convention of depicting animals on maps and charts was well-established and confirmed in instructional works. Thus Richard Eden, in his translation (1561) of Martin Cortes's *Arte de navegar* (Madrid, 1557, f.lviii[r]), gives instructions for "The payntyng of the Carde" in his chapter on "the composition of Cardes for the Sea": "Then shall you drawe or paynte Cities, shyppes, banners, and beastes." Other mathematical practitioners such as the Portuguese Pedro Nuñes thought the profusion of animals inappropriate. In the *Tratado da Sphera* (Lisbon, 1537, p. 131), Nuñes criticized those who decorated their works with too many bears, camels and elephants: "muyto ouro; & muytas bandeyras Ellifantes & Camelos: & outras cousas iluminadas." The much quoted quatrain of Jonathan Swift, published in a poem of 1733, was penned in a similar vein: "So Geographers in Afric-Maps/with Savage-Pictures fill their Gaps;/ and o'er unhabitable Downs/Place Elephants for want of Towns."

Nevertheless, the animals had scientific value, as Wilma George has shown; and as depicted on the printed maps of the day they were popular with the public. Although absent on the maps in Ptolemy's *Geography* published from 1477, they enlivened the modern maps in Sebastian Münster's editions of Ptolemy published at Basel from 1540 onwards. Dutch printed atlases and maps of the second half of the 16th century and the first half of the 17th often showed typical animals, sometimes with explanatory legends. The "Nieuwe Caerte" of Guiana, 1599, by Jodocus Hondius senior, for example, displayed an impressive array of recently discovered animals, including the armadillo and the anteater. Willem Janszoon Blaeu's map "Nova Belgica et Anglia Nova" in the *Novus Atlas* (Amsterdam, 1635) became a prototype for the display of animals on maps of northeastern North America. Appropriate animals were also incorporated into the designs of decorative cartouches, as on Blaeu's map of "Nova Hispania" in the same atlas.

The modern period in zoological mapping begins with E A W von Zimmermann's map "Tabula Mundi Geographico Zoologica sistens Quadrupedes," first published in *Specimen Zoologiae Geographicae* . . . (Leiden, 1777) and then in *Geographische Geschichte des Menchen* . . . (Leipzig, 1778-83). The distributions of mammals are represented on the map by the names of the animals at particular locations; the range of an animal is indicated by the position of the name in relation to temperature lines (primitive isotherms). Carl Ritter used a similar method of representation for a map of domesticated and wild animals in his thematic atlas, *Sechs Karten von Europa* (1804-06). A derivative French version of Ritter's map is the "Carte de la répartition des mamifères terrestres, tant dans l'état de nature que dans l'état domestique," in Maxime Auguste Denaix's *Atlas physique, politique et historique de l'Europe,* Paris, 1829. The zoological maps in Henrich Berghaus's *Physikalischer Atlas* (1845) represent a further major development, portraying the distribution of many species and genera through the use of colour, graded shading and lines.

Several noteworthy zoological maps were published in America in the first half of the 19th century. William C Woodbridge included biological maps in his school atlases of the 1820s and 1830s; Louise Agassiz's map of zoological regions was published in 1848; and Matthew Fontaine Maury's whale charts were among the earliest maps to focus on the worldwide distribution of species of marine mammals. By the middle of the 19th century, thematic maps were an established tool for studying the geographical distribution of animals.

C AGASSIZ, Louis (1848): "Chart of Zoological Regions," in *Principles of Zoology,* by Louis Agassiz and Augustus Addison Gould. Boston, Gould and Lincoln. Also in *Outlines of Comparative Physiology* . . . (1851), edited by Thomas Wright, London, H G Bohn.

BAGROW, Leo (1964): *History of Cartography.* Revised and enlarged by R A Skelton. London, C A Watts; Cambridge, Mass., Harvard University Press.

ECKERT, Max (1921, 1925): *Die Kartenwissenschaft,* 2 vols. Berlin and Leipzig, Walter de Gruyter.

EDEN, Richard (1561): *The arte of Nauigation . . . , Translated out of Spanyshe into Englyshe.* London.

GEORGE, Wilma (1969): *Animals and Maps.* London, Secker and Warburg.

MAGNUS, Olaus (1539): Carta Marina et Descriptio Septemtrionalium Terrarum. Venice.

MAURY, Matthew Fontaine (1851): "Whale Chart (Preliminary Sketch)," Series F of Wind and Current Charts, National Observatory, Washington, DC.

MAURY, Matthew Fontaine (1852): "Whale Chart of the World," Series F of Wind and Current Charts, U.S. Naval Observatory, Washington, DC.

NUÑES, Pedro (1915): *Tratado da Sphera ... Reproduction Facsimilé de l'exemplaire appartenant à la Bibliothèque du Duc de Brunnid à Wolfenbüttel, Edition 1537 Lisbonne.* Munich.

ZIMMERMANN, Eberhardt August Wilhelm von (1777): *Specimen Zoologiae Geographicae, Quadrupedum Domicilia et Migrationes Sistens.* Lugduni Batavorum (Leiden).

GROUP 4

Reference Systems and Geodetic Concepts

4.031 Cardinal Direction

A One of the principle directions which divide the horizontal plan into four equal divisions, one being the direction of the north pole. See also Orientation (4.151).

B The Babylonians divided the heavens as follows: *iltânu* for north (extending from north-northwest to north-northeast), *šûtu* for south (south-southwest to south-southeast), *amurrû* for west (west-southwest to west-northwest), and *šadû* for east (east-northeast to east-southeast). The Greeks at first conceived of only two cardinal directions: *boreas* (N) and *notos* (S); later two more were added: *euros* (E) and *zephyros* (W). The Romans designated the winds coming from the cardinal points as *septentrio* (N), *solanus* (E), *auster* (S), and *favorius* (W), but in the Middle Ages maps lettered in Latin employed the abbreviations of *tramontana* (N), *levante* (E), *ostro* (S) and *ponente* (W) (see Windrose, 5.231). On later maps lettered in modern languages initials were employed: N for *north* (*nord, norden, noord, norte*); E for *east* (*est, este*) except for the Germanic languages which use O for *ost* (*oost*); S for *south* (*sud, süden, sur*) and Z for *zuid* (in Dutch); and W for *west* (except Spain, France, and Italy, which use O for west: *oeste, ouest, ovest*). After the 15th century the cardinal directions of European maps were often indicated by the Latin terms *septentrio* (N), *meridies* (S), *oriens* (E), and *occidens* (W).

One of the cardinal points was usually located at the upper part of the map. On Greek maps north was put at the top; the same appears on the Peutinger map of

163

Roman origin. Maps of the Christian Middle Ages, including circular maps of Roman derivation (see Mappa mundi, 1.2320b), usually placed east at the top. On the Ebstorf world map, *c.* 1235, the head, feet, and hands of Christ indicate the cardinal directions, east, west, south, and north. Notable exceptions to the eastern orientation are the maps of Macrobius (399-423) found in manuscripts dating from 1200 to 1500. These have north at the top, preserving the Hellenistic tradition. Muslim maps placed south at the top, and this placement also appears on Christian maps with elements of Arab or Jewish influences. Examples of this orientation are the world maps of Asaph Judaeus (11th century) and Johannes Utinensis (*c.* 1350), the Borgia world map (*c.* 1430), Fra Mauro's world map (1459), and the map of Vesconte Maggiolo (1512).

The maps of Great Britain by Matthew Paris of St Albans, *c.* 1250, are the first maps of Northern Europe with north at the top and display in Latin the names of the four cardinal points. A local plan of the source (at Wormley) of the water supply of Waltham Abbey, preserved in a mid-13th century copy of the plan of the 1220s, has at the top of the drawing a cross with a splayed foot marking the east. This is believed to be the earliest direction pointer found on any map (Harvey, p.90). The fact that such an innovation should be found on a local plan suggests (with other evidence) that the original plan was probably associated with Matthew Paris (Skelton and Harvey, p.38).

The earliest portolan charts, which date from the end of the 13th and the 14th centuries, have south at the top or are multidirectional. In the atlases of Pietro Vesconte, 1313–*c.* 1327, the charts (with south at the top) have directions indicated by a simple *marteloio* (the rhumb line network). The Catalan maps display discs with signs; for example, on Guillelmus Soleri's chart of the Mediterranean, *c.* 1385, the polestar (N), a red sun (E), a moon with a human face (S). The first compass rose appears on the Catalan map of *c.* 1375. Its north pointer is distinctive, and the decoration of the compass rose also picks out east (levante). The chart itself is multidirectional. The convention of marking east by a cross was followed by many European chartmakers, such as Nicolaus de Caverio on his world chart, *c.* 1505. The Turkish admiral Piri Re'is in his atlas *Kitab-i-Bahriye*, 1521 and 1525, drew his charts with simple rhumb lines with only an arrow in black ink to show the direction of north (Mollat du Jourdin, p.223). The device of depicting the north point as a fleur-de-lys surmounting the compass rose was devised by the Portuguese and first found on Jorge de Aguiar's chart of 1492 and Pedro Reinel's Atlantic chart, *c.* 1504. It was henceforward adopted by most chartmakers (see Windrose, 5.231).

The discovery of Ptolemy's *Geography* and its publication with maps from 1477 led to general acceptance of the convention of placing north at the top for world and regional maps; the network of lines of latitude and longitude made the marking of the cardinal points superfluous. Chartmakers from the late 15th century also increasingly chose north for the orientation of world charts, but the use of other cardinal directions continued into the 18th century, and the compass rose remained a major feature of most charts.

Among exceptions to the general rule of north being at the top are maps whose purpose required a special orientation such as those of Erhard Etzlaub of Nuremberg, a maker of pocket sundials incorporating compasses. He arranged his map "Das ist der Romweg . . ." (*c.* 1500) with south at the top for the traveller from Germany to

Rome, and this is indicated by a dial at the foot of the map. A later mathematician and astronomer of Nuremberg Franz Ritter made maps for the tables of sundials and also issued them on paper in his *Speculum Solis,* 1610. The maps are projected from the north pole with land masses drawn to scale in proportion to their distances from this point. The cardinal directions are marked in Latin round the borders (Shirley, pp. 290-91).

For the Chinese and the Japanese cardinal directions were significant in traditional map-making. By the 5th century AD, geomancy (*feng-shui*), the art of adapting residences of the living and the tombs of the dead so as to cooperate and harmonize with the local currents of cosmic breath, had led to the development of the magnetic compass in China. Geomancers' handbooks included local maps in which cardinal directions were indicated. In Japan the cardinal directions north, south, or east (more rarely, west) were used for orientating maps (see Orientation, 4.151). It was normal for north, south, east, and west to be written in the margins of maps. Maps of Edo during the middle part of the Edo period represented directions by the twelve signs of the eastern zodiac. The compass rose appeared in the later part of the Edo period as a new feature from Europe (Nanba, p.155).

C ALMAGIÀ, Roberto (1960): *Storia della geografiá.* Torino.

BAGROW, Leo (1964): *History of Cartography.* Revised and enlarged by R A Skelton. London, C A Watts; Cambridge, Mass., Harvard University Press.

CRONE, G R (1978): *Maps and Their Makers.* Folkestone, W Dawson and Son; Hamden, Conn., Archon Books.

HARVEY, Paul D A (1980): *The History of Topographical Maps.* London, Thames and Hudson.

MOLLAT DU JOURDIN, Michel, and LA RONCIÈRE, Monique de (1984): *Sea Charts of the Ancient Explorers, 13th to 17th Century.* New York, Thames and Hudson.

NANBA Matsutaro, MUROGA Nobuo, and UNNO Kazutaka (1973): *Old Maps in Japan.* Osaka, Sogensha.

NAVARRETE, Martin Fernandez de (1846): *Disertación sobre la historia de la náutica.* Madrid, La Real Academia de la Historia.

NEUGEBAUER, Paul V, and WEIDNER, Ernst F (1931-32): "Die Himmelsrichtungen bei den Babyloniern," *Archiv für Orientforschung, 17,* 269-71.

SHIRLEY, Rodney W (1983): *The Mapping of the World.* London, Holland Press.

SKELTON, R A, and HARVEY, Paul D A (1986): *Local Maps and Plans from Medieval England.* Oxford, Clarendon Press.

TAYLOR, Eva G R (1956): *The Haven-Finding Art: A History of Navigation from Odysseus to Captain Cook.* London, Hollis and Carter. (Republished 1971, New York, American Elsevier Publishing Co.)

4.032 Compass Direction

A Direction as designated by reference to the divisions of the compass card or a compass rose.

B Compass points were originally wind directions (see Windrose, 5.231). The first reference to a compass card is by Francesco de Buti in 1380, and the first illustration is in Gregorio Dati's manuscript "La Sfera," *c.* 1422. In the 13th century the compass or circuit of the horizon was divided into sixty-four points by the sailors in the Mediterranean, though there were still only eight wind directions with individual names (Taylor, 1956, p.98). In northern countries the compass card was divided into thirty-two parts or points, one point being equal to $11\frac{1}{4}°$ of a circle, and into sixty-four half points. The earliest known compass card marked with the initials of names of winds is believed to be that of Giacomo Giroldi, 1426, in the Biblioteca Marciana in Venice (Brown, p.130). It is clear that the rhumb line networks found from the late 13th century on portolan charts (the earliest example of which is the Carte Pisane, *c.* 1290) were to be used with a compass, but whether the compass featured in their construction is not known.

From the points of the compass and the compass rose the navigator could work out his sailing directions, as explained by the Dieppe hydrographer Jean Rotz, with the help of a diagram in his "Boke of Idrography," 1542 (British Library, Royal MS. 20.E.IX,f.3v): "The maner for to knowe the Wyndis of all the points of the sey compas."

The network of compass lines remained a major element on charts well after the introduction of lines of latitude and longitude from the late 15th century onwards. The straight-line compass course would not, however, lead the navigator to his destination because of the spherical surface of the globe. Gerard Mercator in 1569 invented the projection named after him to answer this problem (see Loxodrome, 4.123).

As late as the 18th century many regional charts and local charts used compass lines as their main reference system. Thus Jacques Nicolas Bellin, first Ingénieur hydrographe de la Marine in France, displayed compass lines without latitude and longitude on many of the local charts in his *Petit Atlas Maritime,* Paris, 1764. In the 19th century naval surveyors still used the compass directions as a major feature of their local charts, giving the co-ordinates of the geographical position in a note. Examples from British surveys are F W Beechey's Plan of Port Lloyd, 1827, A T E Vidal's Dixcove, 1837, and Thomas Graves's Naxia, 1842 (Blewitt). It was common also for 19th century British town plans to display compass roses as their main reference system.

When azimuth compasses were introduced in the 17th century scales of degrees were added, running from 0° at north and south to 90° at east and west. In the 19th century British Admiralty charts displayed compass roses with a magnetic-north arrowhead. In 1863 a magnetic compass rose with the annual change of variation was introduced, and in 1894 degrees were shown as well as points. The 360° card was first proposed for the United States Navy in 1901 and was adopted a few years later, and the compass rose on charts was altered accordingly. A further change came in

1911 with a new compass rose showing an exterior true circle of degrees and an interior magnetic circle (Day, pp. 49-50, 75, 126, 259).

Commercial cartographers in the 19th century, however, remained more conservative than official chartmakers. The charts of William Heather, revised by John William Norie, which came to be called the "blue-back charts," still displayed in the 1830s the network of rhumbs typical of the earlier portolan charts (Howse and Sanderson, p.123).

In Chinese chartmaking the maps derived from the voyages of Chêng Ho (1405-1433), printed in 1621, give elaborate details of compass directions in legends written along the lines of travel (Needham, pp.559-60, fig.236). The compass rose was introduced into Japan from Europe during the later part of the Edo period, in the early 17th century. It differed, however, from the western rose in being divided not into 32 directions but twelve representing the signs of the zodiac, as illustrated by a mid 17th century manuscript portolan chart (Takejirō Akioka's collection, Tokyo; Nanba et al, p.155, pl. 22).

C BLEWITT, Mary (1957): *Surveys of the Seas: A Brief History of British Hydrography.* [London], Macgibbon & Kee.

BROWN, Lloyd A (1949): *The Story of Maps.* Boston, Little, Brown and Co.

CORTAZZI, Hugh (1983): *Isles of Gold: Antique Maps of Japan.* New York and Tokyo, Weatherhill.

DAY, Archibald (1967): *The Admiralty Hydrographic Service 1795-1919.* London, HMSO.

HOWSE, Derek, and SANDERSON, Michael (1973): *The Sea Chart.* Newton Abbot, David & Charles.

NANBA Matsutara, MUROGA Nobuo, and UNNO Kazutaka (1973): *Old Maps in Japan.* Osaka, Sogensha.

NEEDHAM, Joseph (1959): *Science and Civilisation in China,* vol.3. Cambridge, At the University Press.

TAYLOR, Eva G R (1951): "The South-Pointing Needle," *Imago Mundi, 8,* 1-7.

TAYLOR, Eva G R (1956): *The Haven-Finding Art: A History of Navigation from Odysseus to Captain Cook.* London, Hollis & Carter. (Republished 1971, New York, American Elsevier Publishing Co.)

WINTER, Heinrich (1948): "The True Position of Hermann Wagner in the Controversy of the Compass Chart," *Imago Mundi, 5,* 21-26.

4.041 Date Line

A The line (theoretically coincident with the meridian of 180° from Greenwich) at which the calendar day is reckoned to begin and end, so that places immediately to the west of it are one day later than places immediately to the east. For local convenience and by international agreement the line passes to the west of the Aleutian Islands and to the east of Asia as well as to the east of Chatham Island and the Tonga group.

B The need to establish the date line as an anti-prime meridian (see 4.161) was perceived when European powers colonized the borders and islands of the Pacific and as trade in the Pacific developed in the 19th century. Before 1845 there were discrepancies in date according to whether colonizers had come from the east or the west. The Philippines, colonized from America, kept "American date," while the East Indies, colonized from Europe, kept "Asiatic date" (Howse, pp. 161-62). Alaska, as part of the Russian empire also kept Asiatic date. Its purchase by the United States on 1st January 1845 brought it into American time, thus straightening the line in the North Pacific. In 1879 the British governor of the Fiji Islands, which lie astride the 180° meridian, decreed that all the islands should keep the same time, Antipodean Time.

The problem of the date line was corollary to the establishment of a prime meridian for longitude (Howse, p. 160). In 1884 the government of the United States called an International Conference at Washington "for the purpose of fixing a Prime Meridian and a Universal Day." It was proposed to adopt the meridian passing through the observatory of Greenwich and to adopt "a universal day for all purposes for which it may be found convenient, and which shall not interfere with the use of local or other standard time where desirable." This universal day was to be a mean solar day (International Conference, pp. 3, 76-7, 83-4, 112). The fixing of the date line itself was not the result of a formal international agreement. As stated by the Hydrographer of the Navy in Great Britain, the date line was established as a matter of convenience, "merely as a method of expressing graphically . . . the differences of date which exist among some of the island groups in the Pacific" (Howse, p. 163).

German commercial mapmakers appear to have been the first, some 20 years earlier, to introduce the date line on general maps. On the "Chart of the World on Mercators Projection," by Hermann Berghaus and Fr V Stülnagel, third edition, published by Perthes at Gotha, [1865-66], a small double hemispheric inset map, "The world on polar projection" (which is lacking in the first and second editions of 1863 and 1863-64) displays the "present" and "former" portions of the Antipodal Datum Line. Starting from the eastern border of Alaska the "former" line takes a wide sweep across the Pacific westward to include the Philippines which the Manila authorities in 1844-45 had agreed should keep American date. The line then sweeps eastward (in about 20°S) to 144°W before turning westward to 180° within the Antarctic Circle. This alignment brought into the same time zone of Asiatic date Samoa and the archipelago off New Guinea (later the Bismarck Archipelago), both areas of German trade and interest. The "present" line on the map runs from eastern Alaska to the meridian of 180° and holds this position to 80°S. The insets also mark routes for steam navigation round the world and projected and operational telegraph lines, developments which explain why the date line had become of practical importance for map publishers and their public.

In the *Richard Andreas allgemeiner Handatlas*, second improved and revised edition, published at Bielefeld and Leipzig by Velhagen and Klasing in 1887, a "Karte des Weltverkehrs und der Meeresströmungen" (pp. 6-7) shows the "Linie des Datumwechsels" as a chain-dot line marked with "Montag" on the west side and

"Sonntag" on the east. From its northern end in 168°W and 75°N, the line runs south through the middle of the Bering Strait, diverges southwest to pass west of the Aleutian Islands, thence proceeds west and south of the Philippines, turns eastward and runs north of the Solomon Islands, passes west of the Tonga Islands and southeast past Chatham Island, ending in 165°W at 52°S.

Adolf Stieler's Handatlas, published by Perthes at Gotha, [1891], also set out a divergent position from the generally agreed assimilation of the date line to the 180° meridian (Downing, pp. 415-17). On plate 5, "Weltkarte in Mercators Projection," a line named "Wirtschaftliche Datumgrenze oder Tageswende" (Administrative date limit or change of day) runs south to 15°N in about 172°E. This arrangement included in the American date Morrell and Byers islands which appeared on the charts at the western end of the Hawaiian chain (other sources brought them in by giving the 180° line a kink to the west), but when it was proved that the islands did not exist, the line was straightened (Howse, p.162). Stieler's line continued due east in 15°N to 159°W.

J C Bartholomew's "Travellers' Route Chart of the World on Mercators Projection," published at Edinburgh, [1896], displays a "Conventional Division Line for Time." This conforms to the accepted demarcation, except that by following the 180° meridian the line divided the Aleutian Islands. One side of the line is marked "Monday PM" the other "Sunday AM." The 12th edition of Hermann Berghaus's "Chart of the World," revised by Hermann Habenicht and Bruno Domann and published by Justus Perthes in 1897, includes a line annotated only as "Date Boundary," which is the same as the Bartholomew line except in the Aleutian Islands.

The currently generally accepted authority for the position of the line is the British Admiralty Chart 5006, "Time Zone Chart of the World," the earliest version of which was published as chart no. X301 in 1918, with an accompanying memoir, "Memorandum on Time Keeping at Sea." On the chart the date line is marked but not named. In May 1919 a uniform system of one hour time zones with all subdivisions measured as Mean Solar Units was adopted in the British Navy and was identical to systems in use in the French and Italian navies. "The World Time Zone Chart," published by the Admiralty in London in 1919, shows "Time now in use in various countries etc." A system of colours distinguishes the zones.

C BURPEE, Laurence J (1915): *Sandford Fleming, Empire Builder.* London, 220-24,

DOWNING, A M W (1900): "The Date Line in the Pacific," *Geographical Journal, 15,* 415-17.

FIJI, Governor of the Colony of (1879): *Ordinance No.14.*

HYDROGRAPHER, London (1930): "Notes on the History of the Date or Calendar Line." Printed as *Bulletin 78, Dominion Observatory, Wellington, New Zealand.*

HOWSE, Derek (1980): *Greenwich Time and the Discovery of the Longitude.* Oxford, Oxford University Press.

INTERNATIONAL Conference Held at Washington for the Purposes of Fixing a Prime Meridian and a Universal Day (1884): "Protocols of the Proceedings," *Forty-eighth Congress, 2nd Session, House of Representatives, Executive Document No. 14* (Vol. 21).

SMITH, Benjamin E (1899): "Where a Day Is Lost or Gained," *Century Magazine, 58,* no. 5, 742-45.

WOOLHOUSE, W S B (1910): "The Calendar," *Encyclopedia Britannica,* 11th ed., *4,* 987-1004.

4.051 Ellipsoid

A Sometimes called a spheroid ("resembling, but not a true sphere"), the ellipsoid (earth) is the figure generated by the rotation of an ellipse around its semi-minor axis. (See also Geoid, 4.071.)

B From the time of the Greek philosopher Pythagoras (fl. *c.* 525 BC) until the end of the 17th century, the earth was considered to be a sphere. The Dutch physicist and astronomer Christiaan Huygens (1629-1695), who joined the newly founded Académie Royale des Sciences, Paris, in 1666, advanced the understanding of mechanics with his studies of the descent of heavy bodies, and using the pendulum timekeeper which he had invented, he worked with other members of the Académie on the establishment of a standard meridian of longitude (running through the middle of their observatory) and towards the solution of the linear value of a degree of longitude. Newton (1642-1723) developed the subject further with his work on the theoretical shape of a rotating body. He went on to deduce the shape of the earth to be an oblate spheroid, that is flattened at the poles. He determined that the equatorial semi-axis would be 1/230 longer than the polar semi-axis, somewhat larger than the true value (about 1/298).

Measurements along the arc of the Paris meridian from Dunkirk to Spain by Jean Dominique Cassini (1625-1712), continued by his son Jacques (1677-1756), which in 1718 were reported to the Académie and published by Jacques in that year, suggested that the length of a degree of the meridian decreased toward the pole, contradicting Newton's hypothesis and leading to the idea that the earth was a prolate spheroid, flattened at the equator. To settle the question, the French Académie Royale in the mid-1730s sent two expeditions to measure arcs of the meridian: one to Peru in 1734 led by Louis Godin, Pierre Bouguer, and Charles Marie de la Condamine, and the other to Lapland in 1736 led by Pierre Louis Moreau de Maupertuis, Alexis Claude Clairaut, Charles Étienne Louis Camus, Pierre Charles Le Monnier, and Reginaud Outhier. Their measurements confirmed that the earth is indeed flattened at the poles and is therefore an oblate spheroid. The results were published in the *Mémoires de l'Académie Royale des Sciences*; for example, "Triangles de la Meridiene de Quito," *Mémoires . . . 1744,* p.296, pl. XIV (Paris 1748); and "Carte de triangles de la meridienne de Quito," in Pierre Bouguer's *La Figure de la Terre*, Paris 1749, pl.VII. The French geographer Philippe Buache made a map recording the observations world wide, "Carte des Lieux, où les differentes longueurs du Pendule à Secondes ont eté observées," Paris, 1740.

Large-scale topographic mapping requires a regular geometric datum even though the shape of the earth is an irregular geoid (4.071). For this an ellipsoid of revolution is assumed and defined by a semi-major axis, *a*, a semi-minor axis, *b*, and the flattening, *f*, where $f = 1/[(a - b)/a]$. Since 1800 there have been numerous determinations of ellipsoids for mapping and geodetic purposes.

According to Stephen P Rigaud the term spheroid was first used by Isaac Barrow in 1664. John Keill defined it in 1698 as a figure generated by the rotation of an ellipse around its semi-minor axis. Since that time the term spheroid has often been used in geodetic literature.

C BOMFORD, Guy (1971): *Geodesy*. London, Oxford University Press, 109-11.

BROWN, Lloyd A (1949): *The Story of Maps*. Boston, Little, Brown and Company.

KEILL, John (1698): *An Examination of Dr. Burnet's Theory of the Earth*. Oxford.

LA CONDAMINE, Charles Marie de (1751): *Journal du voyage fait . . . à l'équator . . .* Paris, Académie Royale des Sciences.

MAUPERTUIS, Pierre Louis Moreau de (1738): *La figure de la terre déterminée par les observations . . . faites par ordre du Roy au Cercle Polaire*. Paris.

RIGAUD, Stephen Peter (1841, 1862): *Correspondence of Scientific Men of the Seventeenth Century*, 2 vols. Oxford.

SMITH, J R (1986): "The Pear-Shaped Earth," *Geographical Magazine, 58*, 572-77.

4.071 Geoid

A An irregular equipotential surface approximating the shape of the earth (as an ellipsoid of revolution) at every point of which the direction of gravity is normal to the surface. (See also Ellipsoid, 4.051.)

B The shape of the geoid is determined by the uneven distribution and densities of masses comprising the Earth, and in general it corresponds to the surface of seas and oceans at rest (with constant salinity, temperature and atmospheric pressure) and to the levels that water would assume in a system of sea-level canals in the continents. The shape of the geoid is determined with respect to the earth ellipsoid (taken as datum) on the basis of astronomic, geodetic, gravimetric, and satellite observations.

The direction of gravity does not coincide with the ellipsoidal normal at every point. The angle between the plumb line and the ellipsoidal normal is called the deflection of the vertical, and these deflections, when determined at a sufficient number of points, allow computation of the departure (undulation) of the geoid from the ellipsoid. These undulations do not exceed 100 meters.

The uneven distribution of masses on the Earth's surface was considered the cause of the deflections of the vertical. Surveys done by John H Pratt, the British geophysicist, in the mid-19th century did not confirm these suppositions, and in consequence led to the formulation of the hypothesis of isostasy – compensation of masses of mountains and oceans. The hypothesis of Pratt was corroborated by later, 20th century, investigations (John Filmore Hayford, Sir George Biddel Airy, and Veikko Aleksanteri Heiskanen).

The mathematical problem of the connection between the anomalies of gravity and the geoid undulations was investigated by George Stokes, who solved it in 1849. The solution was supplemented and confirmed by Paolo Pizzetti (1896) and by others in the 20th century.

The term geoid for the physical shape of the Earth, as revealed from investigations undertaken in the first half of the 19th century, was introduced in 1873 by Johann Listing, a German mathematician and geodesist.

C BOMFORD, Guy (1971): *Geodesy*. London, Oxford University Press, 109-11.

ENCYCLOPAEDIA BRITANNICA, 15th ed. (1978): "Earth, Figure of."

JORDAN, Eggert Kneissl (1958): *Handbuch der Vermessungskunde,* vol. IV. Stuttgart, 7B, Metzlersche Verlagsbuchhandlung, 21-2.

KAMELA, C Z (1955): *Geodezja dynamiczna* (Dynamic Geodesy), Vol.II. Warszawa, PPWK (in Polish).

LEVALLOIS, J J (1970): *Géodésie générale,* vol. III. Paris, Éditions Eyrolles, 247-73.

MITCHELL, Hugh C (1948): Definitions of Terms Used in Geodetic and Other Surveys. U.S. Coast and Geodetic Survey, *Special Publication, 242.* Washington, DC, 34.

PERRIER, G G (1950): *Kurze Geschichte der Geodësie.* Bamberg, Bamberger Verlagshaus Meisenbach, 82-5.

4.072 Graticule

A (MDTT 322.1/322.2.) a) An imaginary network of meridians and parallels on the surface of the Earth or other celestial body; (i) division of a circle into 360 degrees. (ii) division of a circle into 400 grades. b) A network of lines, on the face of a map, which represents meridians and parallels.

The word graticule so defined clearly distinguishes the latitude and longitude co-ordinate system from the Cartesian co-ordinate system which is known by the term grid (see 4.073). In some parts of the world, however, both systems are referred to as grids.

B The representation of the graticule in the Western world dates from the time of Dicearchus of Messina (350-290 BC) who introduced the idea of a "central parallel," Eratosthenes (c. 275-194 BC) who divided the map into a system of quadrilaterals, and Hipparchus of Rhodes (fl. 146-127 BC) who was the first to specify the positions of places by their latitudes and longitudes. By the time of Claudius Ptolemy of Alexandria (2nd century), the concept of the geographic graticule was firmly established (Fig. 14). Peter Apian (1495-1552), professor of mathematics at Vienna and Ingolstadt, illustrated in a diagram how to use the co-ordinates which were compiled as a geographical index to Ptolemy's *Geography* from 1486 onwards (Apian, *Cosmographicus Liber* . . . , Landshut, 1524, fol. 60; and later works, see Map index, 6.0920b) (Fig. 13).

The meaning of graticule as a system of parallels and meridians dates from about 1875. Previously the word was used in the sense of a division of a design or plan into squares with the object of reproducing it accurately on a different scale. Its extension to involve rectangular and spherical forms in addition to squares and its application to topographical mapping was advocated by General James Thomas Walker (1826-1896): "The sheets of paper on which the details of the survey of any large area of country are to be laid down must be furnished with a system of conventional lines, drawn with a view to assimilate the margins of contiguous sheets and to form a graticulation within which the details may be accurately inserted. The graticule is sometimes rectangular, sometimes spherical, sometimes a combination of both . . . Spherical graticules are constructed in various ways, usually in accordance with some specific method of projection" (Walker, 1887). Examples are the Indian Survey's trans-frontier maps made under Walker's direction 1872-1882 and his "Map of Turkestan" in four sheets, Survey of India, 1882.

In the Eastern world, the Japanese learned of the Chinese square grid (4.191) more than a thousand years ago, but it was not until the late 16th century, when Europeans began to visit Japan, that the Japanese came across the network of meridians and parallels on maps and globes brought from the Western countries. Scholars at that time, however, seem to have paid little attention to it. From the end of the 16th century to the 17th, a number of world maps based on European models appeared in Japan, but with a few exceptions, only the Equator, the Tropics of Cancer and Capricorn, and the Arctic and Antarctic Circles were represented on them. If any meridians appeared at all, they were drawn at large intervals for decoration rather than for use.

Since the middle of the 18th century, when the study of Western sciences expanded, better understanding of European cartography gradually spread among the educated. Nonetheless, the map of Japan published by Nagakubo Sekisui in 1779, which has long been believed to be the first map of Japan with meridians and parallels, was actually based on an early portolan chart and its north-south lines are not proper meridians.

The first Japanese-produced map of Japan to have meridians and parallels based upon original survey is "Dainihon Enkai Yochi Zenzu" (Coastal Map of Japan) which was completed by Inō Tadataka in 1821. In this epoch-making map, the prime meridian goes through Kyoto where the emperor resided then. Although Inō committed some errors, especially in measuring longitudes, this map later proved to be of immense value to the Meiji government in its work to make accurate maps of the whole land of Japan.

C AKIOKA Takejiro (1955): *Nihon Chizushi* (History of Japanese Cartography). Tokyo, Kawade Shobo.

HOYANAGI Mutsumi (1974): *Inō Tadataka no Kagakuteki Gyoseki* (A New Appreciation of the Scientific Achievement of Inō Tadataka). Tokyo, Kokon Shoin.

NANBA Matsutaro, MUROGA Nobuo, and UNNO Kazutaka (1973): *Old Maps in Japan*. Osaka, Sogensha.

WALKER, J T (1887): "Surveying," *Encyclopaedia Britannica*, 9th ed., *22*, 714-15.

WALKER, J T (1896): "India's contribution to Geodesy," *Philosophical Transactions of the Royal Society of London for 1895* (A), *186*, pt 2, 745-816.

4.073 Grid

A A referencing system on a map in which points are defined by their distances from two perpendicular axes, usually with the aid of marginal graduations. This type of grid is the graphic representation of a plane rectangular co-ordinate system. Such a referencing grid system is not to be confused with the grid-like appearance on a map of the lines established by a rectangular land division system, such as the Roman centuriation or the Public Land Survey system of the United States. (See also Graticule, 4.072; Square grid, 4.191.)

B A grid thus defined is a more precise form of reference system than the non-co-ordinate (see 4.141); instead of merely locating a point somewhere within a quadrangle, it may be located precisely by means of Cartesian co-ordinates. This branch of geometry, known as analytical geometry, was created and first used by Réné Descartes in 1637, but the conceptions inherent in Cartesian geometry were appreciated centuries before it was made into an instrument which introduced a new epoch of mathematical invention.

In cartography, the Chinese practice of drawing maps on silk, at least as early as 227 BC and probably earlier, invites the suggestive idea that the position of a place could be fixed by following a warp thread and a weft thread to their crossing point. In European cartography there is more than a hint of the concept present in the index compiled by Conrad Wolfhart (Lycosthenes) for Aegidius Tschudi's map of Switzerland in 1560. The map bears a fine marginal graduation of 80 divisions (gradus) each way, and Lycosthenes provided a long note explaining the use of the system. Implicit is the idea of a square getting smaller and smaller until, at the limit, it becomes a point. Whether Tschudi's first edition of 1538, apparently not extant, had marginal graduations has not been ascertained.

As more large-scale maps were produced in the 17th century and particularly when these were drawn on projections with circular or curved parallels and meridians, the use of plane rectangular co-ordinates for their construction became inevitable. The use of projection-based, precise rectangular co-ordinate systems is a feature of the post-1900 era, and particularly after the period of the First World War (1914-1918).

C BRITISH PARLIAMENTARY PAPERS (1893-4): Report of the Departmental Committee appointed by the Board of Agriculture to Inquire into the Present Condition of the Ordnance Survey [The Dorington Committee], Vol. 72, C. 6895-1, *passim*.

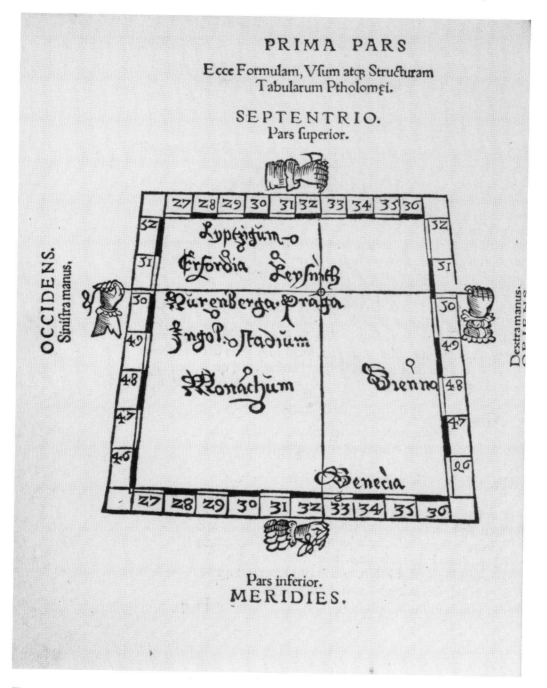

Fig. 13 – Diagram to illustrate the use of Ptolemy's tables of co-ordinates for places. In Peter Apian, Cosmographicus Liber . . . studiose correctus, ac erroribus vindicatus per Gemmam Phrysium, Antwerp, 1529, fol. XXX^v. By permission of the British Library. (Graticule, 4.072; Map index, 6.0920b)

HEAWOOD, Edward (1932): "Early Map Indexing," *Geographical Journal, 80,* 247-9.

NEEDHAM, Joseph (1959): *Science and Civilisation in China,* vol.3. Cambridge, At the University Press, 541.

4.121 Latitude

A The latitude of a place is its angular distance north or south from the equator. There are two major concepts of latitude: astronomic latitude and geodetic latitude. Astronomic latitude is the angle between the axis of the earth and the tangent to the meridional section of the geoid (4.071). For geodetic latitude substitute "ellipsoid" (4.051) for "geoid." Astronomic latitude may be reckoned with the aid of an ephemeris by observation of the angular altitude above the horizontal of the sun or a selected star. Geodetic latitude is computed from astronomic latitude and assumptions relating the geoid and the ellipsoid.

B Astronomers of the ancient world realized at an early date that geographical latitude was related to the appearance of the heavens. Use of a simple instrument resembling a sundial enabled them to become experts in finding the latitudes of places. Babylonian mathematicians had divided the circle into 360 degrees, thus providing a framework for computation. The sexagesimal method of computation dates from at least *c.* 1800 BC, although the earliest cuneiform text with zodiac and equator divided into 360 degrees is found much later, on a horoscope dated 410 BC. The Babylonians had, however, no conception of geographical latitude or of the sphericity of the earth, and their idea of astronomical latitude was limited. They divided the sky into three bands or "paths" of stars roughly parallel to the equator, with *Enlil* extending north from about +17°, *Anu* between +17° and −17°, and *Ea* south from about −17° (Stephenson and Walker, p.20). Distances between the moon and stars were given in cubits *(KÙŠ)* and fingers *(SI)* above (north) or below (south), in front (west) or behind (east). Only in the mathematical lunar ephemerides was there a concept of absolute ecliptic latitude. The cubit seems to have been about 2 or 2.5 degrees, the finger, 5 or 6 minutes, but how the distances were measured is not known (Stephenson and Walker, pp. 14, 20). An important computation in Babylonian astronomy was also the length of daylight, as lunar theory depended on prediction of the first visibility of the new crescent of the moon after conjunction. This was one of the astronomical concepts which passed to the Greeks, together with knowledge of the Babylonian numerical system.

The earliest Greek astronomer to make use of the division of the circle into 360 degrees was Hipparchus of Rhodes (fl.146-127 BC), who worked mainly at Rhodes (Dicks, pp. 148-9). Hipparchus is now believed to have been responsible for the scientific concept of *climata* as narrow belts of latitude within which the astronomical phenomena, such as the length of the longest day, gnomon ratio, and the height of the celestial pole, did not much change (see Mappa mundi, 1.2320b, and Zone map, 3.261). He established a mathematical geography based on scientific determinations of latitude and longitude (4.122), which he recommended as essential for scientific mapping. He drew up a list of latitudes which he believed to be reliable and constructed what appears to have been a table of parallels between the equator and the north pole on the theoretical globe of the earth (Dicks, pp. 35, 148-9, 154, 160-3).

Hipparchus's observations of latitude and longitude were incorporated by Claudius Ptolemy (c. AD 90 to c. 168) into his astronomical treatise the *Mathematical Syntaxis*, written between AD 141 and 147, later known as the *Almagest*. This work with its tables of co-ordinates represents the culmination of efforts by early Greek astronomers to establish the relationship between rising times, seasons, and geographical locations (Neugebauer, p.114). Ptolemy lists 39 parallels of latitude, including the equator, for each of which he gives the length of the longest day and the latitude in degrees north of the equator. Only 29 of the parallels, however, were connected with named localities on earth for which information was available (Dicks, p.161).

Some years later Ptolemy in his *Geography*, c. 150, restricted himself to 21 parallels (representing half hour intervals) north of the equator, extending to 63 degrees through Thule, and to one parallel south of the equator. He gave the co-ordinates of places, for which the latitude is expressed in multiples of 10 (sometimes 5) minutes. Most of these coordinates were derived from calculations of varying accuracy based on travellers' reports; and only those from Hipparchus had a sound scientific basis. The result was an illusory impression of accuracy (Dicks, p.37). The latitudes and longitudes in Ptolemy's *Geography* were recorded in list form in Greek numerals, but none of the extant manuscripts of the *Geography* is earlier than 1200, and by then those figures had undergone much change. The only Ptolemy map which evidently originates earlier than 1200 is that to which the statement of Agathodaemon refers: "I, Agathos Daimon, a technician of Alexandria, drew a map from the *Geography* of Ptolemy." As he could not have lived later than the 6th century AD, this map, presumably the world map, would appear to preserve the latitudes and longitudes as set out by Ptolemy.

The Romans used data taken mainly from Ptolemy in recording the latitudes of provinces on portable sundials. Their land surveyors were taught some cosmology, as recorded in the Corpus Agrimensorum, a collection of surveyors' manuals dating from the 6th and 9th centuries. The texts record five latitudinal zones (Dilke, 1971, pp.61-3). Roman methods of survey, notably centuriation, normally did not involve determinations of latitude. The remarkable legacy of Ptolemy was the transmission of the Hellenistic geographical science to Arabic culture and thence to medieval Europe. In copied and re-copied form Ptolemy's *Almagest* and *Geography* became part of the Byzantine heritage. When Caliph al-Ma'mun established a scientific academy in Baghdad early in the 9th century, among the first works to be translated into Arabic were those of Ptolemy.

The advance of the Arabs westward led to a wide diffusion of Arabic and therefore Hellenistic science. Within a century of the death of the Prophet in 632, the Arabs had occupied most of the centres of ancient learning in Southwest Asia, North Africa, and part of the Iberian peninsula. The re-conquest of Moorish Spain made possible intellectual contacts between Christendom and Western Islam, and Spanish-Arabic learning permeated Western Europe, diffused from places like Toledo and Cordoba. In 1126 Adelard of Bath translated from Arabic into Latin the stellar and solar methods of obtaining latitude set out in the *Khorazmian Tables* of al-Khwārizmī (c. 820). From 1175 Gerard of Cremona's translation of the *Almagest* was available, giving an explanation of parallels of latitude and of places to be found on each parallel. The Toledo Tables, incorporating the revisions of al-Khwārizmī and of al-Zarquali of Toledo (c. 1080), were translated into Latin and used for compiling tables for places in France, England, Italy, and perhaps elsewhere. The popular *De Sphaera* of John of Holywood (known as Sacrobosco), written in Paris c. 1220, and including a zone map, provided instruction in the spherical geometry of the earth.

The maps of the Arabic scholar al-Idrisi (born 1099), who had been a student at the University of Cordoba, were the most notable result of collaboration between Muslim and Christian culture. At the request of the Norman king Roger II at Palermo, Sicily, al-Idrisi completed in 1154 a description of the world accompanied by a map drawn on 70 sheets together with a small circular world map, which displays climates with curved parallels derived from Ptolemy (Bodleian Library, Oxford).

On the continent of Europe there was no comparable attempt to revise the world map (Wright, pp. 96-8). Nevertheless, a scholastic type of scientific geography, which included knowledge of the co-ordinate principle, grew up at various centres, such as the university of Paris in the second half of the 13th century. Roger Bacon (c. 1214-1294), a scholar at the universities of Oxford and Paris, made a world map for presentation to the Pope (as he recorded in his "Opus Majus," written in 1266-67); and this map (no longer extant) was evidently constructed on a co-ordinate system with climates and longitudes, probably on a square grid, with towns located from the Toledo Tables, the *Almagest* and al-Farghani's Elements of Astronomy (Durand, pp. 94-5; Bridges, vol. 1, pp. 294-98; Burke, vol. 1, pp. 315-17). No similar map seems to have been made from co-ordinates between this time and 1424-26. The concept of the *latitudo formarum* which developed in the 14th century, however, was derived from the co-ordinate principle and encouraged a technique of graphic representation (Durand, pp. 96-7).

By the beginning of the 15th century a considerable body of data on latitude and longitude, including co-ordinate tables, was available for mapmakers. Two maps of Germany, 1426, by Conrad of Dyffenbach, a prebendary of the church of St Peter at Heidelberg (fl.1399-1426), use a co-ordinate system and are based on tables preserved in the Vienna-Klosterneuburg corpus of material (the maps are in the Vatican Library, Codex Palat. lat. 1368; Durand, pp. 107-10, 114-22).

These local efforts were forerunners of two revolutionary changes in the later 15th century. The first followed the arrival in Italy of Ptolemy's manuscripts from Byzantium. Ptolemy's *Geography* was translated into Latin in 1406 and circulated in various Latin and Greek manuscripts, some illustrated by maps. From 1477, when the first Ptolemy printed with maps was published at Bologna, knowledge of its contents became widespread. The network of parallels and meridians, combined with the tables, made it possible for places to be laid down by their co-ordinates, thus constituting a major advance on traditional medieval mapmaking. New modern maps were prepared on the same principles. A manuscript of the Latin version of Ptolemy (the Nancy codex), derived from a Byzantine original and prepared by Cardinal Guillaume Fillastre in 1427, contains the first modern map added to a Ptolemy. This map of northern regions, based mainly on that of the Scandinavian cartographer Claudius Clavus, displays a network of lines of latitude and longitude. Another map, which was probably intended as a Tabula Moderna for one of the Rome editions of Ptolemy, 1478 or 1490 (but not included with it), is the map of central Europe by Nicolas of Cusa (1401-1464), inscribed Eystat, 1491, but probably earlier. It is executed in the style of the Rome Ptolemys, drawn in a trapezoid frame and marked with graduations of latitude and longitude (Durand, pp. 252-66). The earliest "modern" world map revised from Ptolemy, with graduations of latitude and longitude, is the large manuscript wall map by Henricus Martellus Germanus, c. 1490 (see Longitude, 4.122). The earliest extant terrestrial globe, by Martin Behaim, which appears to be derived from Martellus or a common source and was made at Nuremberg in 1492, is also graduated in degrees of latitude along the meridian (and also of longitude along the equator).

The second major development in the 15th century resulted from the Portuguese voyages south along the coast of Africa from 1418 onwards. Regular observations of latitude constituted a revolution in navigational method. The first recorded observation from the altitude of the Pole Star made in a European ship, dated 1456, is by Alvise da Cadamosto and appears in his account of his voyage, referring to the mouth of the river Gambia (Albuquerque, p. 227). Later, a latitude scale came to be incorporated in manuscript charts made on or after the voyages. Several undated charts with latitude scales have been assigned to the end of the 15th century, but this dating is contested. As these charts normally also display the traditional network of rhumb lines it is often difficult to determine whether the latitudes marked were superimposed or were a basic element in the construction.

Various innovative features can be identified on the extant charts of the early 16th century. The Cantino world map, 1502, is the first chart of known date marked with lines of latitude in the form of the tropics of Cancer and Capricorn, etc. The King-Hamy map and the planisphere by Nicolaus de Caverio, both c. 1505, are the first world charts to include a latitude scale (Mollat du Jourdin, p. 216). An Atlantic chart of about the same date (c. 1504) by Pedro Reinel (the earliest chart signed by a Portuguese cartographer), is notable in displaying an oblique latitude scale (or "meridian") from 44 to 57 degrees as well as a vertical scale extending through the whole height of the chart (see Magnetic meridian, 4.131). The oldest rutter with latitudes is that of Duarte Pacheco Pereira, c. 1505-08.

Diego Ribeiro (fl.1519, died 1533), a Portuguese serving as Cosmographer-major in the Spanish service, claimed to be the first, on his unsigned chart of 1527 (Thüringische Landesbibliothek, Weimar), to correct the latitudes of the Mediterranean so that its axis was correctly oriented and to explain the principles involved (Cortesão and Teixeira da Mota, I. pp. 92-3). The earlier distortion appears to have been due to failure to correct for magnetic variation. Jean Rotz, hydrographer of Dieppe, in his Boke of Idrography, 1542 (British Library, Royal MS. 20.E.IX), was the first to use different scales of latitude on the left and right hand margins of certain charts, a device designed to adapt them to magnetic variation (Wallis, pp. 32, 37).

In ancient China a square grid was adopted at about the same time that Greek geographers were developing the system of co-ordinates. The earliest reference appears in a biography of the astronomer Chang Heng of the 1st century AD, and Phei Hsiu (224-271) developed it further. Unlike the co-ordinate system which disappeared from use in Europe for almost one thousand years, the square grid (4.191) remained the basic system for Chinese mapping until European missionaries introduced Western principles of mapmaking in the late 16th century.

C ALBUQUERQUE, Luís de (1971): "Two Chapters on History of Astronomical Navigation," in Cortesão, Armando: *History of Portuguese Cartography,* Vol.2. Coimbra, Junta de Investigacioes do Ultramar.

BRIDGES, John Henry, ed. (1897): *The "Opus Majus" of Roger Bacon.* Oxford, Clarendon Press.

BURKE, Robert Belle (1928): *The Opus Majus of Roger Bacon. A Translation.* Philadelphia, University of Pennsylvania Press.

CORTESÃO, Armando, and TEIXEIRA DA MOTA, Avelino (1960): *Portugaliae Monumenta Cartographica.* Lisbon.

CRONE, Gerald Roe (1978): *Maps and Their Makers,* 5th ed. Folkestone, Kent, Dawson; Hamden, Conn., Archon Books.

DICKS, D R (1960): *The Geographical Fragments of Hipparchus.* London, Athlone Press.

DILKE, O A W (1971): *The Roman Land Surveyors.* Newton Abbot, David & Charles.

DILKE, O A W (1985): *Greek and Roman Maps.* London, Thames and Hudson.

DURAND, Dana Bennett (1952): *The Vienna-Klosterneuburg Map Corpus of the Fifteenth Century.* Leiden, E J Brill.

KAMAL, Youssouf (1926-53): *Monumenta Cartographica Africae et Aegypti,* 5 vols. Leiden.

MOLLAT DU JOURDIN, Michel, and LA RONCIÈRE, Monique de (1984): *Sea Charts of the Early Explorers, 13th to 17th Century.* New York, Thames and Hudson.

NEUGEBAUER, Otto E (1983): *Astronomy and History: Selected Essays.* New York and Berlin, Springer-Verlag.

STEPHENSON, F R, and WALKER, C B F, ed. (1985): *Halley's Comet in History.* London, British Museum Publications.

THOMSON, James Oliver (1948): *History of Ancient Geography.* Cambridge, At the University Press.

WALLIS, Helen, ed. (1981): *The Maps and Text of the Boke of Idrography Presented by Jean Rotz to Henry VIII.* Oxford, Roxburghe Club.

WRIGHT, John K (1923): "Notes on the Knowledge of Latitudes and Longitudes in the Middle Ages," *Isis, 5,* 75-98.

4.122 Longitude

A (MDTT 31.12) The angle, measured in a plane parallel to that of the equator, between the plane of the meridian through any point and the plane of the prime meridian or other selected datum meridian. Since every meridian is a semi-great circle terminated by the geographical poles and the plane of a prime meridian bisects the earth, no longitude can be greater than 180°. Longitude is therefore measured along the shorter arc east or west of a prime meridian (see 4.161).

B Longitude is inseparably associated with the rotation of the earth on its axis, with the measurement of this angular value, and thus with the precise measurement of time. The calculation of longitude in antiquity and over the subsequent centuries was more erratic than that of latitude on account of the inherent difficulty. Eratosthenes (c. 275–194 BC) worked out the approximate distance from the Sacred Promontory (Cape St Vincent, the southwest extremity of the Iberian peninsula), to the Ganges and related it to his estimate of the circumference of the earth (see Latitude, 4.121). Hipparchus of Rhodes (fl.146-127 BC) drew attention to the possibility of calculating longitude through accurate simultaneous timing of a lunar or solar eclipse at two given points (Dicks, pp. 42, 65, 121-22), but until the 12th century the longitudinal difference between places was generally determined by converting dead reckoning estimates to angular distance. Roger of Hereford records how the times of the moon's eclipse in 1178 were observed in Toledo, Marseilles, and Hereford and their differences in longitude determined.

Circular medieval maps displayed no latitudes or longitudes, but it has been shown, for example, that the Hereford world map, c. 1285, is based in its western section on itineraries radiating from Rome in directions which are approximately correct. Mappae mundi of this kind may be seen to correspond to an equidistant azimuthal projection (Crone, p. 6; Tobler, pp. 354-55). When the manuscripts of Ptolemy's *Geography* became available in Europe, mappae mundi such as that drawn by the Benedictine Andreas Walsperger in 1448 were adapted to the latitudes and longitudes of Ptolemy without alteration to the basic form of the map.

The first printed maps to display longitude are those in Ptolemy's *Geography,* published at Bologna 1477, Rome 1478, and Ulm 1482. Although the longitudes were grossly in error, in that the Old World landmass was extended over 180 degrees instead of 100 degrees, the maps were regarded as a model of scientific method mainly on account of the graduations of latitude and longitude. A manuscript world map identified as by Henricus Martellus Germanus (fl. 1480-1496), in which Ptolemy's configurations are corrected, ranks as the earliest map outside Ptolemy's *Geography* to show longitude (New Haven, Conn., Yale University Library; Vietor). This or a similar work seems to have been the prototype for Martin Behaim's manuscript globe, made at Nuremberg in 1492, now ranking as the earliest extant terrestrial globe. The equator is divided into 360 degrees, but these are unnumbered, and one meridian, 80° west of Lisbon, is marked. The modern world map by Johann Ruysch, "Universalior Cogniti Orbis Tabula . . . ," printed in the Rome Ptolemy of 1507, is the earliest in a Ptolemy to mark longitudes, showing the whole circuit of the world, 1° to 360°. In the same year Martin Waldseemüller of St Dié, Lorraine, also displayed on his world map the full compass of 360 degrees. Francesco Rosselli's printed world map, Florence, c. 1508, was the first on an oval projection, which became popular through its use by Sebastian Münster, Abraham Ortelius, and others.

Many chartmakers still used the network of loxodromes as the main framework for their maps, marking scales of latitude but not of longitude. Some of the earliest charts to show a longitude scale are the manuscript planispheres, 1525 (anonymous) to 1529, of Diego Ribeiro, the Portuguese emigré who was "cosmographer and maker of charts" to the Spanish hydrographic service, the Casa de Contratacíon (Mantua, Weimar, and the Vatican). They are drawn on an equidistant-cylindrical projection, in which the degrees of latitude and longitude are equal (Cortesão and Teixeira da Mota, I, 93). Another notable work of this period is a world chart probably by the Florentine Girolamo Verrazano, 1529, corrected to 1540 (National Maritime Museum, Greenwich). It is graduated with two sets of meridians, the original in ink with the prime meridian through the west point of Palma in the Canary (Fortunate) Islands; a second set in pencil, with longitude unmarked, based on either the west point of Madeira or the centre of Flores in the Azores (Howse and Sanderson, p.25).

Longitudes in the 16th century continued to be calculated from dead-reckoning. Cartographers were aware, however, that in principle any fast moving astronomical body could be used as a basis for time keeping, provided its motion could be predicted and observed. The earth's satellite, the moon, was an obvious first choice, but to predict and observe its motion presented problems. Johann Werner in his edition of Ptolemy's *Geography,* Nuremberg, 1514, was the first in print to propose the method of lunar distances to establish longitude (Werner, sig. c. vi^v-dii^r). In 1577 the longitudes of different cities of the Spanish Empire were determined by the timing of eclipses (Reparaz Ruiz, pp. 77-8). Galileo Galilei proposed in 1616 a method for determining longitude from the eclipses of the satellites of Jupiter, which he had sighted following his invention of the telescope; as a further aid he had applied the pendulum to the timing of astronomical observations.

Through the supporting observational and theoretical work of Jean Picard and Jean Dominique Cassini in the later 17th and early 18th centuries, it became possible to predict the times of the eclipses of Jupiter's satellites and from 1676 on, this method was extensively used on land for longitude determinations. It was only by the middle of the 18th century that prediction of the moon's motion was sufficiently precise to enable lunar distances to be successfully, if laboriously, employed. At about the same time, the invention of the chronometer, *c.* 1735 by the Englishman John Harrison (1693-1776) and *c.* 1766 by the Frenchman Pierre LeRoy (1717-1785), made it possible to determine longitude at sea without cumbersome calculations.

Captain James Cook on his first voyage to the Pacific, 1768 to 1771, took every opportunity to fix his longitude by observing eclipses of the sun and moon, the transit of fixed stars, and other predictable phenomena (Skelton, 1954, p. 117). By use of the tables in the British *Nautical Almanac,* first published in 1767, he was able to establish longitudes to a remarkable degree of accuracy. On his second voyage, 1772 to 1775, he carried four chronometers, describing the best (the instrument copied by Larcum Kendall from John Harrison's fourth chronometer) as "our faithful guide through all the vicissitudes of climates." On the third voyage, 1776 to 1780 Cook used John Arnold's chronometer. The charts made on these three voyages by Cook and his officers can claim to be the first maps in the history of exploration based on relatively accurate longitudinal observations. Cook's outlines and "fixes" were good enough to be used on British Admiralty charts of the 19th century (Skelton, 1969, p.26).

At the same time that attempts were being made to solve the problem of longitude at sea, some of the more enlightened home-based cartographers were seeking to improve the accuracy of their methods of compilation from co-ordinates of latitude and longitude. The Irish-born cartographer Bradock Mead (*c.* 1685-1757), alias John

Green, was a pioneer in this respect. In his early anonymous work, *The Construction of Maps and Globes*, London, 1717, he enjoined mapmakers to produce tables of all places whose longitude and latitude had been found by observation (p. 151). When later he issued (under the name John Green) his "New Chart of North and South America," London, 1753, and (anonymously) his "New Map of Nova Scotia," London, 1755, he accompanied these by *Remarks* and an *Explanation* respectively, in which he stated and assessed the source of his co-ordinates of places. The commercial chart trade was slow, however, to correct its charts according to recent and improved observations.

In the Eastern world the concept of longitude was introduced in the late 16th century by Jesuit missionaries and European navigators. Father Matteo Ricci, founder of the China mission in 1583, used the oval projection of Abraham Ortelius's "Typus Orbis Terrarum," published in the *Theatrum Orbis Terrarum* (1570), for designing his Chinese world map, 1602. He centred the map on the meridian of 170°E, thus placing China towards the centre. When Ricci's map was introduced into Japan, it became the source of the first western-style world map printed in Japan, "Bankoku Sozu" (Map of the World), Nagasaki Harbour, 1645. The central meridian in the Pacific and the equator made a striking cross, but their "graduations" were purely decorative. It appears, moreover, that some of the artists making world maps for screens had no cartographic knowledge and did not realize that the left and right edges of the map were related (Nanba et al, p. 164).

As the 17th century proceeded the Japanese themselves began to measure longitude. The Shogunate astronomer and globemaker, Shibukawa Shunkai (fl.1690), attempted to measure longitudes of various places out of the necessity to make revisions in the calendar, but he failed to get precise results. The first Japanese to make a map of Japan based on his own measurements of latitude and longitude was Inō Tadataka (1745-1818). During his surveying travels, from 1800 to 1816, he tried to determine geodetically the longitudes of every place he stopped, using specially made instruments such as a pendulum clock and meridian transit. In his early maps he used his house in Kuroe-chō in Edo as meridian. His map of western Japan, "Nishi Nihon," was based on the Sanson-Flamsteed projection, with Nishisan-jō, the location of the calendaric office in Kyoto, as the zero meridian. His small-scale map (1:432,000 in 3 sheets) was published with minor revisions by the Kaiseisho, woodcut and colour printed, in 1865. Although his instruments and techniques were not good enough to get accurate results in longitude (and errors in the projection by Takahashi Kageyasu (1785-1829) had marred the accuracy of the complete map), the published survey was highly regarded by Western authorities (Cortazzi, pp.35-7).

C CORTAZZI, Hugh (1983): *Isles of Gold: Antique Maps of Japan.* New York and Tokyo, Weatherhill.

CORTESÃO, Armando, and TEIXEIRA DA MOTA, Avelino (1960): *Portugaliae Monumenta Cartographica,* Vol. 1. Lisbon.

CRONE, Gerald R (1978): *Maps and Their Makers,* 5th ed. Folkestone, Kent, Dawson; Hamdan, Conn., Archon Books.

DARBY, H C (1935): "A Note on the Early Treatise on the Astrolabe," *Geographical Journal, 85,* 179-81.

DICKS, D R (1960): *The Geographical Fragments of Hipparchus.* London, Athlone Press.

GREGORY, John (1649): "The Description and Use of the Terrestrial Globe," in *Gregorii Posthuma* (1671). London, 274-76.

GUNTHER, R W T (1937): "Astronomy," in *Early Science in Cambridge.* Oxford, 123.

HASKINS, Charles H (1915): "The Reception of Arabic Science in England," *English Historical Review, 30,* 56-69.

HOWSE, Derek (1980): *Greenwich Time and the Discovery of the Longitude.* Oxford, Oxford University Press.

HOWSE, Derek, and SANDERSON, Michael (1973): *The Sea Chart.* Newton Abbot, David & Charles.

HOYANAGI Mutsumi (1974): *Inō Tadataka no Kagakuteko Gyoseki* (A New Appreciation of the Scientific Achievement of Inō Tadataka). Tokyo, Kokon Shoin.

MOLLAT DU JOURDIN, Michel, and LA RONCIÈRE, Monique (1984): *Sea Charts of the Early Explorers, 13th to 17th Century.* New York, Thames and Hudson.

NANBA Matsutaro, MUROGA Nobuo, and UNNO Kazutaka (1973): *Old Maps in Japan.* Osaka, Sogensha.

REPARAZ RUIZ, Gonzalo de (1950): "The Topographical Maps of Portugal and Spain in the 16th Century," *Imago Mundi, 7,* 75-82.

SOKURYO CHIZU HYAKUNENSHI HENSHU IINKAI (1970): *Sokuryo Chizu Hyakunenshi* (A Hundred Years of Surveying and Map-making). Tokyo, Nihon Sokuryo Kyokai.

SKELTON, R A (1954): "Captain James Cook as a Hydrographer," *Mariner's Mirror, 40,* 92-119.

SKELTON, R A (1969): *Captain James Cook – After Two Hundred Years.* London, Trustees of the British Museum.

TAKAGI Kikusaburo (1966): *Nihon niokeru Chizu Sokuryo no Hattatsu ni kansuru Kenkyu* (The Development of Land-Surveying and Map-Making in Japan). Tokyo, Kazama Shobo.

THOMSON, James Oliver (1948): *History of Ancient Geography.* Cambridge, University Press.

TOBLER, W R (1966): "Medieval Distortions: The Projections of Early Maps," *Annals of the Association of American Geographers, 56,* 351-60.

VIETOR, Alexander O (1963): "A Pre-Columbian Map of the World, circa 1489," *Imago Mundi, 17,* 95-6, pl.

WARD, Francis Alan Bennett (1970): *Time Measurement.* London.

WERNER, Johann (1514): *In hoc opere haec continentur Nova translatio primi libri geographiae Cl. Ptolomaei . . .* Nuremberg.

4.123 Loxodrome

A A curve which makes a constant non-right angle with the meridians (a spherical helix), and forms a spiral which approaches the pole as a limit. Also called a rhumb or rhumb line (4.181).

B From the beginning of the 16th century cartographers were concerned about the problem of devising a transformation of the sphere to the plane (see Map projection, 4.134) which would convert loxodromes into straight lines when plotted on a plane chart, a rectangular projection with straight parallel meridians and evenly spaced, perpendicular, straight parallel latitudes. Plane charts of small areas in use at the time created no problem in most areas except in higher latitudes, but for larger areas the convergence of the meridians on the earth caused a rhumb to be curved on the chart. The problem was recognized by mathematicians as early as 1513 when the error of ignoring the convergence of the meridians was pointed out by Johannes Stoeffler. In 1537 Pedro Nuñes (1502-1578), chief cosmographer of Portugal, published two treatises on navigation in which, among other things, he described and showed in a diagram the spiral loxodromes. Gerard Mercator (1512-1594) drew a set of rhumbs on his first terrestrial globe in 1541.

In 1569 Mercator solved the problem caused by using the ordinary plane chart by devising the famous projection in which the meridians are straight parallel lines and the parallels are straight lines perpendicular to the meridians but spaced progressively farther apart poleward. Mercator compensated for the progressive east-west expansion resulting from the parallel meridians by a corresponding exact progressive latitudinal expansion. This is not a simple operation, and although Mercator was trained as a mathematician, he never provided an explanation of how he devised the projection which bears his name. It was not until 1599 that Edward Wright (c. 1558-1615), a Cambridge mathematician, published his *Certaine Errors in Navigation . . . Detected and Corrected* (2nd ed., 1610) in which he explained the theory of Mercator's projection and included tables for its construction.

C BAGROW, Leo (1964): *History of Cartography*. Revised and enlarged by R A Skelton. London, C A Watts; Cambridge, Mass., Harvard University Press.

BROWN, Lloyd A (1949): *The Story of Maps*. Boston, Little, Brown and Company.

NAVARRETE, Martin Fernandez de (1846): *Disertación sobre la historia de la náutica*. Madrid, La Real Academia de la Historia.

NUÑES, Pedro (1537): "Tratado qu o doutor Pero Nunez fez sobre certas duvidas da navegação"; "Tratado quo o doutor Pero Nunez fez em defensam da carta de marear," in Nuñes, P (1940): *Obras*. Lisboa, Academia das cienças de Lisboa.

NUÑES, Pedro (1915): *Tratado da Sphera . . . Reproduction Fac-similé de l'exemplaire appartenant à la Bibliothèque du Duc de Brunnid à Wolfenbüttel, Edition 1537 Lisbonne*. Munich.

REY PASTOR, Julio (1942): *La ciencia y la técnica en el descubrimiento de América*. Buenos Aires, Mexico.

TAYLOR, Eva G R (1956): *The Haven-Finding Art: A History of Navigation from Odysseus to Captain Cook*. London, Hollis and Carter, 175-80. (Republished 1971, New York, American Elsevier Publishing Co.)

4.131 Magnetic Meridian

A The magnetic meridian is defined as a line in the direction of the horizontal component of the total terrestrial magnetic force at any point. On a uniformly magnetized sphere the dip poles are antipodal, and the magnetic meridians are (plane) meridians of these poles. In the 16th and 17th centuries magnetic meridians were believed to be lines of variation coinciding with geographical meridians.

B Once the fact of magnetic declination had become known in the late 15th and 16th centuries, mapmakers attempted, without a satisfactory result, to identify magnetic meridians with lines related to geographical meridians. The Portuguese chartmaker Pedro Reinel marked in the area of Newfoundland on his chart of the North Atlantic, c. 1504, an oblique meridian which is believed to be related to magnetic variation (Albuquerque, pp.415-16). The Nuremberg mathematician Johann Werner in his *Annotationes* to Ptolemy's *Geography*, 1514 (reprinted by Peter Apian in 1533), made what is believed to be the first reference to a magnetic meridian (Propositio XXXVII: "Lineam Meridianum virtute magnetis quomodo perquiras"; Taylor, p.6). In 1525 the Seville physician Felipe Guillen suggested that longitude could be found from local variation. French hydrographers of Dieppe, such as Jean Rotz, called the meridian line of zero variation the "ligne dyametralle" (diametral line) ("Traicte des differences du compas aymante," 1542, British Library, Royal MS.20.B.vii,f.21). Pierre Crignon, the Dieppe cosmographer, using the term "ligne de direction," made it his first meridian, and believed that to establish the pattern of magnetic variation would solve the problem of longitude (Crignon, "La Perle de Cosmographie," MS., 1533; Wallis, pp.6-7, 31, 33-34). The Portuguese navigator João de Castro, on the contrary, concluded in 1538 that magnetic deviation had no connection with differences of longitude ("Roteiro de Lisboa a Goa"; Cortesão and Teixeira da Mota).

Mapmakers and hydrographers continued to search for the link between longitude and magnetic variation. Sebastian Cabot, chief pilot of Spain, marked the isogonic zero line in the Atlantic on his world planisphere published at Antwerp in 1544. His claim on his deathbed that he had solved the problem of longitude by divine revelation was probably connected with his theory of magnetic variation. Gerard Mercator on his world chart of 1569 drew his prime meridian through the Cape Verde Islands, reported by Francis of Dieppe, "a skilful shipmaster," to be a place of no variation. Since other navigators had recorded no variation at Corvo in the Azores, Mercator marked two magnetic poles lying on meridians through the Cape Verde Islands and Corvo (see Magnetic north, 4.132). He evidently believed the magnetic meridian to be a meridian great circle.

Willem Blaeu in a legend on his terrestrial globe published at Amsterdam in 1622 challenged the idea that the prime meridian should be based on magnetic variation, which had been observed to vary along meridians. When Edmond Halley published his maps of magnetic variation in 1701 (the Atlantic) and 1702 (the world), it was possible to contrast the pattern of the isogones with that of the geographical meridians. In his explanation of the Atlantic chart (1703?), Halley pointed out that one of its uses was "in many Cases to estimate the Longitude at Sea, thereby; for where the *Curves* run nearly *North* and *South*, and are thick together, as about Cape *Bona Esperance*, it gives a very good Indication of the Distance of the Land to Ships come from far; for there the Variations alter a Degree to each two Degrees of Longitude nearly . . ." Thus in that area the isogones appeared to coincide with the geographical meridians, whereas in the North Atlantic the curves lay east–west and were no use for establishing longitude (Fig. 16).

C ALBUQUERQUE, Luís de (1971): "Two Chapters on History on Astronomical Navigation," in Cortesão, Armando: *History of Portuguese Cartography,* Vol.2. Coimbra, Junta de Investigaçoes do Ultramar.

CORTESÃO, Armando, and TEIXEIRA DA MOTA, Avelino (1960): *Portugaliae Monumenta Cartographica,* Vol.1. Lisbon, 133-4.

ENCYCLOPAEDIA BRITANNICA, 15th ed. (1978): "Earth, Magnetic Field."

HALLEY, Edmond (1703?): "The Description and Uses of a New and Correct Sea-Chart of the Western and Southern Ocean Shewing the Variations of the Compass." London.

PERRIN, W G (1927): "The Prime Meridian," *The Mariner's Mirror, 13,* 107-24.

TAYLOR, Eva G R (1951): "The South-Pointing Needle," *Imago Mundi, 8,* 1-7.

WALLIS, Helen, ed. (1981): *The Maps and Text of the Boke of Idrography Presented by Jean Rotz to Henry VIII.* Oxford, Roxburghe Club.

4.132 Magnetic North

A Magnetic north is the direction of the north-seeking end of a magnetic compass needle that is not subject to ephemeral or local disturbance. The needle was magnetized by rubbing on a lodestone or by induction through floating it in a bowl in which the lodestone was swinging round in a rapid circular motion.

B Knowledge of the lodestone was widely diffused among ancient peoples, but the polarity of the magnet and the directive property of the earth's magnetic field became known much later. For a long time the discovery of the magnetized needle was attributed to Chinese scholars, but it now seems that it came into use more or less independently by Chinese navigators (year 1100), Arabs (mentioned in 1232), and Europeans (mentioned by Alexander Neckam of St Albans about 1187 in his books *De Utensilibus* and *De rerum naturis*). The first attempt at systematic description of the compass and its application for navigation was given in 1269 in the "Epistola de magnete" by the Frenchman Petrus Peregrinus (Pierre de Maricourt). Records in England show that ships fitted out in 1338 and 1345 were supplied with sailing needles and dials (Mitchell, *37,* 1932, p.128).

At first it was thought that the compass needle pointed exactly to geographic north. Roger Bacon in the late 13th century, however, alluded to the fact that the direction of the magnetic needle and true north did not coincide (Mitchell, *42,* 1937, p.244). In the 15th century it was observed in northwestern and central Europe that the needle pointed some degrees to the northeast. By about 1450 makers of sundials at Nuremberg improved their instruments to take account of magnetic declination, and Flemish and Dutch instrument makers followed suit not long after.

The earliest printed map to mark the variation appears to be that by the Nuremberg mapmaker Erhard Etzlaub, made a little before or in 1500 and entitled "Das ist der Romweg," showing the route to Rome. This displays a compass dial (named "Compast") with a south-pointing needle deviating about ten degrees east of north, which was then and for a long period the correct variation for central Europe (see Compass declination map, 3.031).

The earliest chart which appears to show the direction of the magnetic pole is that of the North Atlantic by the Portuguese cartographer Pedro Reinel, c. 1504. An oblique meridian in the area of Newfoundland provides a supplementary scale of latitudes, believed by some authorities to be designed as a correction to magnetic variation (Albuquerque, pp.415-16). The Dieppe hydrographer Jean Rotz adapted the regional charts in his manuscript "Boke of Idrography," 1542, to show magnetic north by using a different scale of latitudes on each side of the chart (British Library, MS. 20.E.IX; Wallis, pp.32, 37). He describes this device in his manuscript treatise on navigation (BL. Royal MS. 20.B.VII.ff.27v, 42r).

Sebastian Cabot on his world planisphere, Antwerp, 1544, marked a central meridian through the Atlantic and cutting northeast Brazil as a meridian of no variation: "Meridiano adon de el aguia de marear muestra derecamente al norte." In 1551 the Spanish navigational authority Martin Cortes described the magnetic pole as different from the geographical pole, but thought the magnetic one to be a "celestial" pole. Gerard Mercator marked two magnetic poles on his world chart published at Duisburg in 1569. One is a high rocky island in the polar sea north of the Strait of Anian (Bering Strait), as determined by a prime meridian through the Cape Verde Islands; the other lies on an island further to the northwest, on a magnetic meridian through Corvo in the Azores. In an explanatory legend on the origin of geographical longitudes and on the magnetic pole, Mercator states that the true magnetic pole must lie between these two extreme positions (see Magnetic meridian, 4.131). The two magnetic poles are marked similarly on the map of Arctic regions in Mercator's world atlas, 1595, and in various later editions of the Mercator-Hondius atlas from 1606.

English contemporaries of Mercator made major contributions to the theory of terrestrial magnetism. In 1580 Robert Norman, the chart and instrument maker, set out in *The newe Attractive,* London, 1581, the results of his studies of magnetic variation for calculating the position of the magnetic pole. The physician William Gilbert in his academic treatise *De Magnete,* London, 1600, which established the science of terrestrial magnetism, determined that the earth itself was a great magnet and that variation of the compass was due to the attraction of the continents.

Later studies of geomagnetism were carried out by Edmond Halley (1701), Leonhard Euler (1745), and Pierre Charles Le Monnier (1776). For some time a theory of the existence of four magnetic poles was held. A full understanding of the distribution of the magnetic field had to await Carl Friedrich Gauss's publication of his *Allgemeine Theorie des Erdmagnetismus* in 1833.

The orientation of maps to magnetic north remained a common practice. A notable example is William Roy's manuscript survey of Scotland, 1747 to 1755, for which the declination for Scotland was taken by the surveyors to be 19° west (British Library, K.Top. XLVIII.25.). A few of the field sheets (eg. sheet 1, Fort Augustus) have faint guide-lines to help in copying or reduction, and these are aligned true north and south (Skelton, p.5). On charts such as those by the British surveyors Lewis Morris, 1737, 1748, etc., and Murdoch Mackenzie, 1785, wind roses indicate the "Magnetic" and "True" meridians. In the 20th century British Ordnance Survey maps, one-inch to the mile (1:63,360), distinguished by arrows "True North," "Grid North," and "Magnetic North" (Harley, p.23).

C ALBUQUERQUE, Luís de (1971): "Two Chapters on History on Astronomical Navigation," in Cortesão, Armando: *History of Portuguese Cartography,* Vol.2. Coimbra, Junta de Investigaçoes do Ultramar.

CHAPMAN, Sydney, and BARTLES, Julius (1940): *Geomagnetism,* 2 vols. Oxford, Clarendon Press.

FLEMING, John Adam, ed. (1939): *Terrestrial Magnetism and Electricity.* New York and London, McGraw-Hill Book Co.

HARLEY, John Brian (1975): *Ordnance Survey Maps, a Descriptive Manual.* Southampton, Ordnance Survey.

HITCHINS, Henry Luxmoore, and MAY, William Edward (1952): *From Lodestone to Gyro-Compass.* London, Hutchinson's Scientific and Technical Publications.

MITCHELL, A Crichton (1932; 1937; 1939): "Chapters in the History of Terrestrial Magnetism," *Terrestrial Magnetism and Atmospheric Electricity, 37,* 105-46; *42,* 241-80; *44,* 77-80.

SKELTON, R A (1967): *The Military Survey of Scotland 1747-1755.* Edinburgh, Royal Scottish Geographical Society, Special Publication No.1.

WALLIS, Helen, ed. (1981): *The Maps and Text of the Boke of Idrography Presented by Jean Rotz to Henry VIII.* Oxford, Roxburghe Club.

4.133 Magnetic Variation

A (MDTT 321.7) Magnetic variation or declination is the angle between true north and magnetic north.

B The existence of magnetic variation was noted by Roger Bacon in the late 13th century (see Magnetic north, 4.132). Erhard Etzlaub of Nuremberg appears to be the first cartographer to mark variation on a printed map. His "Das ist der Romweg," *c.* 1500, displays a dial with a south-pointing needle deviating ten degrees east of north. Johann Ruysch on his world map of 1507 (one of the new maps in Ptolemy's *Geography,* published at Rome) displays in the North Atlantic a legend referring to variations in the movement of the compass needle in those latitudes: "Hic conpassus navium non tenet, nec naves que ferrum tenent revertere valent." Apart from the Nuremberg maps the earliest known compass rose to indicate variation appears on a map in Peter Apian's *Cosmographiae Introductio,* published at Ingolstadt in 1529 [1532]. Jakob Ziegler's map of Palestine published at Strasbourg in 1532 was the second.

The claim that Christopher Columbus on his Atlantic voyage of 1492 was the first to discover the space-variations in magnetic declination is now contested; his comments indicate that an easterly variation had already been observed in northwest Europe (Mitchell, *42,* 248-68). Information on the world-wide phenomenon of magnetic variation was gradually accumulated on the oceanic voyages of the late 15th and 16th centuries. Ferdinand Magellan used devices for recording variation in the Pacific in 1520 in the course of the *Victoria's* voyage round the world, 1519 to 1522.

The manuscript world planisphere of Diego Ribeiro, cosmographer to the Emperor Charles V, 1525, is said to be the first chart properly corrected for variation. The Mediterranean, for example, is drawn on a correct axis for the first time (Cortesão and Teixeira da Mota, Vol. 1, 92-3). The earliest record of a map showing declination from place to place dates from 1536 in Seville, where at a meeting of Spanish pilots, Alonso de Santa Cruz, Pilot-Major of Spain, displayed a map which he had prepared for this purpose. Only three points were agreed where declination had been satisfactorily determined, a comment perhaps referring specifically to the West Indies (Mitchell, *42,* 270, 280). Jean Rotz, hydrographer of Dieppe, records many observations of variation in his manuscript treatise, 1542 ("Traicte des differences du compas aymante," British Library, Royal MS. 20.B.VII). His main theme in this work was the effect of variation of the compass on the construction and use of charts, and he adjusted the charts of his atlas, 1542, to these observations ("Boke of Idrography," British Library, Royal MS. 20.E.IX; Wallis, 1981, p.32).

Sebastian Cabot, chief pilot of Spain, marked the agonic line on his world planisphere published at Antwerp in 1544. As with Columbus, the claim that he was the discoverer of the space-variation of declination is doubtful, but his assertion on his death bed in 1557 that he had found the secret of longitude was probably connected with magnetic variation. His role (from 1548) as adviser to English explorers seeking the discovery of the northwest and northeast passages partly explains the rapid development of English expertise in this field. William Borough's manuscript chart of the North Atlantic, 1576, with additions to 1578, is marked with five arrows indicating increasing westerly variation as observed by Martin Frobisher on his course to Greenland in 1576. This chart is one of the earliest English records of magnetic variation (Hatfield House, Hertfordshire, CPM I.69; Skelton and Summerson, p.69, pl.6).

The first printed sea charts to represent the declination of the compass were those of Willem Barentszoon in his *Caertboeck vande Midlandtsche Zee,* Amsterdam, 1595. On many of the charts he displays both the Italian and the German compass (Koeman, p.XVI). At about the same time Sir Robert Dudley, the English navigator, included extensive records of variation on his manuscript charts, later published in his sea atlas the *Dell' Arcano del Mare,* Florence, 1646-47.

The first attempt at an "isogonic" map of variation is found on an anonymous manuscript chart of the Pacific Ocean, probably part of a world planisphere by the Portuguese chartmaker Luís Teixeira, *c.* 1585 (Museu de Marinha, Lisbon; Cortesão and Teixeira da Mota, Vol. 3, pp. 71-2, pl.363). Another Portuguese contribution was a map by the Jesuit Father Cristoforo Borri of Milan (fl.1620) who constructed a "mappa geographico-magnetica," as recorded by the Jesuit scientist Athanasias Kircher in 1641 (Wallis, 1973, p.255). Kircher himself obtained extensive data on variation and gave instructions for constructing a map of variation, though he did not make such a map himself. Kircher almost certainly inspired Edmond Halley with ideas for his map of magnetic variation in the Atlantic, published in London in 1701, which ranks as the first isoline map (see Isogones, 5.0910e) (Fig.16). He produced in the following year his magnetic chart for the world, excluding the Pacific.

The fact of secular change in the variation had been observed in 1634 by Henry Gellibrand, professor of mathematics at Gresham College, and published in 1635. Halley referred to this phenomenon in the marginal text to his Atlantic chart, London, 1703?: "it must be noted, that there is a perpetual tho' slow Change in the *Variation* almost every where, which will make it necessary in time to alter the whole System." His charts were revised and republished accordingly. A variety of other isoline maps were produced to record other magnetic phenomena over the years (Robinson, p.86).

C CORTESÃO, Armando, and TEIXEIRA DA MOTA, Avelino (1960): *Portugaliae Monumenta Cartographica,* Vols. 1 and 3. Lisbon.

KOEMAN, Cornelis (1970): "Introduction," in Barentsz, Willem: *Caertboeck vande Midlandtsche Zee* (Amsterdam, 1595). Amsterdam, Theatrum Orbis Terrarum, Ser. 5, Vol.4.

MITCHELL, A Crichton (1937): "Chapters in the History of Terrestrial Magnetism," *Terrestrial Magnetism and Atmospheric Electricity, 42,* 241-80.

ROBINSON, Arthur H (1982): *Early Thematic Mapping in the History of Cartography.* Chicago and London, University of Chicago Press.

SKELTON, R A, and SUMMERSON, John (1971): *A Description of Maps and Architectural Drawings in the Collection Made by William Cecil, First Baron Burghley, Now at Hatfield House.* Oxford, Roxburghe Club.

WALLIS, Helen (1973): "Maps as a Medium of Scientific Communication," *Études d'Histoire de la Géographie et de la Cartographie,* Wroclaw, Warszawa, . . . Poliskiej Akademi Nauk, 251-62.

WALLIS, Helen, ed. (1981): *The Maps and Text of the Boke of Idrography Presented by Jean Rotz to Henry VIII.* Oxford, Roxburghe Club.

4.134 Map Projection

A Any systematic transformation of the curved surface of the sphere or spheroid upon a plane. Every point on the earth's surface, assumed to be a regular geometric surface, is assigned one point in the plane, the system usually being expressed by a mathematical formula. The problem is equivalent to determination of the image of the graticule of meridians and parallels on the flat map, i.e., finding the cartographic co-ordinates using the mathematical relations connecting geographic co-ordinates (latitude and longitude) of a point on the earth's surface and the plane co-ordinates x, y of its image in the projection plane.

As the curved surface of a sphere or an ellipsoid and a plane are not applicable, i.e., one cannot be transformed to the other without differential stretching or shrinking, every projection system causes certain distortions. Practically, only the principal scale is shown on maps. A projection may distort distances, angles or areas. Selection of a particular projection is based on the size, geographical location, and shape of the areas to be represented, and on the scale, character and objective of the map.

B The earliest known Greek cartographers, among them Anaximander of Miletus (early 6th century BC), Hecataeus of Miletus (c. 500 BC), and Aristagoras (c. 500 BC), most probably did not base their images of the earth on a mathematical foundation.

The central gnomonic projection of the celestial dome probably dates from the period of Thales of Miletus (fl. 580 BC). Arbitrary cylindrical type projections were probably used by Dicearchus of Messina (350-290 BC), Eratosthenes (c. 275-194 BC), and Strabo (c. 67 BC-23 AD). The beginnings of systematic plane projections are attributed to Hipparchus (fl. 146-127 BC), who introduced the polar stereographic, gnomonic, and most probably the orthographic projections; the oblique forms were introduced later by the Arabs. Marinus of Tyre (2nd century), who introduced the equidistant cylindrical projection and worked with trapezoidal systems of projection, could be called the founder of mathematical cartography. Rectangular grids, based on the development of a cylinder, were also studied by Claudius Ptolemy of Alexandria (2nd century). In his *Geography,* he also described the simple (illustrated in Fig.14) and the modified conical projections.

During the later Roman period and the Middle Ages, mathematical cartography declined, but the Renaissance saw a growing interest in the improvement of cartography and gave rise to modifications of the projections proposed by Ptolemy. Here should be mentioned the pseudocylindrical projection of Donnus Nicolaus Germanus (c. 1420-c. 1490); the secant conical projections of Marcus Beneventanus (1st quarter of the 16th century) and Johan Cott; the cordiform projections of Bernardus Silvanus (1st quarter of the 16th century), Johannes Stabius/Stöberer (c. 1460-1522), and Johann Werner (1468-1528); the globular projection of Peter Apian (1495-1552) and its modifications by Henricus Glareanus (1488-1563) and Oronce Fine (1494-1555).

Cartography was revolutionized by Gerard Mercator (1512-1594), who modernized it and put it on a firm mathematical ground. To him we owe the revolutionary conformal rectangular projection (Mercator's Projection) and improvements to the conical and equidistant projections.

Johann Heinrich Lambert (1728-1777) investigated the rules governing map projections. He recognized and appreciated the values of equal-area mapping and introduced the azimuthal and cylindrical (isocylindrical) equal-area projections. Another equal-area projection was proposed by Carl Brandan Mollweide (1774-1825).

Among the newer modifications of cylindrical and conical projections are the conformal transverse cylindrical of Cassini-Soldner, the polyhedral and polyconic projections, and the starlike grids of G Jäger, August Petermann (1822-1878), Heinrich Berghaus (1794-1884), and Anton Steinhauser (1802-1890).

Fig. 14 – World map based on the calculations of Claudius Ptolemy of Alexandria, Egypt, c. AD 150. In a manuscript copy, in Greek, of Ptolemy's Geography, late 13th century, probably by Maximus Planudes (c. 1260-1310) from the Vatopedi monastery on Mount Athos. On a conic projection and marked with a graticule. By permission of the British Library, Add. MS. 19391, ff.17b-18. (Map projection, 4.134; see also World map, 1.2320; Wind map, 3.232; Graticule, 4.072)

Of great importance in the development of mathematical cartography is the theory of projection distortions introduced by Nicolas Auguste Tissot (1824-1880).

CAUJAC, Germaine (1966): *Strabon et la science de son temps*. Paris, Collection Bude, 208-11.

BIERNACKI, Franciszek (1973): *Podstawy teorii odwzorowan kartograficznych* (Elements of Theory of Map Projections). Warszawa, PWN.

DAHLBERG, Richard E (1962): "Evolution of Interrupted Map Projections," *International Yearbook of Cartography, 2,* 36-54.

DICKS, D R, ed. (1960): *The Geographical Fragments of Hipparchus*. London, Athlone Press.

DILKE, O A W (1985): *Greek and Roman Maps*. London, Thames and Hudson.

DINSE, P, GELCICH, E, and SAUTER, F (1897): *Kartenkunde geschichtlich dargestellt*. Leipzig.

ECKERT, Max (1921): *Die Kartenwissenschaft,* Vol. 1. Berlin and Leipzig, Walter de Gruyter, 115-31.

KISH, George (1965): "The Cosmographic Heart: Cordiform Maps of the 16th Century," *Imago Mundi, 19,* 13-21.

KEUNING, Johannes (1955): "The History of Geographical Map Projections until 1600," *Imago Mundi, 12,* 1-24.

ODLANICKI-POCZOBUTT, Mical (1977): *Geodezja. Podrecznik dla studiow inzynieryjno-budowlanych* (Geodesy. Handbook for Engineering Studies). Warszawa, PPWK, 372-74.

RICHARDUS, Peter, and ADLER, Ron K (1972): *Map Projections for Geodesists, Cartographers and Geographers*. Amsterdam and Oxford, North-Holland Publishing Company.

RÓŻYCKI, Jan (1973): *Kartografia matematyczna* (Mathematical Cartography). Warszawa, PWN.

THOMSON, James Oliver (1948): *History of Ancient Geography*. Cambridge, 343-5.

4.141 Non-co-ordinate Reference Systems

A A system employing squares or rectangles in which places may be located by referring to the appropriate square or rectangle, normally by a combination of identifying letters and numbers in the map margins.

B The employment of a regular matrix of rectangles or squares to aid in the construction of a map or to indicate scale must be distinguished from, and regarded as different from, the matrix that is similar in appearance which is used to locate places on the map. Possibly the latter is derived from the former, and if so, the innovation would be credited to the famous Chinese astronomer Chang Heng (79-139). A system of squares was certainly used by Phei Hsiu (224-271), and the technique was subsequently employed for the construction of many other Chinese maps (see Square grid, 4.191).

Knowledge of this technique may have passed from Chinese sources by way of Arab scholars to the Middle East and thence to the West, but it seems more likely that the use of the technique in the Western world derives from the Roman land division and referencing practice of centuriation, a map of which would give the impression of regular geometric rectilinear figures. Roman centuriation was influenced by Greek rectangular systems such as can be seen at some Greek colonies in southern Italy and Sicily. The commonest Roman form was one of squares aligned on two axes at right angles, the *decumanus* and the *kardo*. A reference system indicated the number of "centuries" (squares or rectangles) beyond or on the near side of the *kardo maximus* and to the left or right of the *decumanus maximus*. Of the Roman cadasters carved on stone at Orange (Arausio), c. AD 77 and after, that designated Cadaster A was rectangular, while Cadaster B was square.

One of the chief sources of information about Roman land surveying and mapping is the Corpus Agrimensorum, a collection of short works, including illustrations of maps with squares, which survived in copied form until 1493 in the north Italian monastery of Bobbio. The presence of a framework of squares for construction purposes on maps of the Holy Land c. 1320 (Paris, Bibliothèque Nationale) and c. 1327 (British Library) by the Venetian Pietro Vesconte is therefore not difficult to explain (Fig. 17). Other medieval examples of square grids are known, however, on Islamic maps and even on portolan charts of the 15th century.

The modern use of grids for reference systems seems to have begun in the mid-16th century, for at that time grids came into use both on maps and in book illustrations to enable readers to pinpoint specific locations. Edgerton hypothesizes that such uses stemmed from the co-ordinate grids on Ptolemaic maps. On maps the earliest known use in modern times of a matrix of squares for a non-co-ordinate reference system has been traced to Conrad Wolfhart (known also by the Grecized form of the name, Lycosthenes), who employed it in Sebastian Münster's fourth edition of Ptolemy's *Geography* (Basle, 1552). Numbered border graduations have been added to the modern maps in that edition, 24 for the length and 18 for the height, evidently printed from separately engraved strips. In 1560 he did the same for Aegidius Tschudi's map of Switzerland and, he claimed, for others no longer extant (Heawood; see also Map index, 6.0920b). A more satisfactory method of addressing the squares by means of an alpha-numeric notation, southings by letters and eastings by numbers, was used in Philipp Apian's *Bairische Landtaflen* (1568) and by John Norden on his map of Middlesex in his *Speculum Britanniae, the First Parte*, London, 1593.

The map of Bohemia by Paul Aretin, published in Prague in 1619 with later editions 1632, 1645 and 1665, is divided into Czech miles in the margins. An index (still preserved) records some 1150 localities with their rectangular co-ordinates in Czech miles (Kuchař). In England the now recognized cartographic revolution of the 18th century began with the publication of the large-scale map of Cornwall in 1700(?) on which the mapmaker, Joel Gascoyne, placed a non-co-ordinate reference system consisting of squares with sides of two miles to scale and addressed by lower case and capital letters.

C BRADFORD, John (1957): *Ancient Landscapes: Studies in Field Archaeology.* London, Bell, 145-216.

CHEN, Cheng Siang (1978): "The Historical Development of Cartography in China," *Progress in Human Geography, 2,* 101-20.

DILKE, O A W (1971): *The Roman Land Surveyors: An Introduction of the Agrimensores.* Newton Abbot, David & Charles.

DILKE, O A W (1974): "Archaeological and Epigraphic Evidence of Roman Land Survey," in Temporini, H, ed., *Aufstieg und Niedergang der römischen Welt,* II, 1. Berlin and New York, Walter de Gruyter.

DILKE, O A W (1985): *Greek and Roman Maps.* London, Thames and Hudson.

EDGERTON, Samuel Y (in press): "From Mental Matrix to Mappamundi to Christian Empire: The Heritage of Ptolemaic Cartography in the Renaissance," in Woodward, David, ed.: *Art in Cartography: Six Historical Essays.* (6th Kenneth Nebenzahl, Jr., Lectures in the History of Cartography.) Chicago, University of Chicago Press.

FRACCARO, Plinio (1957): *Opuscula,* iii. Pavia, Athenaeum.

GASCOYNE, Joel (1700?): A Map of the County of Cornwall. See "An Explanation of the Severall remarques in this Mapp with the use of the Tables" on the Land's End sheet. British Library, Map Library, Maps 146.e.20.

HEAWOOD, Edward (1932): "Early Map Indexing," *Geographical Journal, 80,* 247-9.

INSTITUT GÉOGRAPHIQUE NATIONAL (edited for) (1954): *Atlas des centuriations romaines de Tunisie,* Paris.

KUCHAŘ, Karel (1937): "A Map of Bohemia of the Time of the Thirty Year's (*sic*) War," *Imago Mundi, 2,* 75-77.

LYCOSTHENES, Conrad Wolfhart (1552): "De utilitate tabularum geographicarum." In *Geographiae Claudii Ptolemaei . . . Libri VIII,* ed. Sebastian Münster. Basle.

SKELTON, R A (not named) (1966): "Bibliographical Note" to facsimile of Claudius Ptolemaeus, *Geographia,* ed. Sebastian Münster. Basle, 1540. Amsterdam, Theatrum Orbis Terrarum.

4.151 Orientation

A (MDTT 931.1) In cartography the term orientation refers to the arrangement of the cardinal directions (4.031) with respect to the normal reading position of a map, specified by the direction of the upper centre of the map. It should not be confused with the meaning of the term in map use, namely, the verb "to orient," wherein it means to put into correct position or relation, that is, to rotate a map (in hand or on a plane table) so that the lines between features on the map are parallel with the corresponding lines in nature.

Both terms derive from the Latin *oriens,* the quarter where the sun rises, the east. In medieval Europe, the ancient significance of the direction of sunrise was reinforced by that being the direction of the Holy Land and the Earthly Paradise, and by the construction of churches and temples with the long axis east–west and the chief altar at the eastern end.

B There has been some argument about the orientation of ancient Bablyonian maps, but the evidence seems to suggest that the east was favoured. Babylonian "north" was actually northwest and thus their "east" was more precisely northeast, which accords with the orientation of the well-known Nuzi map showing an estate in the region of northeast Iraq, dating from *c.* 2500–2300 BC. In Greco-Roman small-scale mapping north was usually at the top, but not unfailingly (Dilke, p. 177). More scientific maps of larger areas or the whole earth involved problems of dealing with the co-ordinate system. In that case, either southern or northern orientation would allow simple arrangements of the graticule. We have no direct evidence of the orientation of Marinus of Tyre's maps (2nd century) which were on a rectangular projection, but Ptolemy in his *Geography* and on the maps that apparently accompanied it clearly opted for a northern orientation. It seems that the practice of orienting maps and plans of local areas was a matter of custom or convenience. Town plans were sometimes orientated to face the way the public was looking, as in the example of the plan of Rome, "Forma Urbis Romae," AD 203-208, and perhaps also in that of the Orange cadasters, *c.* AD 77 (see Plan, 1.1630; Town plan, 1.1630b). The Han topographical and military maps of China, 2nd century BC, are oriented approximately with south at the top.

Arabic maps from early times through the time of al-Idrisi in the 12th century were oriented with south at the top. There are various theories to explain this. One is that south was a sacred direction for Zoroastrians; another, that from Baghdad and the early cultural centres of Islam the faithful looked south to Mecca. It has also been suggested that dynastic Egypt may have used this orientation (Dilke, p.177).

Christian world maps of the Middle Ages usually gave pride of place to the Earthly Paradise (the Garden of Eden), traditionally situated in the orient, and thus east was at the top. Climate maps of the 12th century drawn by Petrus Alphonsus and others and derived from Arabic sources had south at the top. For regional maps, on the other hand, various orientations were employed. The maps of Great Britain, *c.* 1250, by Matthew Paris, the monk of St Albans, have north at the top and are the first maps of northern Europe so oriented (Crone, pp.14-15). The Gough map of the British Isles, made *c.* 1360 apparently as an official road map, has west at the top. That this remained a standard work for two centuries explains why the first separately printed map of the British Isles, 1546, by G L A (George Lily), a derivation of the Gough, has a similar orientation. With the publication of maps in Ptolemy's *Geography* from 1477, it became a general practice to orientate world and small-scale regional maps with north at the top, unless there was a practical need for a different orientation. For example, Erhard Etzlaub's "Das ist der Romweg . . ." (*c.* 1500), made for the traveller from Germany to Rome, had south at the top.

In the 16th century the growing interest in terrestrial globes, on which the north pole stood at the top, also encouraged the northern orientation. Geographers, who used globes as their chief means of instruction, not only regarded the north-south axis as representing the position of the earth in the universe, but also ascribed to the northern hemisphere greater importance than to the southern. Nathaniel Carpenter, the Oxford geographer, wrote "The Northern Hemisphere is the Masculine, the Southerne the Foemenine part of the earth," finding a metaphor taken from the sexes in living creatures appropriate to describe the "great and noble disparity" between the hemispheres. The principle of putting the most important feature or features at the top was thus still being observed. Eighteenth- and nineteenth-century maps of the world which showed south at the top were an inverse reflection of this, illustrating a fashionable satirical theme, "the world turned upside down" (Salviati).

From the 17th century onwards an increasing number of scientific maps on carefully drawn projections were made, and in line with the Ptolemaic tradition and later the Copernicus system, the northern orientation became still more common. The development of large scale topographical mapping, which also followed the practice of northern orientation, gave further authority to this arrangement. Except for specialized maps with a particular need for some other arrangement, north orientation had become the standard by the 19th century in European cartography. Among the exceptions were maps with variable orientations, notably strip maps for roads, of which those in John Ogilby's *Britannia* (London, 1675) were the earliest. Each strip of the road carries a compass to indicate directions.

Chartmakers over the centuries have followed various conventions. The earliest portolan charts (from the end of the 13th and the 14th centuries) had south at the top or were multidirectional. In fact, the user had to turn all portolan charts to follow the names along the coasts. Legends were also often multidirectional. A variety of orientations is found on 16th and also 17th century charts. Before a full appreciation of compass variation, some maps were oriented in the direction of compass north rather than true north; Eckert cites the "Carta Marina" of Olaus Magnus (1539) as one of these. The Dieppe hydrographer Jean Rotz in his "Boke of Idrography," 1542, arranged some of the regional charts (all of which had south at the top) according to magnetic north, as indicated by the differential scales of latitude (British Library, Royal MS. 20.E.IX).

For other chartmakers the short axis of charted sea coast regions often was made the orienting factor. The French sea atlas *Le Neptune françois* (Paris, 1693) has maps oriented in all the cardinal directions. The *Atlantic Neptune* (London, 1779 and 1781) has orientations which are not cardinal directions.

Japanese mapmakers used a variety of orientations, and how to determine what is the top of a Japanese map may itself pose problems, since the characters are written in different directions. Often the only guide is the position of the title, author, or publisher, or the direction of the prominent characters. Many maps had north at the top, as indicated by the symbol *hoku* (Beans). This choice was determined by such factors as the reverence attached to the Pole Star and the use of European models. Another direction used was south, in deference to the theory of opposites, *Yin* and *Yang*. The rising of the sun inspired an eastern orientation, associated with good fortune, whereas the word "setting" had negative connotations, and western oriented maps are rare, with the exception of plans of the city of Edo. Prominent structures such as a castle were often used to orient city maps, as found in some maps of Osaka with eastern orientation. Major topographical features also influenced the arrangement (Nanba et al, pp.155-56). An example is *Fujima Jūsanshū Yochi Zenza* (Complete map of the thirteen provinces commanding a view of Mt. Fuji), by Akiyama Einen, Edo, 1843.

Practical considerations such as the format and size of the paper and the decorative purpose of a map might also influence the arrangement. The diagonal alignment of Japan from southwest to northeast favoured a northwest orientation, with Japan shown horizontally, to conserve space and save paper (Nanba et al, pp.155-56). The first printed world map of Western style, the Shōhō world map, published at Nagasaki Harbour, 1645, was made for a folding screen and has east at the top.

C BEANS, George H (1954): "The Orientation of Japanese Maps," *Imago Mundi,* *11,* 146.

CRONE, G R, ed. (1961): *Early Maps of the British Isles, A.D. 1000–A.D. 1579.* London, Royal Geographical Society.

DILKE, O A W (1985): *Greek and Roman Maps.* London, Thames and Hudson.

MEEK, Theophile James (1936): "The Orientation of Babylonian Maps," in Notes and News, *Antiquity, 10,* 223-26.

NANBA Matsutaro, MUROGA Nobuo, and UNNO Kazutaka (1973): *Old Maps in Japan.* Osaka, Sogensha.

NORDENSKIÖLD, A E (1889): *Facsimile-Atlas to the Early History of Cartography.* Stockholm. (Reprinted, New York, Kraus Reprint, 1961; Dover Publications, 1973.)

SALVIATI, Giuseppi (1822): *The World Turned Upside Down.* London.

SCHNELBÖGL, Fritz (1966): "Life and Work of the Nuremberg Cartographer Erhard Eztlaub († 1532)," *Imago Mundi, 20,* 11-26.

4.161 Prime Meridian

A The meridian from which longitude (4.122) is measured. Since 1884 the meridian of Greenwich has, by general consent, been recognized as the Prime Meridian, i.e., 0°0′0″ longitude.

B The spherical co-ordinate system of latitude and longitude must have a point of origin. In respect of latitude, there is a natural line from which to reckon north–south distances, namely the equator, the great circle lying midway between the poles. With longitude there is no such unique great circle, and one meridian must therefore be chosen as the origin for east–west reckoning. The choice of an arbitrary point, and thus the meridian through that point, from which to begin counting goes back to the innovation of the co-ordinate system itself. The Greek mathematician Eratosthenes selected one of the earliest known prime meridians. It passed through Alexandria and, he assumed, albeit incorrectly, through Rhodes. Hipparchus made use of the meridian of Rhodes, but Ptolemy, not surprisingly, reverted to Alexandria as his base for computational purposes. On the other hand, there was much sense in selecting a meridian at what was then conceived as the most westerly known point. Hence, the Fortunate Islands (Canary Islands) emerged as the zero meridian. Arab geographers favoured a meridian drawn midway between the furthest east and west known points, which for them was at the mythical city of Arin (or Arim) located on the equator and assumed to be 10° east of Baghdad.

With the introduction of Ptolemy's *Geography* into Europe in the 15th century, the prime meridian tended to be located in the Canaries, and more particularly through Ferro, the most westerly of the islands. A geo-political prime meridian assumed significance in 1493 when the Spanish-born Pope Alexander VI decreed by the Bull of May 4 that Spain had a right to all lands not already held by a Christian prince lying west of a line drawn 100 leagues beyond the Azores or Cape Verde Islands. These terms were modified in the Hispano-Portuguese Treaty of Tordesillas of 1494, which placed the line 370 leagues west of the Cape Verde Islands. The earliest extant chart marking this line is the manuscript Cantino world map of 1502 (Biblioteca Estense, Modena), which shows it as the only meridian. The line with its counterpart (the anti-meridian) in the Far East also appears on the anonymous globe gores, *c.* 1530, published probably at Nuremberg, sometimes known as the "Ambassadors' globe." The line is sometimes named "Meridian della demarcacion," for example, by Andreas Garciá de Céspedes on his globe gores in *Regimientio de Navegación,* Madrid, 1606.

From the 1530s attempts were made to place the prime meridian in coincidence with the zero isogonic line (see Magnetic meridian, 4.131). Mercator for this reason changed from the Ptolemaic prime meridian through the Canary Islands to one through the Cape Verde Islands on his world chart of 1569. In 1622, however, Willem Blaeu in a legend on his terrestrial globe of 69 cm diameter exposed the fallacy of so doing and advocated El Pico in Tenerife as an appropriate spot. This was the first attempt to fix the meridian through a point rather than vaguely through an island or group of islands.

On 1 July 1634 Louis XIII issued a decree (Bibliothèque Nationale, Paris, MS. Fr. 12222) fixing the prime meridian at the westernmost point in the Canary Islands, which again encouraged the use of Ferro as a universal prime meridian. Within a few decades, however, most nations had forsaken the idea of a universal prime meridian for one, or more than one, meridian of origin within their borders. This was the period when many of the national surveys began, and a local point of origin seemed more practical than Ferro, whose position in relation to their own point of origin was unknown, owing to the problem of determining longitude exactly. In France, for example, the start of a systematic national survey and interest in the problem of the length of a degree of latitude made it desirable to reckon positions from the meridian of Paris.

By 1700 British cartographers were beginning to use London as their prime meridian. Edward Wells, mathematician and teacher of geography at Christ Church, Oxford, on the two general maps in his instructional atlas, *A New Sett of Maps of Ancient and Present Geography* (Oxford, 1700-1701), displayed the "First Meridian or Meridian of London." British chartmakers in the early and middle years of the 18th century used as their zero either the Lizard Point of Cornwall or St Paul's Cathedral in London. The introduction of the Greenwich meridian on maps followed the establishment of the Royal Observatory at Greenwich in 1675, with the object of undertaking accurate meridian observations and thereby solving the problem of establishing longitude at sea.

Two of the earliest maps with the Greenwich meridian as the origin of longitude are "A Description of the Sea Coast of England and Wales" by Samuel Fearon and John Eyes, 1738, and "The County of Oxford, surveyed by Thomas Jefferys, 1769." When Nevil Maskelyne, the Astronomer Royal, published in 1767 the first Nautical Almanac, seamen were able for the first time to find out their approximate longitudes out of sight of land. This gave a further boost to acceptance of the Greenwich meridian by Great Britain and other nations. Joseph Frederick Wallet Des Barres used it for the charts in *The Atlantic Neptune,* published from 1777. Although commercial publishers were slow to change their practices, Greenwich in due course superseded all other prime meridians on English maps and charts.

The proliferation of national prime meridians caused great inconvenience, particularly among seafarers who were using charts produced by various nations. Finally, in 1884, the United States government called together an international conference "for the purpose of fixing a Prime Meridian and a Universal Day." After lengthy and animated debate it was decided to adopt the meridian passing through the transit instrument at the Observatory of Greenwich as the prime meridian of longitude and as the starting point for a worldwide system of time zones (see Date line, 4.041).

The old practice of circumnavigators to reckon their longitudes from 0° to 360° either east or west continued at least until 1806. After the 1750s examples are also found with divisions east and west of the prime meridian; for instance, Joseph Gilbert's charts of Captain Cook's second voyage (1772-1775) are calibrated from 0° to 180° east and west of Greenwich. Gradually this division became common, and the International Meridian Conference, 1884, merely confirmed it.

C ARH [HINKS, Arthur R] (1935): "Nautical Time and Civil Date," *Geographical Journal, 86,* 153-7.

HAAG, Heinrich (1913): *Die Geschichte des Nullmeridians.* Leipzig.

PERRIN, W G (1927): "The Prime Meridian," *Mariner's Mirror, 13,* 109-24.

WASHBURN, Wilcomb E (1984): "The Canary Islands and the Question of the Prime Meridian: the Search for Precision in the Measurement of the Earth," *American Neptune, 44,* 77-81.

4.181 Rhumb Line

A A line of constant true course making a non-right angle with the meridians. An extended rhumb line forms a spherical helix which approaches the poles as a limit. Also called loxodrome (4.123).

B The straight lines on portolan charts (1.0320d) which radiate from windrose centres in the directions of winds, or later of the compass, were presumed to be rhumbs. Thus the straight line between an origin and destination, the direction of which could be ascertained by reference to the nearest windrose, was thought to be the shortest course to be followed during a sea voyage. The Portuguese cartographer and cosmographer Pedro Nuñes (1502-1578) gave the name rhumb to loxodromes because of their property of "conserving the rhumb," a point of the compass being called a rhumb. Nuñes studied and wrote about the geometric properties of rhumb lines (*Tratado em defensam da carta de marear com o Regimento da altura,* Lisbon, 1537). In this he demonstrated that rhumb lines were not great circles but curves, part of a double spiral cutting the meridians at a constant angle and terminating at the poles. Nuñes showed the spiral configuration of loxodromes in a diagram.

Gerard Mercator, who must have known Nuñes's work, was the first, in 1541, to introduce spherical rhumb lines on a terrestrial globe. On his world map drawn on the projection named after him, Duisburg, 1569, the spherical rhumbs were projected as straight lines. Edward Wright, the English mathematician, also marked the lines on his world chart on Mercator's projection, London, 1599 (see Loxodrome, 4.123).

C BAGROW, Leo (1964): *History of Cartography.* Revised and enlarged by R A Skelton. London, C A Watts; Cambridge, Mass., Harvard University Press.

BROWN, Lloyd A (1949): *The Story of Maps.* Boston, Little, Brown and Company.

NAVARRETE, Martin Fernandez de (1846): *Disertación sobre la historia de la náutica.* Madrid, La Real Academia de la Historia.

REY PASTOR, Julio (1942): *La ciencia y la técnica en el descubrimiento de América.* Buenos Aires, Mexico.

TAYLOR, Eva G R (1956): *The Haven-Finding Art: A History of Navigation from Odysseus to Captain Cook.* London, Hollis and Carter. (Republished 1971, New York, American Elsevier Publishing Co.)

4.191 Square Grid

A A network of straight lines on a map arranged horizontally and vertically at regular intervals to form squares. It has been a prominent feature on Chinese maps since the 3rd century.

B Although the origin of the square grid is not certain, it is generally credited to Chang Heng, a famous astronomer (79-139). The earliest recorded evidence of the square grid dates back to Phei Hsiu (224-271). Although his maps do not survive, records of his mapping principles clearly indicate the use of the square grid on his maps. Following Phei Hsiu, numerous grid maps were made, including Phei Chu's grid map of 605; Chia Tan's "Hai Nei Hua I Thu" (Map of both Chinese and Barbarian Peoples within the [Four] Seas) of 801; "Yü Chi Thu" (Map of the Tracks of Yü the Great), the famous grid map inscribed on stone in 1137 by an unknown cartographer; Chu Ssu-Pên's map of China "Yü Thu," of c. 1315; the Mongolian grid map of the northwestern counties of 1329; and Lo Hung-Hsien's "Kuang Yu Thu," the Chinese atlas of 1555.

The late 16th century marked the first recorded Western influence on Chinese cartography by Jesuit missionaries. At that time the square grid tradition seems to have given way to map projections identifying longitudes and latitudes and to surveying and mapping. By the mid 1800s, the square grid tradition had been revived; maps in the various editions of the *Ta Ching I Tung Yu Thu* (The Ching Dynasty Atlas) and many other maps employed it. The latest evidence of the grid is dated 1896.

All extant Chinese maps with the grid include a statement to indicate the linear distance represented by each square. Thus the primary function of the grid is to indicate scale on the map. In Lo Hung-Hsien's world atlas, "Kuang Yu Thu," 1555, the dimensions of the squares on different maps remain relatively the same, but the linear distances they represent vary. This clearly indicates that Chinese cartographers used the grid to alter map scales by the well-known method of "similar squares." A third possible use of the grid is as a referencing system. Any point on the map can be located from any other point simply by counting off squares.

The square grid also appears on maps of non-Chinese origin. Pietro Vesconte's grid on his maps of the Holy Land which accompany Marino Sanudo's "Liber secretorum fidelium crucis . . ." (A Crusader's Geography) (*c.* 1320) is identical to the Chinese grid (Fig.17). A similar grid is also found on two Islamic maps by al-Mustaufi al-Qazwini (early 14th century). Perhaps the most interesting occurrence of the square grid is on a map of the Mediterranean in a 15th century nautical atlas (Destombes). This sea chart has all the other characteristics of a portolan chart, but the rhumb lines have been replaced by the square grid. It has been suggested that the Chinese square grid stimulated 13th and 14th century Mediterranean navigators and cartographers to create portolan charts. It is more probable, however, that the grid derives from the classical period. A theory that the 15th-century map of the Mediterranean is the lost sea-chart of Marinus of Tyre (2nd century) has been advanced in recent years (Laguarda Trías, pp.24-5).

Martin Cortes, the Spanish navigational authority, in his *Breve Compendio de la sphera y de la arte de navegar*, Seville, 1551, fol.lxvi^v, illustrates the use of the square grid in the compilation of maps and their reduction to a smaller scale. Richard Eden in *The Arte of Navigation,* London, 1561, provides an English version. This use of the square grid has remained a generally accepted procedure.

C CHU, Gregory Hoi-Yuen (1974): "The Rectangular Grid in Chinese Cartography," unpublished Master's Thesis, Univ. of Wisconsin-Madison.

DESTOMES, Marcel (1955): "A Venetian Nautical Atlas of the Late 15th Century," *Imago Mundi, 12,* 30.

FUCHS, Walter (1946): "The 'Mongol Atlas' of China by Chu Ssu-pen, and the Kuang-yü-t'u," Monumenta Serica, Monograph VIII, *Journal of Oriental Studies of the Catholic University of Peking.*

HSU, Mei-Ling (1978): "The Han Maps and Early Chinese Cartography," *Annals of the Association of American Geographers, 68,* 45-60.

LAGUARDA TRÍAS, Rolando A (1981): *Estudios de Cartologia.* Madrid, R A Laguarda Trías.

NEEDHAM, Joseph (1959): *Science and Civilisation in China.* Cambridge, At the University Press, Vol. 3, Chapter 22, 497-590.

WANG, Yung (1958): *Chung Kuo Ti Thu Shih Kang* (History of Chinese Cartography). Peking, San Luen Press.

4.231 Wave Directions

A The azimuth, including convergence and divergence, of wind-generated movements of the surface of the ocean or other water body.

B Various methods of representing wave directions have been devised. The best known are the stick charts of the Marshall Island navigators, first reported in 1862, which show the pattern of swells or wave masses caused by winds. Principal wave systems and local convergent and refracted waves are shown by the arrangement of sticks. Stick charts are fashioned from the centre ribs of palm leaves lashed together with cord made from locally grown fibre.

C BOWDITCH, Nathaniel (1977): *American Practical Navigator: An Epitome of Navigation.* Washington, Defense Mapping Agency Hydrographic Center, 787-800.

DAVENPORT, William (1960): "Marshall Islands Navigational Charts," *Imago Mundi, 15,* 19-26.

THROWER, Norman J W (1972): *Maps and Man: An Examination of Cartography in Relation to Culture and Civilization.* Englewood Cliffs, New Jersey, Prentice-Hall, 5-9.

GROUP 5

Symbolism

5.011 Area Symbol

A Any uniform tone, colour, or pattern of marks on a map which imparts an aspect of homogeneity and distinctiveness to an area.

B The concept of homogeneity and distinctiveness in one area in contrast with the character of another area is as old as cartography. The walls of a city on a map, even though nothing else be symbolized, suggest that there is a difference between the inside and outside.

One of the oldest extant topographical maps, a representation of a region in Mesopotamia on a small clay tablet from Nuzi in northeast Iraq, *c.* 2500-2300 BC, shows mountain regions with area symbols rather like fish scales. The equivalent from the other side of the world, a relief map of southern Changsha, in Hunan province, China (one of the three Han maps dated prior to 168 BC), has blue as an area symbol for the sea. The use of area symbols was not confined to physical characteristics of the earth; the Moorish kingdom of Granada is coloured green on a portolan chart of 1456 by Jacobo Bertran and Berenguer Ripol (on which the Red Sea is characteristically red).

As cartographers became more versatile and as knowledge increased, so did the character and use of area symbols. Further, with the introduction of printing, area symbols were limited by woodcut and engraving techniques. Oceans and shoal waters were stippled; forests were represented by patterns of little trees; fields were lined. Almost anything of a qualitative character, from geological formations to language regions, could be shown by area symbols. The portrayal of quantitative data required increased attention to the problem of clear representation of classes on choropleth maps (1.034), dasymetric maps (1.041), and hypsometric maps (3.081).

205

By the end of the 19th century colour printing (7.031), photographic techniques, and transfer methods such as Ben Day (7.021) had made it relatively easy to employ area symbols.

C BAGROW, L (1964): *History of Cartography.* Revised and enlarged by R A Skelton. London, C A Watts and Co.; Cambridge, Mass., Harvard University Press, 31.

CHANG, Kuei-sheng (1979): "The Han Maps: New Light on Cartography," *Imago Mundi, 31,* 9-17.

HOWSE, Derek, and SANDERSON, Michael (1973): *The Sea Chart.* Newton Abbot, David & Charles; New York, McGraw-Hill Book Company, 19.

5.031 Coastal Depiction

A Details of coastal features.

B Some of the earliest examples of coastal depiction are found in European medieval manuscripts. The portrayal of St John on Patmos, where he wrote his Apocalypse, was a standard feature. Thus the "Douce Apocalypse" of the mid-13th century, which is characteristic of a group of manuscripts assigned to Canterbury, includes a recognizable picture of the island with surrounding islets (Bodleian Library, Douce 180; A G and W O Hassall, pl. 1). As the tradition of portolan chart making developed from the end of the 13th century, coastal features assumed special significance. Rocks and shoals were carefully drawn, and places of importance, such as Venice and the Nile delta, were treated in greater detail within the limitations of scale. Coastal profiles are believed to have been included on early medieval charts, although no such chart survives. The coastal scenes in the manuscripts of *La Sfera* by Leonardo Dati (1360-1425) and his brother Gregorio are presumably drawn from an earlier prototype (see Profile, 5.162). In 1420 Cristoforo Buondelmonte introduced a distinctive style of coastal depiction in his *Liber Insularum Archipelagi,* the first *isolario,* with landmarks and rough coastal profiles sketched in (see Island atlas, 8.0110c). Bartolommeo dalli Sonetti in the first printed *isolario, c.*1485, marks offshore rocks by symbols (cross and dotted cross) which are still in use today (Howse and Sanderson, p.21).

From the second half of the 15th century profiles of coasts were included in the Low German "sea-book," or pilot's guide, and were used likewise in Dutch and French navigational texts, such as Pierre Garcie's *Le grant routtier* (1st ed printed 1520). Lucas Janszoon Waghenaer in his *Spieghel der Zeevaerdt,* Leiden, 1584-85, solved the problem of combining the general chart for use in navigation with the coastal chart for pilotage. In his explanatory text he drew attention to the importance of the appearance of the coasts. He drew them in bird's eye view or, more accurately, bird's flight view, with details of landmarks and offshore rocks and shoals and with soundings at half-tide. He also varied the scale of the coastline in order to treat in greater detail harbours and river mouths. Profiles, usually placed at the top of the charts, give added visual effect.

Sixteenth-century manuscript coastal charts also incorporated panoramic views of coasts as an intrinsic part of the work. The *roteiros* of João de Castro, 1538 to 1542, are masterpieces of the genre (see Profile, 5.162). An early example in England is the long view of the North Kent coast together with the river Swale, *c.* 1514 (British Library, Cotton Charter XIII. 12). Another anonymous chart, depicting the southwest coast of England from Land's End to Exmouth (British Library, Cotton MS. Augustus I.35, 36, 38, 39), emphasizes the features which sailors would look out for by foreshortening of the depth and protraction of the width of the area surveyed.

Skills in coastal depiction improved as more reliable methods of land survey developed in the later 17th and 18th centuries. It may be claimed that the army engineer Joseph Frederick Wallet Des Barres achieved a degree of excellence never surpassed in *The Atlantic Neptune* (1777 onwards). He depicted in remarkable detail, as the title page of volume one, *The Sea Coast of Nova Scotia* (1781), announced, "the Diversities of the Coast, and the Face of the Country near it: The Banks, Rocks, Shoals, Soundings, &c." He described this "beneficial Work" as the result of sixteen years of unremitting exertions.

Coastal depiction was a notable feature of Chinese charts. The famous Mao K'un map recording the voyages of Chêng Ho, the Chinese admiral, 1415-1433, printed in the 16th century, represents in a long narrow strip 7465 miles of coastline, schematic in style, with sketches of mountains and towns (Coastal chart, 1.0320a). Chinese skills in landscape depiction are reflected in some of the fine panoramic scroll maps dating from the late 16th to the early 19th centuries.

In Japan screen paintings of the 17th and 18th centuries portrayed with attractive detail maritime scenes and activity along the bustling coasts of the Seto Inland Sea. A chart of western Japan by Ebi Gaishi, manuscript in 4 volumes, 1680, records the results of an official inspection of the coastlines of the Inland Sea, with information ranging from physical features to the height of the tides along the shore (Nanba et al, pl.34, pp.186-7).

C HASSALL, A G and W O (1961): *The Douce Apocalypse.* London, Faber and Faber.

HOWSE, Derek, and SANDERSON, Michael (1973): *The Sea Chart.* Newton Abbot, David & Charles.

NANBA Matsutaro, MUROGA Nobuo, and UNNO Kazutaka (1973): *Old Maps in Japan.* Osaka, Sogensha.

THOMSON, Don W (1966): *Men and Meridians, The History of Surveying and Mapping in Canada,* Vol. 1. Ottawa, Department of Mines and Technical Surveys.

5.032 Colour

A A complex graphic variable employed in mapmaking through the use of a wide variety of pigments.

B Because colour appears on extant early Egyptian maps, on the Chinese Han maps of the second century BC, and on many medieval European maps, it may be assumed that colour was a normal component of maps from very early times, but there is very little evidence to indicate how colour was used. The fact that on the military Han map colour was employed both to differentiate phenomena and to signify their relative importance (red for military-related features, roads, etc.; light blue-green for water bodies; black for other features and lettering) suggests that the utility of colour for identification and emphasis probably occurred to the earliest mapmakers.

It is not possible to cite many colour conventions on maps in the early Christian era. Colours used on the Madaba mosaic map (560-565) are rather inconsistent although lines of black cubes represent valleys, flat areas are white, and human habitations are in red and brown. The majority of churches are represented by small buildings with sloping red-tiled roofs, probably a derivative of the standard vignette of temples. On the surviving early medieval copies of the *Agrimensores Veteres Romanorum* (Wolfenbüttel, 6th or 7th century; Vatican, 9th century), mountains are brown, mauve or, if wooded, green; rivers are blue; buildings mostly brown, grey or yellow with red roofs; and roads are mostly red or brown, but sometimes green.

The association of blue or green with water and red with human habitation, especially the cities and towns, was current throughout the medieval period. The late 12th century Greek manuscript of Ptolemy's *Geography* from Mt. Athos (the Vatopedi manuscript, British Library, Add. MS 19391) employs red boxes around names on the maps, and on portolan charts significant ports were named in red. By the mid-16th century, careful instructions for colouring charts were available. Martin Cortés in his *Breue compendio de la sphera y de la arte de nauegar* (Seville, 1551, f. lxiiii^r), translated into English by Richard Eden in *The Arte of Nauigation* (London, 1561), states that ports, principal capes, and famous cities were to be depicted in red, the rest in black:

> Then with colours and golde shall you garnyshe and beautifie the Cities, Compasses, Shyppes, and other partes of the Carde. Then shall you set forth the coastes with greene, . . . and make them fayre to syght with a little saffron . . . (Eden, fol. lviii^r).

In general, during the manuscript period of mapmaking, colour was simply one of the available media, and it could be employed in whatever way was desirable to identify, emphasize or clarify the representation. Most pigments were opaque and were applied to the usual parchment or vellum by pen or brush in the same way as in manuscript illustration and illumination.

A notable innovation occurred in the use of colour in cartography when the printed map became common after the 15th century. In the pre-printing era, colour was a basic option in the graphic design. After printing developed, technical problems, particularly registration, effectively prevented any significant amount of multicolour printing for the next 350 years. Nevertheless, colour was desirable as a means of clarification and emphasis, and so it was often added to the printed map by hand. Although there were occasional attempts to print coloured maps, such as the three-colour map of Lorraine in the 1513 Strassburg edition of Ptolemy's *Geography*, maps were generally uncoloured. Colour, if desired, could be added either by an illuminator or by the purchaser himself (see Hand Colouring, 6.081).

The ascendency in the 16th century of copperplate engraving as the preferred technique for the more elaborate and larger separate maps called for the use of lighter, more transparent colours compared with those that had been used up to that time for both manuscript and woodcut maps. The images of engraved maps were more intricate and the linework more delicate than those of woodcut maps, and lighter water colours were desirable when pigments were used primarily to enhance and clarify an already complete black line image.

From the 17th century well into the 19th the hand colouring of printed maps was something of a genteel pastime, but more importantly it developed into a craft represented by artists' guilds, and map colouring became an important adjunct activity to the map publishing industry. The well-known Dutch atlas compiler, Abraham Ortelius, began his career as a map colourist.

In England the use of colour on estate plans would appear from the first to have been symbolic rather than ornamental and was designed to aid the better representation of the plan as a scale drawing of the ground. An early guide to conventions in the use of colour on plans is given in William Folkingham's *Feudigraphia: The Synopsis or Epitome of Surveying Methodized* (London, 1610). Folkingham detailed the colours in general use and indicated how they were to be mixed. Arable land was painted a pale straw colour, meadows a light green, pasture a deeper green, heather and fern a still deeper green, and seas were a "green skie colour." George Atwell, the author of *The Faithfull Surveyour* (London, 1658), advocated the use of similar colour conventions and added to the features to be represented in his chapter entitled "Of colouring and beautifying of plots."

The choice of colours on estate plans may have influenced the colours in the representation of types of land-use, as on the earliest English example in Thomas Milne's "Plan of the Cities of London and Westminster, circumjacent Towns and Parishes, etc., as laid down from a Trigonometrical Survey taken in the Years 1795-1799" in six sheets (1800). On the engraved sheets the several categories of land use were distinguished by letters and subsequently hand-coloured.

The development of thematic mapping and large-scale topographical mapping in the 18th century put increasing demands on the data-carrying capacity of monochromatic maps, mostly black and white. Although mechanical techniques for patterned shading and stippling had been developed, they were insufficient to provide the differentiation and clarity needed for such things as geological maps and detailed topographical portrayals. The advent of lithography and the perfection of transfer techniques and registration systems by the mid-19th century made possible the shift from hand colour to printed colour (7.031). The change was especially helpful in the use of colour as an area symbol (5.011) since the application of colour over large areas was difficult to do by hand.

The change to printing also promoted changes in the way colour was employed. Instead of being primarily a means of enhancing a finished black and white map, colour began to be used as a basic option in map design, as it had been in the manuscript era. Colour again performed a separate information-carrying role in the sense that if the coloured symbols were removed, the map would be incomplete, a simple example being line symbols on topographical maps, such as rivers in blue. In thematic mapping the use of colour to distinguish areas of like composition or value became widespread.

The late 19th century saw continued technical innovations toward increasing versatility of colour in mapmaking. Among the more important were the introduction of photographic techniques and methods of combining tints, such as the Ben Day process (7.021), which greatly increased the range of colours.

From the 17th century Japanese mapmakers developed the use of colour to a fine art. Throughout the Edo period (1600-1868) the principal medium was the woodcut map, which up to about 1765 was normally hand coloured. The most famous world map, "Bankoku Sōzu," published at Nagasaki Harbour in 1645, with its accompanying sheet illustrating the people of the world, was produced as a modified version of the gorgeous sixfold and eightfold Nanban screens of the nobility, and was beautifully coloured to distinguish the countries of the world. When the art of *ukiyo-e* (*ukiyo* means "floating world"; *e,* "picture") became popular under the Tokugawa Shogunate, a tradition of printing in many colours from woodblocks (*nishiki-e*) was established in the early 17th century, with Edo the main centre. A key block (known as the *han*) giving black and white detail was made first. Thereafter each colour was printed from a separate block, as many as twelve being used to make a single map. The publisher could vary the colours used in printing, and thus maps printed from the same key block could show different colours. By 1772 graduation printing, in which the colours were shaded from dark to light, had been invented. This technique was used on a number of panoramic maps. Colour printing of maps continued in this way through the 19th century well after the *ukiyo-e* colour print movement had declined. The panoramic map of Japan by Kuwagati Keisei, early 19th century, and that of China, 1840, by the celebrated artist Hokusai, made in his 81st year, are fine examples of the woodcut, colour-printed Japanese map.

C AVI-YONAH, Michael (1954): *The Madaba Mosaic Map.* Jerusalem, Israel Exploration Society.

DILKE, O A W (1967): "Illustrations from Roman Surveyors' Manuals," *Imago Mundi, 21,* 9-29.

EHRENSVÄRD, Ulla, supplemented by PEARSON, Karen S (in press): "Color in Cartography: An Historical Survey," in David Woodward, ed., *Art in Cartography: Six Historical Essays.* (6th Kenneth Nebenzahl, Jr., Lectures in the History of Cartography.) Chicago, University of Chicago Press.

HSU, Mei-Ling (1978): "The Han Maps and Early Chinese Cartography," *Annals of the Association of American Geographers, 68,* 45-60.

LANGLOIS, Victor (1867): *Géographie de Ptolémée: Reproduction photo-lithographique du manuscrit grec du Monastère de Vatopedi au Mont Athos.* Paris, Firmin Didot.

PEACHAM, Henry (1612): *The Gentlemans Exercise. Or an exquisite practise, as well for drawing all manner of Beasts in their true Portraitures: as also the making of all kinds of colours, to be used in Lymming, [etc].* London.

PEARSON, Karen S (1980): "The Nineteenth-Century Colour Revolution: Maps in Geographical Journals," *Imago Mundi, 32,* 9-20.

SKELTON, R A (1960): "Colour in Mapmaking," *Geographical Magazine, 32,* 544-53.

SMITH, John (1701): *The Art of Painting in Oyl . . . to Which is Added the Whole Art and Mystery of Colouring Maps and Other Prints with Water Colours.* 3rd ed. London.

WALLIS, Helen, and others (1974): *Chinese & Japanese Maps*. British Museum Publications Ltd.

5.033 Conventional Sign

A (MDTT 43.2) "A symbol used to represent information on the face of a map or chart. (By implication a symbol of a kind in common use.)" (MDTT 43.1) A cartographic symbol may be "a letter, character or other graphic device." Cartographic symbols fall into two broad categories, pictographic and ideographic. The former may be defined as a pictorial symbol or sign closely resembling the feature portrayed; the latter is a character or figure symbolizing the idea of the feature depicted.

B The use of conventional signs on maps dates back to at least the third millennium BC as is indicated by those employed on one of the oldest topographical maps extant, a small clay tablet found in the neighbourhood of Nuzi in northeast Iraq. This map shows two mountain ranges in profile by a stylized "running" mountain symbol and several rivers by a series of parallel or flow lines. Both symbols are simple solutions to the problems of representing relief features and water courses. Variants of these two pictographic symbols are common on maps produced both by the peoples of early times and by advanced practitioners of the mapmaker's craft. Thus the mountains and rivers on the Hereford world map, *c.* 1285, and on the engraved maps in Ptolemy's *Cosmographia*, Rome, 1478, are similar to those on the Nuzi map, apart from the addition (on the later examples) of shading on slopes, which was a characteristic of the representation of relief from the 14th to the mid-19th century in European mapping (Lynam, pp. 38-9).

Modern knowledge of symbols used on maps and plans produced in the pre-Christian era is limited because of the fragmentary evidence. The symbols used on the cartographic materials which survive certainly partake of the nature of conventional signs in the form of consistently executed symbols. The rock carvings in the Val Camonica, province of Brescia, northern Italy, dating from *c.* 1500 BC, display rectangles, lines, circles, and stipples representing settlement and fields. A change in style dated *c.* 700 BC distinguishes later carvings, which are more naturalistic in form (Harvey, pp. 45-6; Skelton and Harvey, p. 21).

Signs employed on the few surviving maps and plans from Upper Egypt in the second millennium BC suggest that the Egyptians had a recognized system of symbols for distinguishing different types of trees as well as one for rivers; the latter were marked by a wavy or zigzag pattern of lines (Harvey, pp. 50-1). Another way of delineating a river is found on a map of Madaktu, embodied in a bas-relief of the sixth century BC from Nineveh. The river is lined and decorated by pictographs of fish, a representation employed a millennium later in western Christendom on mappae mundi in the tradition inspired by Beatus Liebanensis (*c.* 730-798) of Spain. The Beatus maps, drawn on vellum from the 10th to the 13th centuries, display symbols of fish in the encircling ocean.

Fragments of Roman maps of a first century cadaster inscribed in marble (found at

Arausio, or Orange, in Provence) show that Roman surveyors had established conventions in plan drawing. The Orange cadaster depicts by double straight lines bridges, ditches, and roads (the principal roads are exaggerated in width), and by curving double lines rivers. Symbols are used to denote rents (Dilke, pp. 188-9). No buildings are marked on this cadaster. On the marble plan of Rome, "Forma Urbis Romae," of the 3rd century AD, which depicts buildings in remarkable detail, the conventional signs for features are nearly all shown strictly in plan. The most notable Roman map for its signs and symbols is the Peutinger, a road map of the Roman world of the 4th or 5th century (probably revised from a first-century map, and known in an 11th- or 12th-century copy). Of the 555 signs, 429 show double towers probably to represent villas, 44 are identified as temples, and 52 are "aquae," watering places, notably spas, shown as hollow squares or rectangles (Levi and Levi, pp. 65-6). The six cities indicated by fortifications on the Peutinger map are depicted in a similar style to illustrations of fortified places in the military instructional manuals, the Corpus Agrimensorum, written from the late first century and known in manuscripts of c. 500 and the 9th century (Dilke, pp. 91-100, 117). The 12th century copy of the Corpus Agrimensorum (British Library, Add. MS. 47679), in contrast, shows the buildings in plan.

A picture-map of Roman descent, the Madaba mosaic map, 560-565, in Jordan, south of Amman, portraying the lands of the Bible, also displays conventional signs similar to those in the Peutinger map. The representation of human habitations is of two main types. Large cities like Jerusalem, Gaza, and Pelusium are represented by pictographs of exaggerated size, drawn from a high view-point in order to show both the city walls and the buildings within. Small towns and villages are marked by conventional signs in the form of towers and walls.

The large medieval mappae mundi (1.2320b) such as the Hereford world map, c. 1285, were predominantly pictorial, with city ideograms which seem derived from the styles of the ancient world. These styles may also have influenced the long-established tradition of picture maps of Italian towns and the plans of towns in the Holy Land which date from the time of the first crusade in the 11th century onwards (Harvey, pp. 70-2).

More abstract signs were used by Arab mapmakers. The Turk Mahmud al-Kashgarī uses circles for settlements on his world map, 1076, and al-Idrisi on his world map of 1154 has segmented circles (see Circle, 5.1610a). The maps made to illustrate Ptolemy's Geography, c. 150, also employ abstract signs for settlements as commonly as pictorial signs. The Byzantine Maximus Planudes (c. 1260-1310), author of the prototype of the earliest manuscript Ptolemy maps known (e.g. Vatican Library, Vaticanus Urbinas graecus 82), employs a system of conventional signs originating in astronomical writings to indicate the distribution of tribes (see Letter symbol, 5.122).

By the end of the 15th century bird's eye views instead of conventional symbols were used for towns and cities on regional maps. At the same time the conventions of manuscript mapmaking were adapted to the European printed map which developed from 1473. These included the circle with a dot or cross for settlements. Erhard Etzlaub on his map of the environs of Nuremberg, 1492, shows towns as circles and marks political limits. Ranking as the first political map (see Jurisdictional map, 2.0210d), this work is notable for its lack of pictorial signs. An early 16th century unsigned and undated copy of another map (also nonpictorial) of the same region by Etzlaub shows roads as dots, each representing a German mile (see Road map, 1.1810d) (Fig. 18).

Various devices were used to explain the signs. The inclusion of a legend or key is found as early as c.1300 BC on the map of the Nubian gold mines, and other examples appear on medieval manuscript maps (see Legend, 6.121). The makers of printed maps, addressing themselves to a wider public, were quick to see the need to enlighten their readers. Bartolommeo dalli Sonetti in his *Isolario*, published at Venice in 1485, the first isolario to be printed, opened his verses by listing (without identifying) the features depicted: "quante insule vi son picole e grande e scogli e seche e citate e castella" (islands . . . small and great and reefs and shoals and cities and castles) (Delano Smith, 1985, p. 16). The earliest explanation on a printed map appears on Etzlaub's environs of Nuremberg, 1492, and the earliest separate sheet of explanation is attached to Etzlaub's Romweg, c. 1500 (see Characteristic sheet, 6.032). Philipp Apian's *Bairische Landtaflen*, Ingolstadt, 1568, is notable for its table of 14 signs entitled "Erklärung nach volgender Zaichen," the first to carry a heading (Legend, 6.121). Gerard Mercator in the preface to his *Galliae tabulae geographicae* (Duisburg, 1585), the first volume of his world atlas, provides a guide to the signs used, "In usum tabularum admonitio," explaining that the small circles indicate the true sites of places and that their distances should be measured from these; however, signs are attached so that places may be distinguished. For these distinctions he explains that he is using easily made signs, so that anyone could supply what has been omitted (Delano Smith, 1985, p. 26). In this way Mercator solved the problem of how to mark the exact position of places as well as classifying them by their appropriate sign (see Circle, 5.1610a).

Pictures were increasingly superseded by signs on European topographical maps from the 16th to the 19th centuries, but the signs themselves were often pictorial in style. This pictorial representation of features makes it difficult sometimes to decide what is a picture and what a conventional sign (Skelton and Harvey, p. 31). A move towards the horizontal plan instead of, or in addition to, pictorial representation is illustrated by John Ogilby in his *Britannia*, London, 1675, as he explained in his preface: "*Capital towns* are described *Ichnographically* according to their Form and Extent, but the *Lesser Towns* and *Villages*, with the *Mansion Houses, Castles, Churches, Mills, Beacons, Woods*, &c. *Scenographically*, or in Prospect". In France the use of signs in perspective detail ("la perspective cavalière") which was characteristic even of the Cassini map of France, c. 1789, was one of the old established traditions swept away by the commission set up in 1802 at the Dépot de la Guerre under general (comte) Sanson to improve topographical maps and plans. The commission decreed that the signs for inhabited places and all human construc-tions must be drawn exclusively in horizontal projection (Dainville, p. 220; Berthaut, vol. I, pp. 137-43).

In maritime charting standardized conventions such as stipple for shoals and crosses for rocks became generally accepted. Lucas Janszoon Waghenaer in the first printed European sea atlas, *Spieghel der Zeevaerdt*, Leyden, 1584, gave instructions in his introductory text on the form of buoys and beacons and other features, which he illustrated by diagrams and by the signs used on the charts.

As the content of maps became more elaborate, a wider range of signs was devised. Since classicial times the major elements depicted on general and topographical maps were mountains, forests, towns, seas, and seashores. To these were added in the 16th and early 17th centuries natural resources and human occupations. Thus maps of Bohemia from 1619 onwards used a variety of symbols for precious metals and minerals (see Mineral map, 3.131). The Polish Jesuit missionary Michael Boym on his manuscript map of China, *c.* 1652, shows minerals by planetary signs, with a key in Latin and Chinese (Legend, 6.121). Military mapping commanded an extensive set of signs, reaching its fullest expression in the "Carte Chorographique des Pays Bas Autrichiens" of the Comte de Ferraris, Brussels, 1777 (see Legend, 6.121; Characteristic sheet, 6.032).

Interest in geographical features and conventional signs prompted the publication of illustrative maps. One of the earliest is that by Matthias Quad, "Vocabulorum geographicorum topica significatio" in his *Geographisch Handtbuch,* Cologne, 1600 (Fig.1). A well-known later example is the "Mappa Geographiae Naturalis sive Tabella Synoptica," by Matthias Seutter, Augsburg, *c.* 1730 (see Imaginary map, 1.091).

The maps of non-European nonliterate peoples reveal many similar styles of sign, indicating the universality of symbolism applied to terrestrial features. The *churinga* of the Australian aborigines, for examples, comprise stone plaques or wooden tablets inscribed with abstract line designs, recording the doings of a tribe's distant ancestors and the lands where they lived (Gaur, pp. 25-6) (Fig.19). In western Australia, the spear-throwers of the Bindibu people are carved on the back with geometrical designs which may be seen as highly conventionalized maps, their concentric circles and spirals representing the waters in their territories (Thomson, p. 274 and plates between pp. 272-3; Skelton and Harvey, pp. 20-21). The Aztec maps of the early 16th century use similar pictorial styles to those of the Old World. A notable combination of symbols, Aztec and Spanish, appears in the representation of roads by double lines with footprints, as on a map of the valley of Tepetlaoztoc in Mexico, 1583 (British Museum, Add. MS. 13964, f.209). A territorial map showing the division of lands between two chiefs in the state of Pueblo, *c.* 1600, is schematic (British Museum, Egerton MS. 2896). In North America an Indian birchbark map of a route by land and water (see Route map, 1.1810, and Map surface, birchbark, 6.1310a) provides a topological sketch comparable to the London Underground plan of today (Figs. 8 and 21).

C BERTHAUT, Henry Marie Auguste (1898-99): *La Carte de France, 1750-1898: étude historique,* 2 vols. Paris, Service géographique de l'armée.

DAINVILLE, François de (1964): *Le langage des géographes.* Paris, Éditions A et J Picard.

DELANO SMITH, Catherine (1982): "The Emergence of 'Maps' in European Rock Art," *Imago Mundi, 34,* 9-25.

DELANO SMITH, Catherine (1985): "Cartographic Signs on European Maps and Their Explanation before 1700," *Imago Mundi, 37,* 9-29.

DILKE, O A W (1985): *Greek and Roman Maps.* London, Thames and Hudson.

GAUR, Albertine (1984): *A History of Writing.* London, The British Library.

HARVEY, Paul D A (1980): *The History of Topographical Maps.* London, Thames and Hudson.

LEVI, Annalina, and LEVI, Mario (1967): *Itineraria Picta*. Rome.

LYNAM, Edward (1953): *The Mapmaker's Art*. London, Batchworth Press.

SKELTON, R A (1965): *Decorative Printed Maps of the 15th to 18th Centuries*. London, Spring Books.

SKELTON, R A, and HARVEY, Paul D A, ed. (1986): *Local Maps and Plans from Medieval England*. Oxford, Clarendon Press.

THOMSON, Donald F (1962): "The Bindibu Expedition: Exploration among the Desert Aborigines of Western Australia," *Geographical Journal, 128*, 262-78.

5.041 Density Symbol

A A marking on a map intended to convey directly the idea that in one place there is a greater or lesser number of some phenomena per unit of area than there is in another place. As so defined, density symbolism does not include digits, dots or isarithms.

B The concept of geographical density was known at least in the 18th century, especially as it related to numbers of persons per unit of area, when it was often called specific population. The first attempt to symbolize density values on a map apparently was by Baron Charles Dupin who made a choropleth map in 1826 (published in 1827) of the number of persons per male child in the elementary schools in each département of France, "Carte figurative de l'instruction populaire de la France" (Plate 1 in Vol. 2 of *Forces productives et commerciales de la France*, Paris, Bachelier). Dupin attempted to shade each département according to its value, i.e., employing no class categories, with the darker shades assigned to the départements having the lesser instruction (larger number of persons per student). Adrien Balbi and André Michel Guerry used the same technique in 1829 on their set of maps "Statistique comparée de l'état de l'instruction et du nombre des crimes dans les divers Arrondissements des Académies et des cours R.les de France," and similar but much less rigorous and systematic shading was employed from the early 1830s on to show variations in the occurrence of cholera in various parts of cities.

In 1831 Adolphe Quetelet symbolized variations in density (instruction and incidence of crimes) by smooth shading without reference to enumeration boundaries, with the darker symbolizing greater density (*Recherches sur le penchant au crime aux différens âges*, Bruxelles, M. Hayez). This form of symbolization was used in the 1840s by Heinrich Berghaus in his *Physikalischer Atlas* (1845) to show variations in the density of such things as animal life, as well as by August Petermann in his population and cholera maps of the British Isles, published in 1849 and 1852.

The first attempt to symbolize classes of density was by George Poulett Scrope in 1833 in a small, crude world map of population density categories (Frontispiece, *Principles of Political Economy, deduced from the Natural Laws of Social Welfare . . . .*, London, Longman, *et al.*). Symbolization of density classes was employed by

Adolphe d'Angeville in 1836 for maps of France, by Henry Drury Harness in 1837 for a map of Ireland, and in the Irish Census of 1841, generally with the greater densities as the darker. Joseph Fletcher in 1849 used a system of classes with seven tones according to proportions below and above the average (Fig.15) (see Moral statistics map, 2.132).

In 1866 Charles Joseph Minard, in a first attempt to relate the symbolic shadings to the density in a commensurable way, prepared a map of population density in Spain, "Carte figurative et approximative des Populations specifique des provinces d'Espagne" (Paris, Régnier et Dourdet). In it he employed horizontal rulings spaced in strict proportion to the density, the greater the density the closer the lines, with the number of lines per centimetre being one-fifth the number of persons per square kilometer in that area. Minard's system seems to have been either ignored by or unknown to others.

After the 1830s, if statistics referring to enumeration units were available, density symbolization was by classes; if such data were not available, density variations were shown by smooth shading, with the darker usually being the more in both systems.

C ANDREWS, John H (1966): "An Early World Population Map," *Geographical Review, 56,* 447-48.

DU BUS, Charles (1931): *"Démocartographie de la France, des origines à nos jours.* Paris, Librairie Félix Alcan.

JARCHO, Saul (1973): "Some Early Demographic Maps," *Bulletin of the New York Academy of Medicine, 49,* 837-44.

ROBINSON, Arthur H (1955): "The 1837 Maps of Henry Drury Harness," *Geographical Journal, 121,* 440-50.

ROBINSON, Arthur H (1967): "The Thematic Maps of Charles Joseph Minard," *Imago Mundi, 21,* 95-108.

ROBINSON, Arthur H (1982): *Early Thematic Mapping in the History of Cartography.* Chicago, University of Chicago Press.

5.061 Flow Line

A A line of varying width which symbolizes movement along a route, the width of the line at any place being made proportional to the volume of movement.

B Rivers portrayed as lines tapering more or less in proportion to the volume of flow are a quite old, non-rigorous form of flow line, as are the representations of tidal and ocean currents by Athanasius Kircher in 1665 and Eberhard Werner Happel in 1687. It is probable that manuscript maps were made before 1837 which show the flow of goods and are attached to official reports in various archives. None are now known, however, and the first systematic and sophisticated use of flow lines was by Henry Drury Harness in 1837. He employed them on two maps in the *Atlas to Accompany the Second Report of the Commissioners Appointed to Consider and*

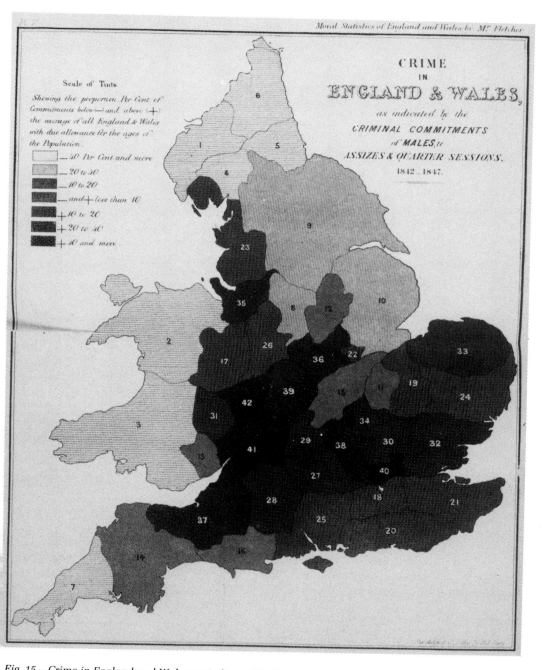

Fig. 15 – Crime in England and Wales, as indicated by the criminal commitments of males to assizes and quarter sessions, 1842-1847, by Joseph Fletcher. In "Moral Statistics of England and Wales, by Mr. Fletcher," Journal of the Statistical Society of London, 12 (1849), map after p.236. By permission of the British Library. (Density symbol, 5.041; see also Choropleth map, 1.034; Moral statistics map, 2.132)

Recommend a General System of Railways for Ireland (Dublin, HMSO, 1838). One map portrayed passenger traffic, the other the flow of goods. Harness's maps seem to have been quickly forgotten.

In 1845 Alphonse Belpaire in Belgium and Charles Joseph Minard in France employed flow lines to show passenger and goods traffic. Whether one or both knew of Harness's maps is not known, and the priority between Belpaire and Minard has not been established. Minard employed flow lines of great variety and complexity, and it was his use of them that established flow lines as a standard symbol.

C HAPPEL, Eberhard Werner (1687-89): *Mundus mirabilis tripartitus.* Ulm.

KIRCHER, Athanasius (1665): *Mundus subterraneus.* Amsterdam.

ROBINSON, Arthur H (1955): "The 1837 Maps of Henry Drury Harness," *Geographical Journal, 121,* 440-50.

ROBINSON, Arthur H (1967): "The Thematic Maps of Charles Joseph Minard," *Imago Mundi, 21,* 95-108.

5.081 Hachures

A (MDTT 432.21: Schraffen) Lines, often short, following the direction of maximum slope (*lignes de plus grande pente*) which in series indicate relief. By 1730 the French were using the terms *hachures diagonales, hachures horizontales,* etc., for symbols depicting terrain (e.g. Jean de la Grive, in the text to his map of the environs of Paris, 1730). A later development was the use of lines, the thickness and spacing of which indicate the degree of slope (MDTT 432.24: Böschungsschraffen).

B Simple hachuring was introduced in the middle years of the 17th century when cartographers modified the pictographic system of hatching, which employed oblique shading as though the object were illuminated from the northwest. Hachuring assumes the light to fall vertically from above. One of the earliest to employ the method was the Maltese Giovanni Francesco Abela on a map of the archipelago of Malta, 1647, published in his *Della Descrittione di Malta isola . . . Libri Quattro* (Malta, 1647), reproduced by Roberto Almagià, *Monumenta Italiae cartographiae* (1929). The earliest map in France was David Vivier's "Carte particulière des environs de Paris," prepared by Vivier and others under the auspices of the French Académie Royale des Sciences in 1674 and engraved by F de la Pointe in 1678. The crude hachuring adopted was to be the characteristic feature of relief depiction on the "Carte Géometrique de la France" ("Carte de Cassini") (1750–), of which Vivier's map was the forerunner. The style was also copied by French, Dutch, and German mapmakers, such as Alexis H Jaillot, Claes J Visscher III, and Johann B Homann. An original attempt to show the pattern of valley systems was that of the English physician Christopher Packe (1686-1749) on his ". . . Philosophico-Chorographical Chart of East Kent . . ." (1743). He emphasized the valleys by hatching their slopes, leaving the ridgetops white. The first world map of mountain systems using hachuring appears to be that of the French geographer Philippe Buache (1700–1773), published in the *Mémoires de l'Académie Royale des Sciences, Année 1752* (Paris, 1756, pp. 399-416, plate 13).

The earliest proponent of hachuring as a means of portraying degrees of slope was an officer in the Saxon Army, Major Johann Georg Lehmann (1765-1811). His method of employing equally spaced lines of varying thickness was first published in *Darstellung einer neuen Theorie der Bergzeichnung der schiefen Flächen* (Leipzig, 1799). The technique supposed vertical illumination of the surface, the return reflection varying with the angle of slope. Vertical rays falling on a slope inclined at an angle of 45° from horizontal would be reflected horizontally, and such a slope was represented by complete black on the "scale of shade." A horizontal plane, reflecting all incident light upwards, was represented by white. The intermediate slopes were divided into nine categories, each comprising a 5° range of slope, thus the proportion of white to black varied according to the slope and was obtained by varying the width of the lines. Other scales were proposed later, for example by Friedrich Carl Ferdinand von Müffling (1775-1851). The first map to show relief by this system of hachuring appears to be one of the Hemsdorf district drawn by Lehmann in 1798, and reproduced in *Die Lehre der Situation-Zeichnung, 1820.*

Lehmann's method was modified and adopted by 1815 by the armies of the Austro-Hungarian Empire, Prussia and Imperial Russia. Napoleon is reported to have had the book translated (but not published) for use in the French Army. An early German modification of the system was produced by J A Schienert, whose own book, published in 1806, was first translated into English in 1812. Lehmann's original system was translated into English by Colonel William Siborne and published in 1822, but Lehmann's technique of hachuring was never adopted by the British Army. A modification of it by Colonel Jan Egbertus van Gorkum was used in the Netherlands in the 1820s, and in the early 1830s was proposed by Major-General Sir James Carmichael-Smyth (1779-1838) for use by the British Royal Engineers. Hachuring was extensively employed on the engraved topographical maps of the 19th century, especially on published national surveys.

The "Topographische Karte der Schweiz 1:100,000," 1842-64, made under the direction of General Guillaume Henri Dufour (1787-1875) and known as the Dufour-karte, is regarded as one of the finest examples of hachuring. It represents a further innovation by modifying Lehmann's system so as to give an impression of relief by shadow obtained by oblique lighting (MDTT 432.25, Schattenschraffen).

C BERTHAUT, Henry Marie Auguste (1898-99): *La Carte de France 1750-1898: étude historique,* 2 vols. Paris, Service géographique de l'armée, I, 57-58.

BROWN, Lloyd A (1949): *The Story of Maps.* Boston, Little, Brown and Company.

CAMPBELL, Eila M J (1949): "An English Philosophico-Chorographical Chart," *Imago Mundi, 6,* 79-84.

CARMICHAEL-SMYTH, Sir James (Bart.) (1828): *Memoir upon the Topographical System of Col. Van Gorkum.* London, T. Egerton.

CUÉNIN, René (1972): *Cartographie générale: Tome 1, Notions générales et principes d'élaboration.* Paris, Éditions Eyrolles.

DAINVILLE, François de (1964): *Le langage des géographes.* Paris, Éditions A et J Picard, 167-71.

GALLOIS, L (1809): "L'Académie des Sciences et les origines de la carte de Cassini. Premier article," *Annales de Géographie, 18,* 193-204.

IMHOF, Eduard (1965): *Kartographische Geländedarstellung.* Berlin, Walter de Gruyter, 237-54.

IMHOF, Eduard (1982): *Cartographic Relief Representation.* English language version translated and edited by H J Steward. Berlin, New York, Walter de Gruyter.

JONES (afterwards O'DONOGHUE), Yolande (1974): "Aspects of Relief Portrayal on 19th Century British Military Maps," *Cartographic Journal, 11,* 19-33.

LEHMANN, Johann Georg (1820): *Die Lehre der Situation-Zeichnung.* (Ed. by G A Fischer). Dresden.

RAVENSTEIN, E G (1911): "Maps," *Encyclopaedia Britannica,* 11th ed. Cambridge, *17,* 630.

SCHIENERT, J A (1806): *Das Situationszeichnen für Soldaten.* Berlin. (Translated into English (1812): *Essay on Military Drawing.* London.)

SCOTT, H Y D (1863): "On the Representation of Ground," *Professional Papers of the Royal Engineers,* New Series, *12,* 144-62.

SIBORNE, William (1822): *Instructions for Civil and Military Surveyors in Topographical Plan-Drawing . . . founded upon the System of J G Lehmann.* London, G and W B Whittaker.

5.0910 Isarithm

A (MDTT 433.1: isoline) A line on a map along which some quantity or value is, or is assumed to be, constant. Also called isogram.

The generic term "isarithm" derives from Greek *is(os)*, equal, plus *arithm(os)*, number, and encompasses the lines used to portray the distribution of a wide variety of commensurable phenomena. Isarithms are employed for two main classes of data: (1) to show values which can exist at points, e.g., departure from a datum (elevation, depth), average annual precipitation, flowering dates of a plant species; (2) to show ratios which cannot exist at points because some surface measure (area, distance) is included in their derivation, e.g., density, average slope, potential. The first is sometimes referred to generically as "isometric lines," the second often as "isopleths" (German: *Pseudoisolinien*), but usage is not consistent.

B The isarithm developed as a map symbol along two unrelated paths. The one with a somewhat longer history was its use for representing variations of the compass, the earliest employment apparently being in 1536 when compass declinations (3.031) began to be mapped in manuscript. The Portuguese chartmaker Luís Teixeira seems to have been the first to draw, *c.* 1585, curved lines of equal declination on a map (see Compass declination, 3.031). The first printed map with isogones (5.0910e) by Edmond Halley is dated 1701. Development of the isarithm along the other path, the representation of depths in navigable waters, also began in the 16th century. The earliest known manuscript depth isarithm is dated 1584 (see Isobath, 5.0910d); the first printed map with an isobath, by L F Marsigli, is dated 1725.

Isarithmic maps to show distributions of various elements of geomagnetism continued development during the 18th century. The use of isarithms to show depth was extended to the open sea during the same century, but so far as is known no one observed the conceptual similarity between geomagnetic isarithms and depth isarithms. The depth isarithm "climbed out of the sea like Aphrodite" and was applied to elevations of the land surface in the late 18th century (Dainville; see Contour, 5.0910a).

An important derivative of the geomagnetic isarithm was introduced in 1817 by Alexander von Humboldt who applied the concept to show the distribution of surface temperatures (see Isotherm, 5.0910g). The concept of the isotherm in turn spawned numerous other applications of the isarithm in meteorology.

The conceptual union of the two mainlines of development, the isobath-contour on the one hand and the isogone-isotherm on the other, occurred in 1845 when it was suggested by Léon Lalanne that the variable statistical surface of population density could be represented by isarithms. Such a map was made in 1857 by N F Ravn (see Isopleth, 5.0910f). Since that time innumerable applications of the isarithm have been made in a wide variety of fields.

Those who have applied the isarithmic concept, especially since the early 1820s, have generally endowed the symbolism with excessive jargon by naming the lines according to the phenomena being mapped. Thus we have such terms as "isanther" (simultaneous blooming time), "isostalact" (equal density of planktonic organisms), "isophygm" (equal earthquake frequency) and so on, almost without restraint; the list is well over a hundred. Max Eckert called the practice an "isodisease."

C GULLEY, J L M, and SINNHUBER, K A (1961): "Isokartographie: Eine Termino-logische Studie," *Kartographische Nachrichten, 11,* 89-99.

HORN, Werner (1959): "Die Geschichte der Isarithmenkarten," *Petermanns Geographische Mitteilungen, 103,* 225-32.

ROBINSON, Arthur H. (1971): "The Genealogy of the Isopleth," *Cartographic Journal, 8,* 49-53. (Also in *Surveying and Mapping, 32* (1972), 331-38. Translated into Dutch, "De Afstammung van de Isopleth," *Kaartbulletin, 34* (1973), 5-14.)

5.0910A Contour

A (MDTT 432.5) A line joining points of equal vertical distance above or below a datum.

B The earliest use of a line of equal vertical distance was to show the depth below water level of the bed of a river, the first now known being by a Dutch surveyor,

Pieter Bruinsz., in 1584. The employment of the contour as an isobath (5.0910d) grew rapidly after the late 17th century.

The application of the principle of the contour to distance above a datum seems to have derived from its use as an isobath. The earliest development took place in France where, by the mid-18th century, it was the practice of the military engineers to determine and mark on maps the distances of points on the land below the highest elevation in an area. In 1777 J B Meusnier proposed that all such points with the same numbers be joined by curves.

Meanwhile, in 1771, Marcellin Du Carla presented to the Académie Royale des Sciences his idea for using contours, which was subsequently published in 1782 (see Hypsometric map, 3.081). Du Carla's proposal was accompanied by hypothetical maps, but Jean Louis Dupain-Triel (who published Du Carla's proposal) made an actual contour map of France in 1791. In the first years of the 19th century, the contour was clearly established as a cartographic device, but in spite of its acceptance by military engineers, the contour was not widely accepted for general use until the second quarter of the 19th century.

C BERTHAUT, Henry Marie Auguste (1898-99): *La Carte de France 1750-1898: étude historique,* 2 vols. Paris, Service géographique de l'armée.

CLOSE, Sir Charles (1926): *The Early Years of the Ordnance Survey.* Chatham, Kent, The Institution of Royal Engineers, 141-44. (Reprinted, Newton Abbot, Devon, David & Charles, 1969.)

DAINVILLE, François de (1958): "De la profondeur à l'altitude," in *Le Navire et l'économie maritime du moyen âge au XVIIIᵉ siècle principalement en Méditerranée,* Travaux du Deuxième Colloque international d'histoire maritime, Paris, 195-213. (Reprinted in *International Yearbook of Cartography, 2* (1962), 151-62. Translated into English in *Surveying and Mapping, 30* (1970), 389-403.)

DU CARLA, Marcellin (1782): *Expression des nivellemens, ou méthode nouvelle pour marquer rigoureusement sur les cartes terrestres & marines les hauteurs & les configurations du terrain.* Paris, J L Dupain-Triel.

STEINHAUSER, Anton (1858): "Beiträge zur Geschichte der Entstehung und Ausbildung der Niveaukarten, sowohl See- als Landkarten," *Mitteilungen der k.–k. geographische Gesellschaft* (Wien), *2,* Heft 2, 58-74.

5.0910B Illuminated Contour

A A method for creating an impression of relief by applying white and dark shading to contours as though the contours were the edges of successive, darkened horizontal slabs lit by an oblique light.

B A shading technique of differentially thickening contour lines on the "shadow" side was originally crudely applied to form lines on sketch maps and, more elaborately, to contours (Michaelis, 1843). Illuminated contours require a neutral ground tint such as is found in a specimen for Ireland (Rosenberg, 1846). Illuminated contours were used on the one inch Ordnance Survey map of Scotland for the Edinburgh region (sheet 32, 1858), with "shaded zones of altitude every 200 feet up to 1000 above that every 250." The Ordnance Survey applied the technique especially to maps of investigations of water supply catchment areas. Attempts to popularize its educational value in "oroscopic" maps were largely discredited, but there was a revival of interest among some British troops during World War I in an attempt to improve relief depiction on maps of northern France by manuscript annotation. Illuminated contours are sometimes now used for decorative effect.

C GREAT BRITAIN, ORDNANCE SURVEY (1857): *Report on the Ordnance Survey of the United Kingdom for 1855-56.* (Signed by Lt-Col Henry James.) Parliamentary Papers, House of Commons, *Accounts & Papers,* 1857, Sess 2, vol 27, no 147.

GREAT BRITAIN, ORDNANCE SURVEY (1862): Catalogue of Maps.

GREAT BRITAIN, ORDNANCE SURVEY (1867): One inch map of the Lake District with shaded contours, in 9 sheets.

INSTITUT CARTOGRAPHIQUE MILITAIRE (1904): "Croquis hypsometrique de la Belgique," 1:1,000,000.

MICHAELIS, Ernst Heinrich (1843): "Skizze von der Verbreitung des Cretinismus im Canton Aargau." (Map.)

MORROW, Frederick (1913): *Contours and Maps Explained and Illustrated.* London, Meiklejon & Son. Reviewed in *Geographical Journal, 43,* 1914, 333.

ROSENBERG, C (1846): "Specimen of Contoured Ground in the County of Kilkenny." Published by J Weale.

5.0910C Isobar

A An isarithm which links points of equal values of barometric (atmospheric) pressure. Observations of surface barometric pressure are usually reduced (changed) to equivalent sea-level values to eliminate the complicating effects of differences in the elevations of the observation points.

B Observations of barometric pressure were regularly made after the invention of the barometer by Evangelista Torricelli in the mid-17th century, but variations from place to place were not mapped until 1820. In that year H W Brandes of Breslau made a synoptic map of conditions on 6 March 1783 when an unusually strong storm had occurred in western and central Europe. Brandes plotted the deviations from normal pressure and joined equal deviations with lines (Hellmann). In 1827 L F Kämtz used isobarometric lines, i.e., lines of equal barometric variation, and in 1832 he included a map of isobarometric lines for the earth in the second volume of his three volume *Lehrbuch der Meteorologie.*

Isobars as lines of equal average pressure, conceived as being parallel with latitude, were first used by Heinrich Berghaus in his *Physikalischer Atlas* on a map dated 1839. Elias Loomis in the United States mapped irregular lines of barometric deviation in 1846, but it remained for E Renou to publish in 1864 the first map of irregular isobars, a map of mean barometric pressure in France. The fact that isobars are not regular lines as had been thought was confirmed by Alexander Buchan who published isobaric maps of the earth in 1869.

C BERGHAUS, Heinrich (1845; 1848): *Physikalischer Atlas oder Sammlung von Karten . . .* Gotha, Justus Perthes.

BUCHAN, Alexander (1868): *Handy Book of Meteorology,* 2d ed. Edinburgh.

BUCHAN, Alexander (1866-1869): "The Mean Pressure of the Atmosphere over the Globe for the Months and for the Year. Part I.—January, July, and the Year," *Proceedings of the Royal Society of Edinburgh, 6,* 303-7 (paper presented 16 March 1868).

BUCHAN, Alexander (1869): "The Mean Pressure of the Atmosphere and the Prevailing Winds over the Globe, for the Months and for the Year. Part II," *Transactions of the Royal Society of Edinburgh, 25,* 575-637.

ECKERT, Max (1921; 1925): *Die Kartenwissenschaft,* 2 vols. Berlin and Leipzig, Walter de Gruyter.

HELLMANN, Gustav (1897): "Meteorologische Karten," in his *Neudrucke von Schriften und Karten über Meteorologie und Erdmagnetismus,* No. 8. Berlin, A. Asher.

HORN, Werner (1959): "Die Geschichte der Isarithmenkarten," *Petermanns Geographische Mitteilungen, 103,* 225-32.

KÄMTZ, Ludwig Friedrich (1831-36): *Lehrbuch der Meteorologie,* 3 vols. Halle.

LOOMIS, E (1846): "On Two Storms Which Were Experienced throughout the United States, in the Month of February, 1842," *Transactions of the American Philosophical Society, 9,* 161-84.

RENOU, E (1864): "Hauteurs moyennes du baromètre en France," *Annuaire de la Société météorologie de France,* 240-44.

5.0910D Isobath

A (MDTT 432.15a) A contour (5.0910a) depicting the location of the same vertical distances beneath a given surface datum, such as mean sea level.

B The first cartographer known to employ the device was the Dutch surveyor, Pieter Bruinsz., who drew a single isobath on his manuscript chart of the river Spaarne, 1584. The next extant example is a map of the river Maas (Meuse) and environs (1697) by the Dutch cartographer Pierre Ancelin (fl. 1695-1711). The earliest known printed map to show an isobath is the "Carte du Golfe du Lion," published by the Italian Luigi Ferdinando Marsigli in his *Histoire physique de la mer* (1725).

The Dutch engineer Nicolaas Samuel Cruquius was the first cartographer to apply the technique of submarine contouring systematically to a restricted area. He had proposed a "waterstaatkaart" of Holland in 1725, and in 1726 was commissioned to make a hydrographic survey of the river Merwede. His chart (Caarte . . . van de Rivier Merwede), published at Leiden in 1730 in 3 sheets, contours the bed of the river in the vicinity of Gorichem using a great number of soundings.

In the 18th century the use of multiple isobaths on charts of the open sea was also developed, notably by Philippe Buache in his chart of the English Channel, "Carte . . . du Canal de la Manche," presented to the French Academy of Sciences in 1737 and published in 1756. A claimant for the earliest attempted use of isobaths on a sea chart may be the English naval officer, Nathaniel Blackmore, who drew in 1715 a manuscript chart of the coasts of Nova Scotia which shows what appear to be depth lines (Public Record Office, Kew, C.O. 700, Nova Scotia no. 4).

C ANDREAE, S J Fockema, and HOFF, B van 't (1947): *Geschiedenis der Kartografie van Nederland.* The Hague, Martinus Nijhoff, 75, pl. 15.

BUACHE, Philippe [1756]: "Carte physique et profil du Canal de la Manche et d'une partie de la Mer du Nord . . . ," *Mémoires de l'Academie Royale des Sciences, Année 1752* (Paris, 1756), Pl. 14, facing p. 416.

DAINVILLE, François de (1958): "De la profondeur à l'altitude," in *Le Navire et l'économie maritime au moyen âge au XVIIIe siècle principalement en Méditerranée,* Travaux du Deuxième Colloque international d'histoire maritime. Paris, 195-213. (Reprinted in *International Yearbook of Cartography, 2* (1962), 151-62. Translated into English in *Surveying and Mapping, 30* (1970), 389-403.)

KOEMAN, C (1983): *Geschiedenis van de Kartografie van Nederland.* Alphen aan den Rijn, Canaletto.

RIEL, H F van (1924): "Pierre Ancelin," *Tijdschrift voor Kadaster en Landmeetkunde, 40,* 51-63, 133-44.

ROBINSON, Arthur H (1971): "The Genealogy of the Isopleth," *Cartographic Journal, 8,* 49-53. (Also in *Surveying and Mapping, 32* (1972), 331-38. Translated into Dutch, "De Afstammung van de Isopleth," *Kaartbulletin, 34* (1973), 5-14.)

ROBINSON, Arthur H (1976): "Nathaniel Blackmore's Plaine Chart of Nova Scotia: Isobaths in the Open Sea?" *Imago Mundi, 28,* 137-41.

5.0910E Isogone

A A line (isarithm) that connects all points which have the same angular amount and direction of variation between magnetic north and true north (Compass declination map, 3.031; Magnetic variation, 4.133).

B Several instances of lines of equal compass declination being drawn on manuscript maps are reported in the literature from the 1530s onwards. The first clear use of isogones on a published map is by Edmond Halley in 1701 on a map of the Atlantic Ocean, "A New and Correct Chart shewing the Variations of the Compass in the Western and Southern Oceans" (Fig.16). Halley referred to the lines of equal compass declination simply as "curve lines," that being the standard term to describe a line along which some mathematical relationship remains constant. Later the lines were called Halleyan lines after their originator. According to Hellmann (1895, p. 20), Christopher Hansteen probably introduced the name "isogonic lines" in 1825 or 1826.

C HELLMANN, Gustav (1895): *Neudrucke von Schriften und Karten über Meteorologie und Erdmagnetismus,* No. 4. Berlin, A. Asher.

HELLMANN, Gustav (1909): "Magnetische Kartographie in historisch-kritischer Darstellung," *Veröffentlichungen des Königlich Preussisches Meteorologisches Institut,* No. 215, Bd. III (3), 5-61.

HORN, Werner (1959): "Die Geschichte der Isarithmenkarten," *Petermanns Geographische Mitteilungen, 103,* 225-32.

5.0910F Isopleth

A (MDTT 433.2) An isarithm representing ratios involving area which cannot exist at any particular point; e.g., number of persons per square kilometre.

B The French engineer and mathematician Léon Lalanne suggested the isopleth in 1845 as a modification of Alexander von Humboldt's isotherm line (5.0910g), invented in 1817. Lalanne had used isarithms for the graphical representation of temperature tables in the appendix to L F Kaemtz's *Cours complet de météorologie* (Paris, 1843). Combining the principle of the "curve line" isogone (5.0910e) with that of contours (5.0910a), as developed by Philippe Buache and Marcellin Du Carla, Lalanne recommended using isopleths for maps of population. The technique was first used on a map by an officer of the Danish Hydrographic Department, Nils Frederik Ravn, who probably learned of Lalanne's work from C Andrae, a Danish mathematician and engineer who had studied in France. Ravn's "Populations Kaart over Det Danske Monarki 1845, 1855" were published in *Einleitung zu dem Statistischen Tabellenwerk, neue Reihenfolge, 12* (Köbenhavn, 1857).

Lalanne's proposals for an isopleth map of population density were not followed up in France until 1874, when Louis L Vauthier used isopleths to show the distribution of population in Paris.

C FUNKHOUSER, H Gray (1938): "Historical Development of the Graphical Representation of Statistical Data," *Osiris, 3,* 269-404.

LALANNE, Léon (1843): "Sur la représentation graphique des tableaux météorologiques et des lois naturelles en général," Appendix to C Martins, ed., *Cours complet de météorologie de L F Kaemtz.* Paris.

RAVN, Nils Frederik (1857): "Populations Kaart over Det Danske Monarki 1845,

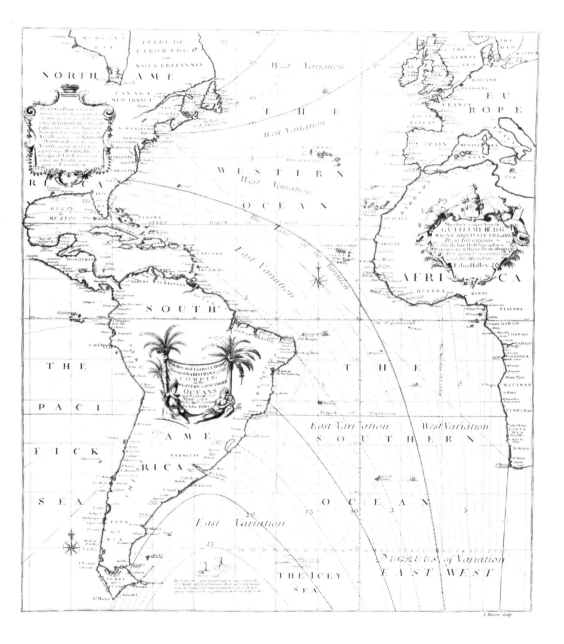

Fig. 16 – Edmond Halley: A New and Correct Chart shewing the variations of the compass in the Western & Southern Oceans, London, 1701. In a cartouche (top left) Halley describes the purpose of the "Curve Lines." By permission of the British Library. (Isogone, 5.0910e; see also Compass declination map, 3.031; Magnetic meridian, 4.131; Magnetic variation, 4.133)

1855," in *Einleitung zu dem Statistischen Tabellenwerk, neue Reihenfolge,* 12 Bd, Köbenhavn, between pp. XVI and XVIII.

ROBINSON, Arthur H (1971): "The Genealogy of the Isopleth," *Cartographic Journal, 8,* 49-53. (Also in *Surveying and Mapping, 32* (1972), 331-38. Translated into Dutch, "De Afstammung van de Isopleth," *Kaartbulletin, 34* (1973), 5-14.)

VAUTHIER, Louis L (1874): "Note sur une carte statistique figurant la répartition de la population de Paris," *Comptes Rendus de l'Académie des Sciences, 78,* 264-67.

5.0910G Isotherm

A An isarithm (5.0910) which links points of equal values of temperature. The term is usually applied, but not restricted, to atmospheric temperatures.

B Isothermal-like climatic lines appeared on maps before true isotherms were devised. Examples are the lines on E A W von Zimmermann's zoological maps (3.262), published in 1777 and 1783, which show the range of animals, and Carl Ritter's "Map of Cultivated Plants of Europe, Geographically and Climatically Represented," in his atlas published in 1806.

True isotherms were first devised by Alexander von Humboldt and appeared on a diagram map published in 1817. He had earlier developed the concept of an isothermal parallel as a "curve [isarithm] drawn through the points on a globe which receive an equal quantity of heat" (Humboldt, 1816). In his paper on isothermal lines he points out the similarity between the concepts of isotherms and the isogonic lines (5.0910e) employed by Edmond Halley. The simple diagram map of seven isotherms for the zone of the earth north of the Equator and extending from North America to eastern Asia has no coast lines and shows the names of only a few places. Curiously, the map was not published with Humboldt's essay on isothermal lines, but appeared with an abstract of that paper published in another journal ([Humboldt], 1817b).

Because of Humboldt's great renown and the significance and applicability of the concept, the scientific community was quick to take up the idea. Humboldt's essay was translated and published without a map in Edinburgh (*Edinburgh Philosophical Journal, 3* (1820), 1-20, 256-74; *4* (1821), 23-37, 262-81). Isotherms appeared on a map in a school atlas by Woodbridge in Connecticut in 1826. Humboldt's map was reproduced in Germany in 1827, and the technique was developed and applied by L F Kämtz (*Lehrbuch der Meteorologie*) in 1832 for the preparation of maps of annual isotherms. It became widely known soon after through the appearance in 1838, separately, of what was to be the first map in Heinrich Berghaus's *Physikalischer Atlas* entitled "Alexander von Humboldt's System der Isotherm-Kurven in Merkator's Projection." By the 1840s the concept of the isotherm was well known.

C ENGELMANN, Gerhard (1964): "Der Physikalische Atlas des Heinrich Berghaus und Alexander Keith Johnstons Physical Atlas," *Petermanns Geographische Mitteilungen, 108,* 133-49.

HELLMANN, Gustav (1897): "Meteorlogische Karten," in his *Neudrucke von Schriften und Karten über Meteorologie und Erdmagnetismus*, No. 8. Berlin, A Asher.

HORN, Werner (1959): "Die Geschichte der Isarithmenkarten," *Petermanns Geographische Mitteilungen, 103*, 225-32.

HUMBOLDT, Alexandre de (1816): "Sur les lois que l'on observe dans le distribution des formes vegetales (1)," *Annales de Chimie et de Physique, 1*, 225-39. Paris, Crochard.

HUMBOLDT, Alexandre de (1817a): "Des lignes isothermes et de la distribution de la Chaleur sur le globe," *Mémoires de Physique et de Chimie, de la Société d'Arcueil* (Paris), *3*, 462-602.

[HUMBOLDT, Alexandre de] (1817b): "Sur les lignes isothermes. Par A. de Humboldt (Extrait)," *Annales de Chimie et de Physique* (Paris), *5*, 102-111. Map folded between pp. 112-13.

KÖRBER, Hans-Günther (1959): "Bemerkungen über die Erstveröffentlichung der schematischen Jahresisothermenkarte Alexander von Humboldt," *Forschungen und Fortschritte, 33*, 355-58.

MEiNARDUS, Wilhelm (1899): "Die Entwicklung des Karten der Jahres-Isothermen von Alexander von Humboldt bis auf Heinrich Wilhelm Dove," in *Wissenschaftliche Beiträge zum Gedächtnis der hundertjährigen Wiederkehr des Antritts von Alexander von Humboldts Reise nach Amerika am 5. Juni 1799* (Humboldt-Centenar-Schrift). Berlin.

RITTER, Carl (1806): *Sechs Karten von Europa mit erklärendem Texte . . .* Schnepfenthal.

ROBINSON, Arthur H, and WALLIS, Helen (1967): Humboldt's Map of Isothermal Lines: A Milestone in Thematic Cartography," *Cartographic Journal, 4*, 119-23.

WOODBRIDGE, W C (1826): *School Atlas to Accompany Woodbridge's Rudiments of Geography/Atlas on a New Plan . . .* Hartford, Connecticut, O D Cooke.

5.121 Layer Tints

A Uniform tones or colours placed between selected isarithms to enhance the graphic quality of an isarithmic map.

B Both Carl Ritter and Johann August Zeune are often credited with originating layer tints in the early 19th century; although neither made contour maps, they did grade the representation of landforms from darker for lower elevations to lighter for the higher. The first employment of layer tints appears to be by Jean Louis Dupain-Triel in 1798-99 (year VII) who added at least four grey tones or shadings between the contours of a map of France. The first use of coloured layer tints was by Göran Wahlenberg, a Swedish botanist, who used them to differentiate altitudinal-vegetational zones on a map of the Tatra Mountains made in 1815 (Szaflarski). Apparently Franz Ritter von Hauslab was the first (1830) to use layer tints to enhance depths on a chart (Steinhauser), and layer tinting between contours on wall maps was employed by Emil von Sydow in 1837.

The first layer tinting of other than contour-isobath maps seems to have been by Berghaus on his 1841 map of precipitation in Europe for his *Physikalischer Atlas.* The first layer tinted isopleth maps seem to be a pair of population maps of Denmark by Nils Frederik Ravn which were published in 1857.

Layer tinting in lieu of plastic shading to portray terrain was included in the 19th-century controversy over whether a representation should be "the higher the darker" or "the higher the lighter." There was a great variety of tints and some went to extremes. For example, the first layer tint map in the British Isles, produced in 1845 for a report on the "Occupation of Land in Ireland" by Thomas A Larcom, had five zones, the lowest being yellow and the highest, black. A clear innovation was put forward by Karl Peucker of Austria in 1898 who proposed arranging the colour tints in a kind of spectral progression from a blue-green at the lowest elevations through yellows and oranges to red at the highest. His proposal was based on the idea of "advancing" and "retreating" hues, that is, the shorter wave lengths (blues) were supposed to appear farther away and the longer (reds) would appear closer, as would be appropriate in the orthogonal view of a map. Although opposed by many, especially Swiss cartographers, and without much other than theoretical justification, the general spectral scheme for layer tinting smaller scale maps became an accepted convention.

C BERGHAUS, Heinrich (1845, 1848): *Physikalischer Atlas oder Sammlung von Karten . . .* Gotha, Justus Perthes.

DUPAIN-TRIEL, J L (1798-99): "Carte de la France où l'on a essayé de donner la configuration de son territoire par une nouvelle méthode de nivellements." Paris.

ECKERT, Max (1921): *Die Kartenwissenschaft.* Berlin and Leipzig, Walter de Gruyter, Vol I, 486-94, 625-34.

EHRENSVÄRD, Ulla, supplemented by PEARSON, Karen (in press): "Color in Cartography: An Historical Survey," in David Woodward, ed., *Art in Cartography.* Chicago, University of Chicago Press.

LARCOM, Thomas A (1845): "Map of Ireland to Accompany the Report of the Land Tenure Commissioners," in "Index to Minutes of Evidence . . . in respect to the Occupation of Land in Ireland," Part 5, of "Occupation of Land (Ireland)," *Reports from Commissioners 1845, No. 9* (Vol. 22 of U.K. *Parliamentary Papers,* Session 1845).

PEUCKER, Karl (1898): *Schattenplastik und Farbenplastik. Beiträge zur Geschichte und Theorie der Geländedarstellung.* Vienna, Artaria.

RAVN, Nils Frederik (1857): "Populations Kaart over Det Danske Monarki 1845, 1855," in *Einleitung zu dem Statistischen Tabellenwerk,* New Series, *12,* Copenhagen, Statistical Bureau.

ROBINSON, Arthur H (1982): *Early Thematic Mapping in the History of Cartography.* Chicago, University of Chicago Press, 210-15.

STEINHAUSER, Anton (1858): "Beiträge zur Geschichte der Entstehung und Ausbildung der Niveaukarten, sowohl See- als Landkarten," *Mitteilungen der k.-k. geographische Gesellschaft* (Wien), *2*, Heft 2, 58-74.

SZAFLARSKI, Józef (1959): "A Map of the Tatra Mountains Drawn by George [sic] Wahlenberg in 1813 as a Prototype of the Contour-Line Map, *Geografiska Annaler, 41,* 74-82.

5.122 Letter Symbol

A (MDTT 433.9) A letter used as a point symbol. Letters may also be used as an indexing device in map reference systems, which from the later years of the 16th century were commonly alpha-numeric (see Map index, 6.0920b).

B Maximus Planudes (*c.* 1260-1310), who revived Ptolemy's Geography of *c.* AD 150 and made the maps for the codex Vaticanus Urbinas graecus 82, late 13th century, used a system of conventional signs for the native tribes of Western Europe. These signs, which attribute *poleis* ("cities", i.e. places) to tribes, are derived from astronomical texts and include letter symbols (Dilke, p. 159). For example, the signs for British tribes include the letter λ, the twelfth letter of the Greek alphabet, for the Silures. These signs are marked both against the regional tribal name and in the rectangular castellated sign bearing the name of the settlement. The same system is followed on the map of the British Isles in the Latin codex Vatican Latin 5698, dating from the first half of the 15th century. The editions of the first printed Ptolemys delineate the regional tribal boundaries instead of using symbols.

Letter symbols are one of the innovations appearing in the map of the Holy Land prepared by Pietro Vesconte for the later version of the book of the Venetian traveller Marino Sanudo, "Liber secretorum fidelium crucis," written *c.* 1327 (British Library, Add. MS. 27376, ff 188-89) (Fig.17). Upper-case letters distinguish the regions of Palestine, and a legend, "sub signo A B C D E," sets out descriptions of each part of the country. When the map reappeared as one of the five modern maps in the codex of Ptolemy's *Geography,* from Francesco Berlinghieri, 1482 (Vatican Library, Cod. Urb. Lat. 273; Almagià, pp. 23, 103-4), it was presumably based on the earlier version and the letter symbols and explanatory text did not feature; they are similarly absent in Berlinghieri's printed edition, Florence, 1482, and in another version of the map in the Ulm Ptolemy of 1482.

In the early 16th century letter symbols came into general use in Europe to supplement the vocabulary of conventional signs. The French cosmographer Oronce Fine used initials for settlements (e.g. M for Paris, L for Lyons) on his diagrammatic map of France which illustrates how to make a map of a region and was published in *Cosmographia sive de Mundi Sphaera,* Lutetiae, 1530 (Dainville, 1970, pp. 49-50). The Swedish mapmaker Olaus Magnus on his pictorial map of northern regions, "Carta Marina . . . ," published in Venice in 1539, deploys a double system of lettering, with large capitals for regional divisions and small capitals for specific features. Pieter van der Beke (Peter de la Beke) of Ghent on his map of Flanders, 1538, classifies ecclesiastical establishments by upper case initials, A for abbeys, P for priories for men (Paternostres), F for priories for women (Femmes). This device was followed on the later derivative maps of Flanders by Gerard Mercator, 1540, and by Domenicus Zenoi, Venice, 1559.

Another early example of the use of letter symbols is found in the plan of Dover harbour drawn in 1581 by Thomas Digges (British Library, Add. MS. 11815a). The locations of fourteen points above sea level are indicated by letters which relate to a key giving heights in fathoms (see Spot height, 5.193).

Matthias Quad identified by capital letters the geographical features on his imaginary map "Vocabulorum geographicorum topica significatio," published in his *Geographisch Handtbuch,* Cologne, 1600 (Fig.1). In a legend below the map written in Latin and German he explained them thus: A. Continens, B. Insula, C. Peninsula, etc.

C ALMAGIÀ, Roberto (1944): *Planisferi, carte nautiche e affini dal secolo XIV al XVII esistenti nella Biblioteca Apostolica Vaticana.* Vatican City.

DAINVILLE, François de (1964): *Le langage des géographes.* Paris, Éditions A et J Picard.

DAINVILLE, François de (1970): "How Did Oronce Fine Draw His Large Map of France?" *Imago Mundi, 24,* 49-55.

DELANO SMITH, Catherine (1985): "Cartographic Signs on European Maps and Their Explanation before 1700," *Imago Mundi, 37,* 9-29.

DILKE, O A W (1985): *Greek and Roman Maps.* London, Thames and Hudson.

5.123 Line Symbol

A Any elongated continuous mark or discontinuous series of marks (as a line of dots) on a map that serves as a sign for some geographical phenomenon or concept.

B The line symbol is probably as old as cartography. The oldest maps known from all cultures show the positions of walls, canals, rivers, coastlines, etc., by means of various kinds of lines. The possibilities of variation in the design of lines is almost unlimited, and there is a general progression of sophistication in their use that parallels the development of cartography.

The earliest use of lines to show the occurrence of distinctive features — rivers, coasts, edges of mountain areas, etc. — was basically qualitative. Also relatively early were lines to represent more abstract concepts such as *klimata* (latitude), the limits of ownership or jurisdiction, or the system of radiating rhumbs on portolan charts (1.0320d), first used to represent wind directions and then the points of the compass. Adapting the design to portray quantitative information came later, as in road distances on Erhard Etzlaub's map, "Das ist der Romweg . . ." (*c.* 1500; see 1.1810d), and in the 19th century hachure (5.081) and flow line (5.061). Varying the design of lines to convey qualitative distinctions by classes of phenomena also came early in road maps and other maps portraying transportation facilities.

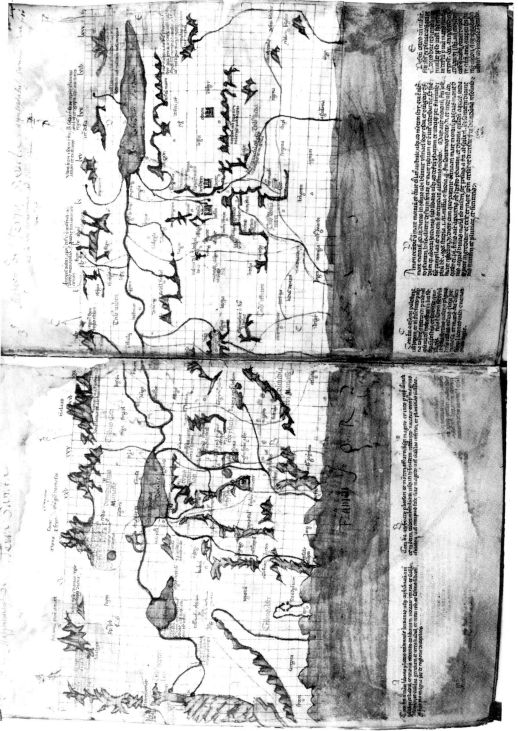

Fig. 17 – Map of the Holy Land, by Pietro Vesconte, in Marino Sanudo's "Liber secretorum fidelium crucis," c. 1327. By permission of the British Library, Add. MS. 27376, ff.188-189. (Letter symbol, 5.122; see also Non-co-ordinate reference systems, 4.141; Square grid, 4.141; Legend, 6.121)

The use of the line as a symbol for showing the distribution of equal values began with the representation of depths in rivers (see Isobath, 5.0910d) and magnitudes of compass variation (5.0910e) from which spawned a great number of such lines (see Isarithm, 5.0910) ranging from the familiar contour (5.0910a) to the isotherm (5.0910g) and isopleth (5.0910f).

C HARVEY, Paul D A (1980): *The History of Topographical Maps.* London, Thames and Hudson.

HODGKISS, A G (1981): *Understanding Maps: A Systematic History of Their Use and Development.* Folkestone, Kent, Dawson.

KEATES, J S (1973): *Cartographic Design and Production.* London, Longman.

ROBINSON, Arthur H, SALE, Randall D, MORRISON, Joel L, and MUEHRCKE, Philip C (1984): *Elements of Cartography,* 5th edn. New York, London, and elsewhere, John Wiley.

5.141 Names

A Geographical names on maps identify natural and manmade features. Names of other things, such as plants, animals or minerals, when placed on maps, show the places where the named items occur.

B The inclusion of geographical names on maps and the identification of symbols probably dates from the time when maps were first made on permanent substances such as clay tablets and cloth. Early Mesopotamian city plans on clay tablets identify with cuneiform characters canals, walls, houses, terraces, etc., and the Han maps from Changsha in the Hunan province in China contain dozens of names of rivers, cities, towns and villages. The oldest printed maps in both western and eastern cultures contain geographical names.

The use of names of things, such as animals or minerals, placed in the appropriate places on a map to show the occurrence of the named items is considerably more recent than the use of geographical names. Although some of the elaborate medieval mappae mundi (1.2320b) contain legendary non-geographical names, the systematic use of names as a form of symbol for showing the distribution of a class of items seems to have been started in the 18th century. Gottfried Hensel published in 1741 four continental maps of language areas on which samples of the written languages are placed in their proper positions. Similar kinds of maps of languages and nationalities are listed in Arnberger. The first map of animal geography by E A W von Zimmermann, dated 1777, distributes the names of large numbers of quadrapeds in the supposedly proper places, a practice followed by Carl Ritter for maps of animals and plants in his early 19th century atlas. By the mid-19th century the technique was quite common, but where precision allowed it some other sort of symbolism seems generally to have been preferred.

C ARNBERGER, Erik (1966): *Handbuch der thematischen Kartographie.* Wien, Franz Deuticke, 94-97.

CHANG, Kuei-sheng (1979): "The Han Maps: New Light on Cartography in Classical China," *Imago Mundi, 31,* 9-17.

HENSEL, Gottfried (1741): *Synopsis universae philologiae, in qua miranda unitas et harmonia linguarum totius orbis terrarum . . . eruitur.* Nuremberg.

RITTER, Carl (1806): *Sechs Karten von Europa mit erklärendem Texte . . .* Schnepfenthal.

ROBINSON, Arthur H (1982): *Early Thematic Mapping in the History of Cartography.* Chicago, University of Chicago Press.

UNGER, Eckhard (1935): "Ancient Babylonian Maps and Plans," *Antiquity, 9,* 311-22.

ZIMMERMANN, Eberhardt August Wilhelm von (1777): *Specimen Zoologiae Geographicae, Quadrupedum Domicilia et Migrationes Sistens.* Lugduni Batavorum (Leiden).

5.1610 Point Symbol

A Any non-linear sign which by its position on a map shows the location of some geographical phenomenon. Point symbols may also portray quantitative information about a place.

B The use of distinctive point symbols is limited only by the media being used and the ingenuity and knowledge of the mapmaker. Among the earlier uses of point symbols were pictorial representations of cities by drawings of buildings or banners and the portrayal of real and fanciful animals and human-like forms. The development of more detailed general topographical maps (1.203) in the 16th and following centuries led to a great increase in point symbols, some maps having many small pictorial and geometric drawings to show the locations of such varied phenomena as mills, quarries, towns with a church, towns the seat of nobility, etc. The mineral maps (3.131) and product maps (2.163) which came later carried on the tradition of using large numbers of point symbols. Mineral symbols were frequently based on planetary and alchemical signs.

The use of point symbols for the quantitative portrayal of ordinal differences (larger, smaller, etc.) is quite old, usually by means of stylized pictorial symbols being drawn larger or smaller. Point symbols for more precise values came later with the use of proportional circles (5.1610a), divided circles (5.1610b), and rectangles (5.1610c).

5.1610A Circle

A (MDTT 431.3(4)) A symbol in the form of a circle.

B The earliest mention of a circle as a conventional sign is that of Roger Bacon (*c.* 1214-1294), in his "Opus Majus," 1266-67. Referring to the world map (now lost) which he made for presentation to the Pope, he writes: "Secundum igitur praedicta praesentem affero descriptionem in albiori parte pellis, ubi civitates notantur per circulos rubros" (I am offering the present description on the whiter part of the parchment, where the cities are marked by red circles) (Bridges, ed., I, 300, 320).

In his use of abstract signs Bacon may well have been influenced by Arab (and through Arab Greek) geography, which he had extensively studied. The schematic maps in the various 10th century versions of the "Atlas of Islam" and the world map by the Turk Mahmud al-Kashgarī, 1076, illustrate the use of circles for settlements, while the Norman-Arabic world map of al-Idrisi, 1154, has segmented circles. The maps in the 15th century Latin (as distinct from Greek) manuscripts of Ptolemy's Geography also use circles, as in the work of Donnus Nicolaus Germanus (for example, Cod. Vaticanus Urbinas Latinus 277, dated 1468).

With the introduction of map printing the circle proved to be one of the most useful conventional signs. The line engraver working in reverse on copper and the block maker of woodcut maps with his preference for simple designs could easily transfer the circle from the manuscript original. Four manuscript maps of English counties (Chester, c. 1602, Hertford, 1602, Warwick, 1603, Worcester, 1602) by William Smith, evidently prepared as fair drafts for the engraver of the "anonymous series" published in 1602 to 1603, illustrate the procedure. The circles for settlements are punched through the paper as a guide for the reversed image, which is sketched in pencil on the dorse (British Library, Maps C.2.cc.2 (12.-15.)).

The circle thus appears as the symbol for settlement on the first printed regional maps. The Bologna Ptolemy, 1477, the first printed Ptolemy with maps, has circles with a central dot or prick. Arnold Buckinck, printer of the Rome Ptolemy, 1478, copper-engraved like the Bologna, uses a simple circle of uniform size, as does the Ulm Ptolemy, 1482, which was based on the work of Donnus Nicolaus Germanus and was the first with woodcut maps. Erhard Etzlaub, adopting a strictly nonpictorial style, uses circles for towns on his woodcut map of the environs of Nuremberg, 1492, and for his road map, "Das ist der Romweg," c. 1500, writes on the explanatory sheet that "all notable cities have been drawn as large points, while the smaller ones as small round points" (translation Krüger, p. 22). Most towns are shown by simple circles as opposed to vignettes for large or capital cities.

The circle was the most convenient device for road maps, on which distances had to be precisely measured. On the early 16th century road map of the region round Nuremberg, copied from one of Etzlaub's, circles for towns have an additional feature, a pointer to indicate the town's name nearby (Fig.18). This was later used by Gerard Mercator in his world atlas to avoid ambiguity, as he explained in his users' guide in the preliminaries to the first volume, *Galliae tabulae geographicae,* Duisburg, 1585, sig.+v^r.

In the early years of the 16th century the circle remained the common symbol for settlements. After about 1520, however, it became general practice to combine the circle with the town and city ideogram or pictograph. The earliest printed example of the double symbol seems to be the map of central Europe by Nicolas of Cusa published at Eystat with the date 1491 (perhaps a misreading of a manuscript date). Nicolas places the circle next to the pictograph. When the circle symbol reappeared

Fig. 18 — Road map of the environs of Nuremberg after Erhard Etzlaub, woodcut, Nuremberg, c. 1545(?) Distances along roads are indicated by dots, and towns by circles. North is at the top. A small map of the environs of Bamberg is pasted on (bottom left). By permission of the British Library. (Circle, 5.1610a; see also Road map, 1.1810d; Conventional sign, 5.033)

237

later, *c.* 1540, it was normal to mark the circle at the church door as a point from which distances could be measured.

Various systems were devised to adapt the circle to different types of settlements. On a manuscript map of Allgäu, 1534 (University Library, Basle), Achilles P Gasser used different sizes of circle to show the relative importance of towns, with a pointer to their names (Siegrist). Philipp Apian on his woodcut map *Bairische Landtaflen,* Ingolstadt, 1568, distinguished places by different types of circles, single, double, divided, etc., as explained in his key, and set the circles in a great variety of pictographs.

Flemish and Dutch mapmakers also developed the use of the circle. Gerard Mercator on his map of Flanders, *c.* 1540, used church symbols and city vignettes round a single or double circle. On his map of the British Isles, Duisburg, 1564, his circle symbols have single or double crosses attached to show ecclesiastical status, explained in a legend. For his world atlas he devised a series of circles for different types of settlement which he defined in the guide to conventional signs in the first volume, 1585 (see above and Conventional sign, 5.033). Jacob van Deventer, author of over 300 Dutch town plans, used to good effect the combined circle and pictograph on his maps of the Dutch provinces, published at Antwerp from 1533 to 1559, providing on his map of Gelderland, 1556, a detailed key. The maps in Abraham Ortelius's *Theatrum Orbis Terrarum,* 1570, derived from many sources, well illustrate the variety of styles then in vogue, from simple to classified circles, with the associated diversity of ideographs and pictographs.

In England John Norden, influenced probably by his friend William Smith who had been resident in Nuremberg, *c.* 1578-84, introduced various conventional signs, including circles. He used seven distinctive circles on his map of Sussex, 1595, to distinguish market towns, parishes, etc, all explained in a key. John Ogilby on his map of Kent, published at London *c.* 1670, further elaborated the circle symbol, using twelve characters explained in a key.

The device of the circle in the pictograph remained standard in European mapmaking from the 16th to the end of the 18th century. Circles in spires and towers overran the map of Europe (Lynam, p. 43). Many engravers and publishers used the signs indiscriminately as a stereotype. Only hamlets with no church were given a simple circle, which was in fact nearer to correct scale than the pictograph.

The circle regained in 1800 its earlier status as a standard symbol. As atlases of general small scale reference maps proliferated in the 19th century, the circle came increasingly to be used to show the locations of settlements, often being distinguished by size or design to show character, such as villages, towns, or capital cities.

The use of circles of varying sizes to show proportional quantities seems to have its origin in the graphic statistics made popular toward the end of the 18th century by August Friedrich Wilhelm Crome in Germany, William Playfair in Scotland, and Denis François Donnant in France. The use of proportional circles on maps came later, the first apparently being in 1838 on a population map of Ireland by Henry Drury Harness. On that map the populations of settlements are shown by circles, the areas of which are strictly proportional to the population numbers. The next uses of proportional circles were on maps designed by Samuel Clark, and drawn by August Petermann, in Clark's *Maps Illustrative of the Physical, Political and Historical Geography of the British Empire,* London, 1852, also to show populations. In 1852

Charles Joseph Minard used proportional circles to show tonnages of shipping at the various ports of France; and W Bone and Petermann used them on maps prepared to accompany the 1851 census of Great Britain, published in 1852 (Robinson, 1982, pp. 125-28). After that time the proportional circles became a common cartographic feature.

C BRIDGES, John Henry, ed. (1897): *The Opus Majus of Roger Bacon.* Oxford, Clarendon Press.

DAINVILLE, François de (1964): *Le langage des géographes.* Paris, Éditions A et J Picard, 324-29, Plates IX-XXII.

KRÜGER, H (1951): "Erhard Etzlaub's *Romweg* Map and Its Dating in the Holy Year of 1500," *Imago Mundi, 8,* 17-26.

LYNAM, Edward (1953): *The Mapmaker's Art.* London, Batchworth Press.

ROBINSON, Arthur H (1955): "The 1837 Maps of Henry Drury Harness," *Geographical Journal, 121,* 440-50.

ROBINSON, Arthur H (1967): "The Thematic Maps of Charles Joseph Minard," *Imago Mundi, 21,* 95-108.

ROBINSON, Arthur H (1982): *Early Thematic Mapping in the History of Cartography.* Chicago, University of Chicago Press, 117-30, 203-8.

SIEGRIST, Werner (1949): "A Map of Allgäu, 1534," *Imago Mundi, 6,* 27-30.

SKELTON, Raleigh A (1952): *Decorative Printed Maps.* London, Staples Press.

5.1610B Divided Circle

A A circle representing a whole quantity which has been divided into sectors by two or more radii so as to indicate the fractions provided by components.

B The divided circle was first used by William Playfair in 1801 in a small volume of graphic statistics entitled *The Statistical Breviary . . .* to show the relative areas of the Turkish Empire in Europe, Africa and Asia, but it does not seem to have appeared on a map until half a century later. Its first cartographic use was probably by Charles Joseph Minard, an innovative French thematic cartographer, who employed the technique on a map dated 1858 to show the quantities of the various kinds of butcher's meat dispatched to Paris from the individual départements. In 1859 he used the technique again to show the comparative amounts of foreign and coastwise trade handled by the ports of France.

Minard pointed out that although it was difficult to compare the areas of different sizes of circles, it was easy to judge the proportional sectors. He also specifically disclaimed being the first to use divided circles pointing out that they had been employed before he used them, but he did not say whether it was by Playfair or on a map.

C MINARD, Charles Joseph (1858): "Carte figurative et approximative des quantités de Viandes de Boucherie envoyées à Paris." Paris.

MINARD, Charles Joseph (1859): "Carte figurative et approximative de l'importance des portes maritimes de l'empire français measurée par les tonnages effectifs des navires entrées et sortis en 1857." Paris.

ROBINSON, Arthur H (1967): "The Thematic Maps of Charles Joseph Minard," *Imago Mundi, 21,* 95-108.

ROBINSON, Arthur H (1982): *Early Thematic Mapping in the History of Cartography.* Chicago, University of Chicago Press, 144-45, 206-8.

5.1610C Quadrangle

A A sign used on a map to indicate the location of something. Quadrangles of varying sizes have also, but infrequently, been employed to symbolize different quantities.

B Quadrangles as point symbols on maps have been used from the earliest times to depict single buildings and aggregates of buildings in settlements, in effect showing the ground plan of built structures. Rectangles filled with regularly spaced points appear in the Bedolina petroglyph at Val Camonica, dating from the second and first millennia BC, and seem to have been drawn as representational signs (Delano Smith).

The use of quadrangles as military symbols or representations can be traced back to the Han period in China. The garrison map of southern Changsha, second century BC, marks by quadrangles the areal extent of each military base (Chang). By the end of the 17th century quadrangles of various sorts had become recognized military symbols in European cartography. They developed from the pictorial representations of groups of tents (camps) and the massing of men and/or cannon for the attack.

A mosaic of quadrangles to give a pictorial impression of the field pattern, without necessarily indicating true boundaries, became a familiar feature of topographical maps from the 16th century onwards and, later, of land-use maps.

With the advent of mineralogical maps in the 18th century quadrangles and rectangles found favour as signs for various rock types and minerals. Philippe Buache used at least six different rectangular symbols on his 1746 map which accompanied an article by Jean-Étienne Guettard (see Geological map, 3.071). In 1849 August Heinrich Petermann (see Population map, 2.161) employed a solid black square to indicate a category of population on his map of the 1841 census of Great Britain.

The first use of the rectangle as a proportional map symbol may be on an 1843 map of the incidence of cretinism and the proportion of deaf mutes in the population in Canton Aargau by E H Michaelis. The symbolism is tortuous but includes red rectangles the lengths of the bases of which represent the relative frequency of deaf mutes in the populations of adjacent villages.

C CHANG, Kuei-sheng (1979): "The Han Maps," *Imago Mundi, 31,* 9-17.

DAINVILLE, François de (1964): *Le langage des géographes.* Paris. Éditions A. & J. Picard, 324-29, Plates IX-XXI.

DELANO SMITH, Catherine (1982): "The Emergence of 'Maps' in European Rock Art," *Imago Mundi, 34,* 14-16.

KISH, George (1976): "Early Thematic Mapping: The Works of Philippe Buache," *Imago Mundi, 28,* 129-36.

MICHAELIS, Ernst Heinrich (1843): "Skizze von der Verbreitung des Cretinismus im Canton Aargau." (Map.)

ROBINSON, Arthur H (1982): *Early Thematic Mapping in the History of Cartography.* Chicago, University of Chicago Press, 52-54, 174-76.

5.162 Profile

A A sketch made along a predetermined line to show vertical elevations or depths.

B In the Middle Ages and early renaissance the art of navigation in northern Europe and the Mediterranean was mainly concerned with pilotage or coasting, described by Michel Coignet as "la navigation comune," as opposed to oceanic navigation, "la navigation grande." One of the aids to navigation was therefore the coastal profile, which was included in sailing directions or added to charts. In the Middle Ages in southern Europe coastal profiles were drawn on portolan charts. Although no portolan of this kind is known from the early Middle Ages, the manuscripts of *La Sfera* by Leonardo Dati (1360-1425) and his brother Gregorio include drawings and scenes derived from an earlier prototype. The *Isolarii* by Cristoforo Buondelmonte (MS., 1420) and Bartolommeo dalli Sonetti (Venice, 1485) include drawings in profile (for example, Sonetti's map of Cyprus). In the early 16th century Portuguese chartmakers began to include coastal sketches in their roteiros. The earliest appear to be in the Book of Francisco Rodrigues *c.* 1512 (Bibliothèque de la Chambre des Députés, Paris), which includes 69 folios of panoramic views in the East Indies (Cortesão and Teixeira da Mota, 1, 79-84). João de Castro, fourth Viceroy of India, developed the technique to a fine art in his three roteiros, 1538-42, of which the most famous is that of the Red Sea (British Library, Cotton MS Tiberius D. IX; Cortesão and Teixeira da Mota, 1, 127-143). Rodrigues and De Castro incorporated the panoramas into the body of the charts, creating the effect of a series of bird's eye views.

In Northern Europe the Low German "sea-book" was the equivalent of the Mediterranean or Portuguese roteiro. The Frenchman Pierre Garcie (1430-1503?) included 59 crude woodcut illustrations of coastal elevations in his *Le grant routtier* (1st ed. printed 1520) and claimed to have made his illustrations as early as 1484. The Dutch *leeskaart* from the 1540s onwards contained woodcuts of coastal elevations by the fine artist and seaman Cornelis Anthoniszoon (1499-1557) of Amsterdam. His "Caerte van die oostersche See" is notable as the first Dutch sea-book accompanied by charts as well as coastal views. Lucas Janszoon Waghenaer of Enkhuizen in his

Spieghel der Zeevaerdt (Leiden, 1584-85) appears to have been the first to place coastal elevations directly on printed maps to supplement the outline of the coasts (Koeman). The clarity of the profiles made them immediately popular, and they were imitated by others in the following years. It is possible, however, that Aelbert Haeyen whose "Amstelredamsche Zee-carten" appeared in 1585 developed the same concept independently.

Spanish chartmakers of the 17th century used the coastal profile in designing the charts in their roteiros of the west coasts of America, and copies were made in England by William Hack, 1682, and Basil Ringrose, *c.* 1682-83 from a Spanish book of sea charts and sailing directions. A typical feature of "Waggoners," as such collections of charts were called (after Waghenaer), was the coastal profile. Henceforward it was common for maritime surveys, both printed and manuscript, to include profiles. Alexander Dalrymple, hydrographer to the East India Company from 1779 until his death in 1808 and first hydrographer to the British Admiralty (1795-1808), was one of the strongest advocates for the use of the device.

C BAGROW, Leo (1964): *History of Cartography.* Revised and enlarged by R A Skelton. London, C A Watts & Co.; Cambridge, Mass., Harvard University Press.

CORTESÃO, Armando, and TEIXEIRA DA MOTA, Avelino (1960): *Portugaliae Monumenta Cartographica.* Lisbon.

KOEMAN, Cornelis (1964): *The History of Lucas Janszoon Waghenaer and his "Spieghel der Zeevaerdt."* Lausanne, Sequoia, 38-39.

VROOM, U E E, and others (1984): *Lucas Jansz. Waghenaer van Enckhuysen.* Enkhuizen, Vrienden van het Zuiderzeemuseum.

WATERS, David W (1967): *The Rutters of the Sea: The Sailing Directions of Pierre Garcie.* New Haven and London, Yale University Press.

5.191 Shading

A The application of continuous tone (smooth gradations from light to dark) to represent variations in numerousness (density) or frequency of some phenomenon, with the lighter usually indicating less and the darker, more. This is not to be confused with the use of continuous tone for hill shading (or plastic shading) to render a pictorial impression of the land form.

B The first use of shading was by Adolphe Quetelet of Belgium in 1831 for three maps of moral statistics (see 2.132). The technique was adopted by Oluf Nikolay Olsen in 1839 for a map of precipitation in Europe and northern Africa (see 3.161) and by Heinrich Berghaus as early as 1841 for numerous maps in his *Physikalischer Atlas.* In the 1840s and 1850s shading was widely used to represent variations when the character of a distribution was known but data were too sparse to allow the use of an isarithmic representation. Other notable maps employing continuous-tone shading are Berghaus's, Johnston's and Petermann's world maps of precipitation (see 3.161) and Petermann's 1849 and 1852 maps of population in the British Isles (see 2.161). The use of shading decreased after the mid-19th century as more data became available which allowed the use of more favoured (but less graphic) commensurable representations.

C QUETELET, Adolphe (1831): *Recherches sur le penchant au crime aux différens âges*. Brussels, Hayez. Also in *Nouveaux mémoires de l'Académie Royale des Sciences et Belle-Lettres de Bruxelles, 7* (1832).

ROBINSON, Arthur H (1982): *Early Thematic Mapping in the History of Cartography*. Chicago, University of Chicago Press.

5.192 Soundings

A The depth of water at a place obtained with a pole or a line weighted by lead and noted by a number on a chart at that point.

B Although depths and bottoms off the French coast are noted in a 15th century German manuscript "seebuch" (Commerz bibliothek, Hamburg), such information does not appear on charts until the 16th century. De Smet has pointed out that Pierre Pourbus "in 1551-52 ... made three maps covering the eastern part of the Franc [de Bruges], showing depths of the sea and of the Zwin" (p. 33). An English coastal chart of the north Kent coast and the river Swale from Faversham to Margate, probably from 1514, includes information about depths written in words (British Library, Cotton MS. Augustus. I.ii.75). The earliest English chart to show soundings by numbers so far determined is a manuscript chart of the river Humber, *c.* 1569 (Tyacke and Huddy). Destombes identifies the earliest Dutch charts to show soundings as *c.* 1565-68, and numerous other charts with soundings date from before 1600 (Denucé and Gernez). In Lucas Janszoon Waghenaer's *Spieghel der Zeevaerdt*, 1584, many of the charts have coastal soundings. This is also true of Aelbert Haeyen's "Amstelredamsche Zee-carten," 1585.

Dainville observes that the marine charts of the Dieppe chartmakers in the first third of the 17th century are distinguished by the minute detail of the soundings entered in red or black numbers along the coasts, around islands, and at the entrances to estuaries, sounding lines being too short for deep waters. Robert Dudley on his manuscript maps drawn in 1636 (Bayerische Staatsbibliothek, Munich) and on the derived printed charts in his *Dell' Arcano del Mare* (1646-47) shows soundings in many areas of the world, including Africa and the Americas. Eventually equal measurements of depth were joined together to form isobaths (see 5.0910d).

C DAINVILLE, François de (1958): "De la profondeur à l'altitude," in *Le Navire et l'économie maritime du Moyen Âge au XVIIIe siècle principalement en Méditeranée*, Travaux du Deuxième Colloque international d'histoire maritime, Paris, 195-213. (Reprinted in *International Yearbook of Cartography, 2* (1962), 151-62. Translated into English in *Surveying and Mapping, 30* (1970), 389-403.)

DE BOER, G, and SKELTON, R A (1969): "The Earliest English Chart with Soundings," *Imago Mundi, 23,* 9-16.

DENUCÉ, Jean, and GERNEZ, G (1936): *Le livre de mer*. Académie de Marine de Belgique, Anvers (facsimile).

DESTOMBES, M (1968): "Les plus anciens sondages portés sur les cartes nautiques aux XVIe et XVIIe siècles," *Bulletin de l'Institut Océanographique* (Monaco), num. spécial *2,* 199-222.

HOWSE, Derek, and SANDERSON, Michael (1973): *The Sea Chart.* Newton Abbot, David & Charles, pl. ix.

SMET, Antoine de (1947): "A Note on the Cartographic Work of Pierre Pourbus, Painter of Bruges," *Imago Mundi, 4,* 33-36.

TYACKE, Sarah, and HUDDY, John (1980): *Christopher Saxton and Tudor Map-Making.* London, British Library Series No. 2.

WAGHENAER, Lucas Janszoon (1584): *Spieghel der Zeevaerdt.* Leiden, Christopher Plantin.

5.193 Spot Height

A (MDTT 432.1) A number on a map which shows the position and the altitude of a point above a given datum.

B The inclusion of written information about specific heights on maps dates from the end of the 16th century. One of the earliest maps to contain reference to measurements of elevation is the plan of Dover harbour, drawn in 1581 by Thomas Digges. The heights of fourteen points above sea level are indicated in fathoms by a key to letters placed in position on the map. Until the 18th century the method of indicating heights by forms of explanatory written information seems to have been general. In 1703 Francis Nevel, a "collector of Her Majesty's revenue" in Ireland, surveyed and levelled an area between Lough Neagh and Newry in preparation for cutting a canal and indicated on the map "depths" along the route, giving the depth in yards below summit level at Poyntzpass.

True spot heights began to appear on maps in the 1740s. In England Christopher Packe published a ". . . Philosophico-Chorographical Chart of East Kent . . ." (1743), which marked some spot heights, mainly in the Stour Valley. They were measured by barometer from a datum level of the low-water mark at a point in Sandwich Bay. In Ireland the surveyor Charles Baylie and a scientific inventor John Mooney surveyed the Demesne of Carton, County Kildare, in 1744 and included some 160 spot heights, probably levelled with the theodolite, from a datum described in the map's explanation as "at the foot of the Rocks of the Riverside marked thus (0)." In France, the principle of representing spot heights on maps was first elaborated by an officer of the Corps of Engineers Louis Marie Antoine Milet de Mureau in his *Mémoire pour faciliter les moyens de projeter dans les pays de montagnes* (1749), which was illustrated by an imaginary example of such a map. The earliest *Plans cotés* (spot height plans) were of Ciutadella and Fort Saint-Philippe à Mahon, Minorca (1761), in which the highest elevation was expressed by 0, all other values of altitude being negative. In Scotland John Laurie's map of the County of Midlothian (1763) was probably the first to show spot heights above sea level. The practice of including spot heights on maps was adopted more generally in France during the

1770s and 1780s, notably by the Corps of Engineers who, at the instigation of Marshal du Muy, prepared a manuscript atlas of *plans cotés* between 1774 and 1789. At about this time spot heights were joined together to form contours (5.0910a).

C DAINVILLE, François de (1958): "De la profondeur à l'altitude," in *Le Navire et l'économie maritime du Moyen Âge au XVIIIᵉ siècle principalement en Méditerranée,"* Travaux du Deuxième Colloque international d'histoire maritime, Paris, 195-213. (Reprinted in *International Yearbook of Cartography, 2* (1962), 151-62. Translated into English in *Surveying and Mapping, 30* (1970), 389-403.)

HORNER, A A (1974): "Some Examples of the Representation of Height Data on Irish Maps before 1750, Including an Early Use of the Spot-Height Method," *Irish Geography, 7,* 68-80.

SKELTON, R A (1957): "Cartography," chap. 20, in Singer, C, *et al, A History of Technology, Vol. 4.* Oxford, 611-12.

5.231 Windrose

A A circle with directions of winds inscribed around its margins. The different directions of the winds were always identified with the names of winds which blow from those directions.

B The division of the circle was made long before the building of the Tower of the Winds at Athens (*c.* 100 BC). The earliest references give only two winds *boreas* (N) and *notos* (S), but shortly after *euros* (E) and *zephyros* (W) were added; intermediate winds were already known in Homeric time, namely the *boreas-euros* (NE), *notus-apeliotes* (SE), *argestes-notos* (SW), and *zephyros-boreas* (NW). These names were maintained during the classical period, but they did not always correspond with the same directions. Two new names appeared: *caecias* (NE), and *libs* (SW). In fact, every wind corresponded to a horizontal angle, and the corresponding name of the wind was given to the direction of its bisection.

A later division was established but it was not uniform. At first twelve winds *(vegetios)* were listed, and afterwards Vitruvius (1st century BC) in his treatise "On Architecture" (Bk 1,vi.10) employed twenty-four: *septentrio* (N); *gallicus, supernas, aquilo* (NE); *caecias, carbas, solanus* (E); *ornithiae, euricircias, eurus* (SE); *volturnus, leuconotus, auster* (S); *altanus, libonotus, africus* (SW); *sudvesperus, argestes, favonius* (W); *etesiae, circias, caurus* (NW); *corus* and *thracias.* A division into sixteen winds was also established. In Chinese maps, the division into twelve winds was maintained.

In the Middle Ages the names of the eight main wind directions were similar in all the North Mediterranean countries: *tramontana* (N), *greco* or *gregal* (NE), *levante* (E), *siroco* or *xaloc* (SE), *ostro* or *austro* (S), *libeccio, lleveig, africo* or *garbino* (SW), *ponente* (W), and *maestro* (NW). The portolans reproduced this windrose, giving the initials of the name of each wind: T, G, L, S, O, L, P, and M. Also the direction north was indicated by a spearhead as well as by the T for *tramontana.* This produced a figure which evolved into a fleur-de-lis. A cross was employed to represent the east, instead of the L of *levante*. The initial M of *maestro* (NW) became successively a sign similar to a divided "omega," an inverted E, and an R. The complete windrose with thirty-two directions was introduced by the navigators of Amalfi in the 14th century.

The first surviving portolan chart, the "Carte pisane," *c.* 1300 (Bibliothèque Nationale, Paris, Rés [Ge.B.1118]), has the directions of the windrose around its edge, as do those of Pietro Vesconte in in his atlas of 1313 (Bibliothèque Nationale, Paris, Rés Ge. DD.687), and in Marino Sanudo's "Liber secretorum fidelium crucis," *c.* 1320 and *c.* 1327 (examples in the Bibliothèque Nationale and the British Library) (Fig. 2). As a separate design, the windrose appears first in the Catalan atlas (1375) by Abraham Cresques (Bibliothèque Nationale, Paris). The nomenclature is Catalan (*tramuntana,* etc.)

From about 1300 the wind (or compass) rose was a prominent and decorative feature of nautical charts, forming the nodal points for the network of radiating lines which provided the frame of the chart. The nomenclatures used for the winds according to the various schools of cartography help with the attribution of anonymous charts, with the proviso that foreign mapmakers usually followed the convention of their land of adoption. The Portuguese introduced a new type of rose with its chief feature the fleur-de-lys and a lozenge-shaped central petal, first used on the charts of Jorge de Aguiar, 1492, and Pedro Reinel, *c.* 1504, and this was destined to become international (Winter).

In the 19th century Matthew Fontaine Maury of the U.S. Navy developed as a meteorological device a new style of wind rose. As officer in charge of the Depot of Charts and Instruments of the Navy Department, Washington, later the National Observatory and Hydrographic Office, he published from 1847 onwards his wind and current charts based on the records in log-books "stored as rubbish" in his office. The winds are denoted by the feather end of an arrow, the butt end pointing in the direction whence the wind blows, the character of the feathering showing the character of the wind, and colours indicating the season. On his pilot charts (1849 to 1851), he recorded by means of windroses "how the winds blow in each month and in every 5° square of the Ocean." The windroses record the total winds (or calms) for each month in the appropriate section.

C ALMAGIÀ, Roberto (1960): *Storia della geografiá.* Torino.

BAGROW, Leo (1964): *History of Cartography.* Revised and enlarged by R A Skelton. London, C A Watts and Co.; Cambridge, Mass., Harvard University Press.

BROWN, Lloyd A (1949): *The Story of Maps.* Boston, Little, Brown and Company.

MAURY, Matthew Fontaine (1848-80): *Wind & Current Chart Series A-B, C-F.* Washington, D.C., United States Hydrographical Office.

NAVARRETE, Martin Fernandez de (1846): *Disertación sobre la historia de la náutica.* Madrid, La Real Academia de la Historia.

TAYLOR, Eva G R (1956): *The Haven-Finding Art: A History of Navigation from Odysseus to Captain Cook.* London, Hollis and Carter. (Republished 1971, New York, American Elsevier Publishing Co.)

WINTER, Heinrich (1947): "On the Real and the Pseudo–Pilestrina Maps and Other Early Portuguese Maps in Munich," *Imago Mundi, 4,* 25-27.

GROUP 6

Techniques and Media

6.031 Cartouche

A (MDTT 21.23.) An enclosed area within a map, often in elaborate baroque or roccoco style, which variously contains the title, legend, dedication, author, figures, etc. From Italian *cartoccio*, an oval to enclose the arms of the Pope or members of royal families.

B Scroll-like figures have been used from Egyptian to modern times to enclose inscriptions on tablets and pillars, but the earliest extant map to bear an embellishment which might be considered a cartouche is a manuscript map of the world (1448) drawn by Giovanni Leardo of Venice (reproduced in the Viscount de Santarem's *Atlas* [1842-53], pl. 49). On that map the title is contained within strapwork. The device is also used on a manuscript map of Italy (1449), preserved in the Museo Civico Correr de Venezia, on which the title and certain geographical names are written within scroll-like entablatures. Cartouches also appear on some of the earliest copper-engraved maps, e.g., those included in the first edition of Ptolemy's *Cosmographia* printed with maps, published at Bologna (1477).

By the middle of the 16th century the title and other legends were sometimes enclosed in a rectangular cartouche, such as the one on the map of the British Isles, "Britanniae insulae . . . descriptio" (Rome 1546), by GLA, the Englishman George Lily. As the 16th century proceeded, cartouches became more ornate with heavy strap-work or roll-work simulating shields with intricate rolled edges. Such designs were available in pattern books and were widely copied by architects and other designers as well as by mapmakers. Many of the Dutch maps of the 17th century have elaborate cartouches incorporating perspective views, hunting and fishing scenes, flora and fauna, human figures, cherubs, battle scenes, and so on. The decorative cartouche persisted into the era of the lithographic map, as illustrated by maps of the settlements in New South Wales, published by the Quarter Master Generals Office at the Horse Guards, Whitehall, 1817 (Fig. 22).

247

C BROWN, Lloyd A (1949): *The Story of Maps.* Boston, Little, Brown and Company, 174-75.

DAINVILLE, François de (1964): *Le langage des géographes.* Paris, Éditions A et J Picard, 64-66.

WELU, James A (in press): "The Sources and Development of Cartographic Ornamentation in the Netherlands," in Woodward, David, ed., *Art and Cartography.* Chicago, University of Chicago Press.

6.032 Characteristic Sheet

A (MDTT 23.4a). A separate sheet bearing a key to the symbols and lettering styles and sizes, which serves as the legend for a map or, more usually, a map series.

B The characteristic sheet developed as a refinement of the practice of printing on a map a legend or explanation of symbols (see Legend, 6.121). An early forerunner was devised by the Nuremberg cartographer Erhard Etzlaub for his Romweg map, *c.* 1500. To supplement the explanatory text at the base of the map he provided an accompanying slip which he called a "register," comprising a verbal description of the symbols displayed (Krüger, 20, 22). As the use and explanation of conventional signs became increasingly more general and elaborate, the printing of a separate sheet devoted to the explanatory legend was introduced as a further improvement, and it became an essential feature of topographical map series from the later years of the 18th century. The earliest characteristic sheet to a topographical map was the *Explication* to "La carte chorographique des Pays-Bas Autrichiens . . ." published at Brussels in 25 sheets in 1777, and based on surveys carried out under Joseph Comte de Ferraris. The *Explication* distinguishes four types of river crossings and includes a *nota* that cultivated ground and gardens are left blank (les terres labourées et les jardins sont en blanc). For the first time the method of representing high and low tide marks is explained. The *Explication* is not complete, however, as it does not include the conventional signs for woods, marshes, sands and swamps.

At about the same time, Joseph F W Des Barres in his maritime atlas *The Atlantic Neptune,* published in London from 1777 for the use of the Royal Navy of Great Britain under the direction of the Lords Commissioners for the Admiralty, provided a sheet entitled *References,* which explained the elaborate symbols used for the physiography of shore and off-shore features.

In response to the need of military commanders for precise knowledge of ground cover the army major Johann Georg Lehmann of Saxony laid down at the end of the 18th century precise methods for the topographical delineation of features, and set out the full range of symbols in the form of a characteristic sheet ("Topogràphische Bezeichnung der einzelnen Gegenstände des Landbodens," 1819, published at Dresden and Leipzig in 1843). His instructions were adopted by the central European military schools and were translated into English by William Siborne (1822): *Instructions for the Civil and Military Surveyors founded upon the System of J G Lehmann,* London (Campbell, 1952, p. 430; ill. 429).

C BRUWIER, Marinette (1978): "Le comte de Ferraris et son oeuvre," in *La cartographie au xviiiᵉ siècle et l'oeuvre du comte de Ferraris* (1726-1814). [Brussels], Crédit Communal de Belgique, 19-26.

CAMPBELL, Eila M J (1952): "The Development of the Characteristic Sheet, 1533-1822," in *Proceedings of the Eighth General Assembly and Seventeenth International Congress*. International Geographical Union, Washington, DC, 426-30.

DELANO SMITH, Catherine (1984): "Cartographic Explanations before the Key," *The Map Collector, 26,* 38-41.

KRÜGER, H (1951): "Erhard Etzlaub's *Romweg* Map and Its Dating in the Holy Year of 1500," *Imago Mundi, 8,* 17-26.

LEHMANN, Johann Georg (1843): *Plane zu J G Lehmann's Lehre der Situation-Zeichnung*. Dresden and Leipzig.

6.081 Hand Colouring

A A method of applying colour to a map by hand with a brush, either freehand, or with a stencil. Since manuscript maps, if coloured, are by definition coloured by hand, the term usually refers to printed maps, thus distinguishing hand colouring from printed colour, which was not commonly used for maps before the 19th century. Also known in the past as tinting, limning, or washing.

B Hand colouring was used by European mapmakers throughout the period of the engraved and woodcut map, both as an adornment and as a supplement to enhance the printed features (see Colour, 5.032). The map engravers of the 15th and early 16th centuries followed the traditions of manuscript cartography in their use of colour. Those employing the woodcut technique with its limited scope for detailed expression saw the value of colour to enrich the map. Thus the Nuremberg cartographer Erhard Etzlaub has blue rivers and red territorial boundaries on his map of the environs of Nuremberg, 1492, and, as he explained in the legend (see 6.121), he marked towns as red dots. His Romweg map, *c.* 1500, also calls for colour, as stated in his *Register* (Characteristic sheet, 6.032). He proposed to show by colour the countries and language areas adjoining Germany, but not all the extant maps are coloured (see Linguistic map, 2.122) (Krüger, pp. 19-20, fig. 3, p. 22).

The Italian copperplate engravers of the 16th century with their skilful linework were content to leave their maps uncoloured. When the leading centres of publishing moved to the Netherlands later in the century, map colouring became fashionable and reached a high degree of artistry. Abraham Ortelius issued his atlases at Antwerp in both coloured and uncoloured versions. The maps were coloured either by illuminators in the map shops, who included Ortelius's sister, or by other craftsmen such as the specialist Bernard van der Putte. With their elaborate strapwork cartouches picked out in a variety of tones and their fine decorative borders, Dutch maps took on the appearance of brilliantly coloured paintings. John Smith (1701) commented that "the only way to colour maps well is by a pattern done by some good Workman, of which the Dutch are esteemed the best." Hand colouring was usually done with a water based paint of either thick or thin consistency. Gilding with liquid gold was a frequent feature of presentation copies. One of the famous practitioners of the 17th century was the master-illuminator Dirch Jansz van Santen (1637/8-1708), who worked for the Amsterdam collector Laurens van der Hem. Van Santen, using precious paint and gold dust, coloured and heightened with gold the engraved maps of Blaeu's Atlas which formed the nucleus of the celebrated "Van der Hem Atlas (National Library, Vienna) (Schilder). Van Santen's works include, in addition to the maps of Blaeu, those of Hondius, Visscher, De Wit, and Jansson.

In the eighteenth century the development of a more detailed geographical vocabulary reduced the effectiveness of colour, which became more restrained on topographical maps and was limited to distinguishing and clarifying boundaries and areas. As thematic mapping began to develop at the end of the 18th century, however, hand colouring came to be employed not as decoration but as an intrinsic element of the map. Thomas Milne's land utilization map of the London region, London, 1800, distinguished by key letters and hand colouring seventeen different types of land use, but it survives in only one fully coloured example (British Library, Map Library, K. Top. VI.95). The first geological map of England, by William Smith, London, 1815, employs a colour scale "assimilated to the colour of each stratum, except chalk, which being colourless, seemed best represented by green." As colour printing developed in the mid-nineteenth century, the hand coloured map became a rarity.

In Japan, where maps were displayed as decorations on screens, hand colouring was an important feature up to 1765, when superseded by colour printing. The earliest printed Japanese world map, which was drawn by a painter of Nagasaki and published at Nagasaki in 1645, is a woodcut map displaying in colour countries and their peoples. Such world maps were inspired by Dutch cartography. Ishikawa Ryūsen, widely known as an *ukiyo-e* painter under the name of Ryūshū, made beautiful woodcut and hand-coloured maps and plans such as the map of Japan, "Honchō Zukan Kōmoku," Edo, 1687, and the plan of Edo, "Edo no Ezu," Edo, 1689.

C DAINVILLE, François de (1964): *Le langage des géographes*. Paris, Éditions A et J Picard, 329-37.

KRÜGER, H (1951): "Erhard Etzlaub's Romweg Map and Its Dating in the Holy Year of 1500," *Imago Mundi, 8*, 17-26.

SCHILDER, Günter (1984): "The Van der Hem Atlas, a Monument of Dutch Culture in Vienna," *The Map Collector, 25*, 22-26.

SKELTON, R A (1952): *Decorative Printed Maps of the 15th to 18th Centuries*. London.

SKELTON, R A (1960): "Colour in Mapmaking," *Geographical Magazine, 32*, 544-53.

SMITH, John (1701): *The Art of Painting in Oyl . . . to which is Added the Whole Art and Mystery of Colouring Maps and Other Prints with Water Colours*. 3rd ed. London.

WALLIS, Helen, and others (1974): *Chinese and Japanese Maps*. London, British Library Board.

6.091 Imprint

A (MDTT 21.52) A brief note in the margin of a map which gives some or all of the following: date of publication, date of printing, name of publisher, name of printer, place of publication, number of copies printed and related information.

B Prior to the invention of printing in Europe the makers of portolan charts often included signatures on their work, together with place and date of manufacture. The earliest signed and dated chart is Pietro Vesconte's of 1311 (Florence, Archivio di Stato). The first to include place of manufacture (Genoa) is Vesconte's atlas of 1313 (Bibliothèque Nationale, Paris) (Fig.2). Some of the earliest cartographic incunabula display imprints. Ptolemy's *Cosmographia,* the first with maps, published at Bologna in 1477, bore an imprint in the form of a colophon, though with missing digits in the date. The woodcut world map in Ptolemy's *Cosmographia* published at Ulm in 1482, gives the blockcutter's imprint, "Insculptum est per Johannes Schnitzer de Armssheim" (cut by Johannes, blockcutter of Armsheim). The undated woodcut world map by Hanns Rust, which is assigned to *c.* 1480 and is certainly pre 1485, has a simple author statement: "Hanns Rust." An apparent imprint (its significance now disputed), is found on Nicolas of Cusa's map of central and northern Europe "Eystat anno salutis 1491. XII Kalendis augusti perfectum" (completed at Eichstatt 21 July 1491). The year, however, may be an engraver's misreading of a manuscript date. The earliest unequivocal imprint on a published map is on Erhard Etzlaub's unsigned and untitled woodcut circular map of the environs of Nuremberg. The map is lettered with the printer's name "Jorg Glogkendon" and dated in the legend, 1492. Glogkendon provided a full imprint on Etzlaub's road map of the Holy Roman Empire (Germany), 1501: "Getruckt von Georg Glogkendon zu Nurnberg 1501" (Schnelbôgl, pp. 18-20, fig. 1).

In the middle of the sixteenth century when the Italians dominated the European map market, publishers and printsellers such as F Bertelli, G F Camocio and Paolo Forlani in Venice, and Antonio Salamanca, Antonio Lafreri and Claudio Duchetti in Rome normally included imprints on their maps, many of which were copies of other men's work. The Latin terms in common use were *excudit (excud., exc.), formis, sumptibus, apud, ex officina.* With the move of the centre of cartographic enterprise to the Netherlands, the Dutch publishing houses with their highly organized industry used the accepted practices of the book trade. Statements of the grant of a privilege from government or sovereign are found on maps from the 16th century onwards, providing a protection of copyright, for example, "Cum privilegio decem annorum" (with a privilege of ten years).

In England the widespread piracy arising from the unacknowledged use of other men's work, with a mere change of imprint, provoked the protests of the artist William Hogarth, who was largely responsible for the Act of Parliament passed in 1734 which gave protection to the owner of copyright in an engraved map, chart or print. Thereafter imprints on maps included the phrase "published according to Act of Parliament," "published as the Act directs," or a similar statement. The Act also required the "day of first publishing" to be stated. Imprints, or at least the statements of date, were frequently left off eighteenth and nineteenth century maps, however, because this enabled publishers to continue reissuing them without the maps becoming visibly out of date. There was also less need for an imprint on maps which were published in atlas form, as were so many in this period. It was common for map series of the eighteenth and nineteenth centuries not to give details of date and publication on the individual sheets, although they might be published and reissued over a period of years.

In China where the invention of printing dates from the 8th century AD and maps were circulating in multiple copies from the 12th century, imprints have been a long established and prominent feature of the map. The earliest extant world map published by a Chinese cartographer after the arrival of the Jesuit missionary Father Matteo Ricci in 1583 is that by Liang Chou of 1593-94, and bears an imprint which translates as follows: Carefully engraved by Liang Chou . . . during the reign Wan-li in the Kuei-ssu year (1593-94) in the autumn. Printed by the Chang-ssu Tang in the 4th ward of the Li District of Nanking. Japanese mapmakers also took care to record details of publication. The Shōhō world map, the earliest published in Japan, 1645, bears the imprint "published at Nagasaki Harbour in the hinoto tori (cock) year of the Shoho era."

C SCHNELBÔGL, Fritz (1966): "Life and Work of the Nuremberg Cartographer Erhard Etzlaub (†1532)," *Imago Mundi, 20*, 11-26.

SKELTON, Raleigh Ashlin (1952): *Decorative Printed Maps of the 15th to 18th Centuries.* London.

WALLIS, Helen, and others (1974): *Chinese & Japanese Maps* (British Library Exhibition Catalogue). London.

6.0920 Index

A An alphabetical list of names recurring in a book or on a map, with indication of the places in which they occur. To index can mean to furnish the parts of a diagram with different symbols to facilitate identification. These two different meanings of index are reflected in two main types of guide to the use of maps, the map index (6.0920b), which is a list of names appearing on the map, and the index map (6.0920a), which is a diagram to the layout of sheets of a multi-sheet map. See also Grid, 4.073; Non-co-ordinate reference systems, 4.141; Square grid, 4.191.

B When maps came into general use in the later years of the 15th century the need for a guide to the location of places became evident. The geographical co-ordinates of places given by Ptolemy provided one way to find their location on the map, if their country was known (since arrangement was regional). Peter Apian (Bennewitz or Bienwitz), born at Leisnig in Saxony, doctor and professor of Ingolstadt University, seems to have been the first (in 1524) to illustrate the procedure in a diagram (Fig. 13). The German grammarian Conrad Wolfhart, known also under the Grecized form of the name, Lycosthenes, provided a map index to the Basel edition of Ptolemy (known as Münster's fourth edition), 1552. In 1560 Lycosthenes further developed the method with his index to Aegidius Tschudi's map of Switzerland.

Philipp Apian, son of Peter, born in 1531 in Ingoldstadt, devised what now ranks as the earliest index map as a key to the 24 sheets of his *Bairische Landtaflen*, 1568.

6.0920A Index Map

A (MDTT 22.21) A small-scale map, usually fitted to an outline base which shows the layout and numbering system of the individual maps of a set or series. For topographical maps, two main types of index maps may be recognized, skeleton and full.

B The skeleton index has usually been employed for wall maps consisting of several sheets of which the inner adjoining ones of the intended assembly have no printed border, hence the index map may be only an outline of sheet shapes with the addition of numbers and/or names. Latitude and longitude may be indicated as an aid to more precise geographical location. This type of index map has not normally been used for a map series, the sheets of which are issued individually, perhaps over a considerable period of time. Each sheet in such a series is surrounded by an outer border which contains essential information.

The full index map may be not only an outline of sheets showing their inter-relationship together with their numbers and/or names but also may include sufficient cartographical detail for it to be considered and used as a map in its own right. The earliest example printed is believed to be "Brevis toti[us] Bavariae Descriptio," compiled by Philipp Apian and engraved by 'HF' (Ingolstadt, 1568) to serve as a key to the 24 sheets of his *Bairische Landtaflen*. In the early 18th century the large scale multi-sheet maps of the Dutch polders required index maps. One of the earliest appears to be the "Caartboeck van Voorne," 1701, by Heyman van Dyck, drawn at a scale of about 1:5000 on 33 sheets. A "Generale Caarte" depicts the whole area, delimiting the polders, and this is followed by 32 sheets of separate polder maps, diversely oriented to fit the format of the page (British Library, Beudeker Collection, Maps C.9.d.3.(85-118).). The map of Delflant by Nicolaas and Jacob Cruquius (Krukius), published at Delft in 1712 and drawn at a scale of 1:10,000 on 25 sheets, has an index map marked with sheet lines and is notable for displaying a miniaturized version of the whole, complete with iconographic detail (British Library, Beudeker Collection, Maps C.9.d.3.(46-71).).

The earliest example of an index map for an official national map series may be "Nouvelle Carte qui comprend les principaux triangles qui servent de fondement à la description géométrique de la France . . ." by Jean Dominique Maraldi and César-François Cassini de Thury, and engraved by Guillaume Dheulland ([Paris], 1744), which served as an index map to the 182 sheets; the intended order of publication was also indicated by a number.

For hydrographical charts, a different kind of index is required in respect to river surveys. The index may be in the form of a simplified route map with toponyms, the sheet lines being indicated by short dividing lines within which are the individual sheet numbers. An example is to be found in the *Atlas chasti reki Volgi . . .* compiled by Captain Pliskov in 1857 and engraved by Kolodeyev (St. Petersburg, 1860).

253

6.0920B Map Index

A A list of place-names on a map or in an atlas, normally in alphabetical order, with an indication of their locations on the map.

B The method of indexing maps by a system of co-ordinates defining the positions of places was not devised until some time after printed maps came into general use in the 15th century. The co-ordinates given in the lists of places in Ptolemy's *Geography* served as an index to the maps only if the map user knew the region in which the place lay. To meet this difficulty Johann Reger, editor of the Ulm Ptolemy of 1486, compiled a geographical index to the text and maps under the title "Registrum alphabeticum super octo libros Ptolemei," with a preface, "Nota ad inveniendum igitur regiones," explaining its purpose and use. He also obtained, or compiled, an anonymous tract entitled "De locis ac mirabilibus mundi." Both these were added in manuscript to the Ptolemy codex at Castle Wolfegg, *c.* 1474; they were printed in 1486 and inserted in some unsold copies of the 1482 Ulm *Cosmographia.* The Ulm edition of 1486 included the "Registrum" (Skeleton, 1963, p.x), which followed the order of the first letter of the names (but with an haphazard arrangement within each letter), giving references to the book and chapter of the text, and to the number of the map in which the places appeared. The "Registrum" and tract "De locis . . ." were then incorporated in the Rome Ptolemy of 1490, printed by Petrus de Turre (Pietro de la Torre), who had acquired the plates from Arnold Buckinck, printer of the edition of 1478. The lists were reprinted in the Rome editions of 1507 and 1508. In 1513 the printer of the Strassburg Ptolemy, Johann Schott, included in his text 15 pages of index, derived from the anonymous tract "De locis . . . ," as printed in the Rome editions of 1507 and 1508. The alphabetical order follows the second as well as the first letter of the names. Peter Apian illustrated how to use Ptolemy's co-ordinates by a diagram in his *Cosmographicus Liber Petri Apiani Mathematici studiose collectus,* Landshut, 1524, fol. 60 (Fig.13).

The earliest extant work published with a true map index is the Basel edition of Ptolemy's Geography (known as Münster's fourth), 1552. The author of the index, as stated on the title page, was the German grammarian Conrad Wolfhart (or Lycosthenes). In the introduction he explains the use of the index, which is arranged in two parts, the first with Ptolemaic names, the second with those on modern maps. For the ancient maps Münster provided geographical co-ordinates in the margins. The modern maps had either no notations or only notations of latitude. Lycosthenes therefore gave an arbitrary graduation "extra omnem poli altitudinem," and explained how places could be found with its help, together with the use of a graduated ruler. The graduations on the modern maps are 24 across, 18 deep, printed from separately engraved slips. The index itself is in strict alphabetical order.

Lycosthenes used the same method for the revised edition of Aegidius Tschudi's large map of Switzerland, "Nova Rhætiæ atque totius Helvetiæ descriptio," 1560. The index to the 1538 edition of the map (Münster's issue, now lost) gave references to the text only. By means of the system of rectangular co-ordinates on the map Lycosthenes defined the location of places listed in the accompanying text *De prisca ac vera Alpina Rhætia . . . ,* Basel, 1560, sig. t1ʳ–x3ᵛ. The marginal graduations comprised 80 divisions (called "gradus") each way, and Lycosthenes explained the system in a long note on the map.

Lycosthenes also refers to other maps with indexes, namely the Ptolemys, and then Oronce Fine's map of the Holy Land (not now known), Heinrich Zell's map of Germany (in its original issue, also unknown); the edition of 1560 is the first known today, which raises the question of priority between the two cartographers in the matter of indexing. The third map mentioned is Caspar Vogel's world map of which the first edition (unknown today) appeared in 1545, with later editions 1549 and 1552. The new version was published at Antwerp in 12 sheets in 1570. The later issues, however, do not display marginal graduations, which may mean that Vogel's indexes gave geographical, not conventional, co-ordinates.

In English cartography John Norden introduced the grid as a map-reference system for use either with a gazetteer of place-names or other descriptive text. This innovation, like others, shows his knowledge of continental map techniques. In his *Speculum Britanniae, the First Parte*, London, 1593, his map of Middlesex displays a grid with numbers along the top and bottom and letters down the sides. Lists of streets and buildings appear in the bottom margin of the map of London. The map of Hertfordshire in *Speculum Britanniae Pars. The Description of Hartfordshire*, London, 1598, has two-mile divisions, lettered and numbered, as on Philipp Apian's map of Bavaria, 1568. The maps of Norden's manuscript "Topographicall and historical description of Cornwall," 1607, also bear a reference grid in two-mile units. The maps are now in the possession of Trinity College, Cambridge, and kept in Cambridge University Library; the accompanying text is in the British Library, Harl.MS.6252.

C HEAWOOD, Edward (1932): "Early Map Indexing," *Geographical Journal, 80,* 247-9.

RAVENHILL, William (1972): *John Norden's Manuscript Maps of Cornwall and Its Nine Hundreds.* Exeter, University of Exeter, 31.

SKELTON, R A (1963): *Claudius Ptolemaeus Cosmographia Ulm 1482.* Amsterdam, Meridian Publishing Co.

SKELTON, R A (1966): *Claudius Ptolemaeus Geographia Strassburg 1513.* Amsterdam, Theatrum Orbis Terrarum Ltd.

6.121 Legend

A (MDTT 21.33) An explanation or amplification of the symbols and conventions used on a map.

B Precursors to the legend or key supplied on European printed maps to explain their use date at least from the later Middle Ages, with one example from ancient Egypt, *c.* 1200 BC. This is a map of the mines in Nubia drawn for a lawsuit (Museo Egizio, Turin). One of its inscriptions reads, "The hills from which the gold is brought are drawn red on the plan," and others refer to hills of silver and gold and to roads (Delano Smith, 1984, 1985). The world map made in the early 14th century by Giovanni da Carignano of Genoa (destroyed in the second world war) has a legend explaining that colours distinguish Christian and Turkish-held towns. Pietro

Vesconte has a long legend on his maps of the Holy Land, *c.* 1320 (Fig.17). Andreas Walsperger, a Benedictine monk of Salzburg, described in a legend on his world map of 1448 the signs and colours used; for example: "the red spots are cities of the Christians." A similar type of written explanation appears as a short legend on Lazarus Secretarius's printed map of Hungary, 1528, explaining the use of colour and the symbol delineating the limits of Turkish occupation. Also of note is the manuscript map of Allgäu, 1534, drawn by the doctor of medicine Achilles Pirum Gasser, and intended as one of his contributions to Sebastian Münster's *Cosmographia*, although never printed (University Library, Basle). In a legend on the back of the map Gasser explains the conventional signs used and the style and colour of lettering for names (Siegrist, 1949).

The earliest explanation of how to use a printed map was given by Erhard Etzlaub on his map of the environs of Nuremberg, 1492, where he describes how to measure distances between towns, which (he notes) are marked as red dots. In the "register" to his Romweg map of *c.* 1500, which ranks as the earliest separate sheet of explanatory text (see Characteristic Sheet, 6.032), he provides a short essay on the meaning of his colouring and symbols. Georg Erlinger followed Etzlaub's example, producing in 1515 an explanation (*Erklärung*) for his revised version of Etzlaub's road map of the Holy Roman Empire (Delano Smith, 1985). A more elaborate explanatory text in the form of a book or pamphlet is found in Martin Waldseemüller's *Instructio Manuductionem Prestans in Cartam Itinerarium* (1511), comprising in its first chapter a guide for the use of his map "Carta Itineraria Europae" of that year. A forerunner of this type of textual explanation had already appeared in the *Isolarii* of the 15th century, where the text contains explanations of features shown on the maps, as illustrated by Bartolommeo dalli Sonetti in the first printed *Isolario, c.*1485 (Delano Smith, 1984).

The use of long explanatory legends sometimes obscures the fact that a key to conventional signs was contained within the text, as on Oronce Fine's "Nova Totius Galliae Descriptio," 1533. A much clearer statement was given by George Lily on his map of Great Britain, Rome, 1546, with its simple key. Gerard Mercator's map of Europe (Duisburg, 1554) and Jacob van Deventer's "Kaart van Gelderland" (1556), in which the symbols are given in lines of text, are other examples of 16th century maps with legends. Only five of the fifty-three maps in Abraham Ortelius's world atlas, *Theatrum Orbis Terrarum* (1570), have a key. Some of the maps also display legends giving other information which the reader could not otherwise obtain, such as special categories of religious establishments.

The first map to employ a specific heading for the explanatory legend was Philipp Apian's *Bairische Landtaflen* (Ingolstadt, 1568), where a table of fourteen signs was entitled 'Erklärung nach volgender Zaichen' (sheets 13 and 17). In England the herald and mapmaker William Smith (*c.* 1550-1618), who had worked in Nuremberg *c.* 1578-84, seems to have introduced the use of a table of signs into English mapmaking (e.g. Smith's MS map of Cheshire, 1585, British Library, Harleian MS.1046, Fol. 132). This practice was evidently copied by his friend John Norden who on the map of Middlesex published in his *Speculum Britanniae, the First Parte,* 1593, gave a title to the table of fourteen signs: "Caracters distinguishing the difference of places." The series of English county maps printed in 1602-03, now identified as by Smith, also gives lists of symbols, some in Latin as well as English.

During the 17th century the practice of including an explanation of signs became more general and the tables were more elaborate, reflecting a fuller map content. The detailed topographical map of Bohemia, "Regni Bohemiae nova et exacta descriptio," 1619, by Paul Aretin, a citizen of Prague, as revised by Daniel Vusin (1665), is notable for its use of symbols for precious metals and mineral resources, *fodine auri, argenti, stanni, ferri, thermae, officina vitriaria,* which are set out in the key (Kuchař, 1937). The map of Bohemia, "Mappa geographica totius regni Bohemiae," Augsburg, 1777, by Johann Christoph Müller, marked a further stage in the development of the legend. In the "Characterum explicatio. Erklärung der Zeichen" (on sheet XXI) he explains some 48 symbols, with emphasis on industrial and mineral works, buildings and settlements. Three types of rural settlement are distinguished, the *Strassendorf* (street village) *(Pagi longo tractu excurrentes, Lange Dörfer),* the isolated farm *(Villae solitariae, Einschichtige Mayerhöfe),* and scattered farms *(villae rusticae passim dispersae, hin und wieder zerstrauete Bauernhöfe)* (Campbell, 1952; Kuchař, 1961).

The special interest of eastern European mapmakers in mineral and other economic resources is reflected also in the innovations which the Polish Jesuit and sinologue Michael Boym, assisted by the Chinese scholar Andreas Chên, introduced on his manuscript map of the Middle Kingdom of China, "Chung kuo t'u," *c.* 1652 (private collection, London). Minerals marked on the map are listed in a legend addressed "Ad lectorem," which is notable as the earliest displaying both Chinese characters and roman script, and gives the planetary signs for the minerals (Wallis and others, C9; Wallis).

The fullest development of the legend in the 18th century was the achievement of the Comte de Ferraris on his "Carte Chorographique des Pays Bas Autrichiens," Brussels, 1777. This was the first map to give the legend on a separate sheet (see Characteristic Sheet, 6.032). A still more detailed system of symbols was recommended by Johann Georg Lehmann for military mapping and printed as a characteristic sheet in 1819, with an English version by William Siborne published in 1822 (6.032).

C BRUWIER, Marinette (1978): "Le comte de Ferraris et son oeuvre," in *La cartographie au xviiiᵉ siècle et l'oeuvre du comte de Ferraris (1726-1814).* [Brussels,] Crédit Communal de Belgique, pp.19-26.

CAMPBELL, Eila M J (1952): "The Development of the Characteristic Sheet, 1532-1822," in *Proceedings of the Eighth General Assembly and Seventeenth International Congress.* International Geographical Union, Washington, DC, 426-30.

CAMPBELL, Eila M J (1962): "The Beginnings of the Characteristic Sheet to English Maps," Part II of Crone, G R, Campbell, E M J, and Skelton, R A, "Landmarks in British Geography," *Geographical Journal, 128,* 411-15.

DAINVILLE, François de (1964): *Le langage des géographes.* Paris, Éditions A et J Picard.

DELANO SMITH, Catherine (1984): "Cartographic Explanations before the Key," *The Map Collector, 26,* 38-41.

DELANO SMITH, C (1985): "Cartographic Signs on European Maps and Their Explanation before 1700," *Imago Mundi, 37,* 9-29.

KRÜGER, Herbert (1951): "Erhard Etzlaub's *Romweg* Map and Its Dating in the Holy Year of 1500" *Imago Mundi, 8,* 17-26 (Fig. 1).

KRÜGER, Herbert (1958): "Des Nürnberge Meisters Erhard Etzlaub älteste Strassenkarten von Deutschland," *Jahrbuch für Fränkische Landesforschung, 18.*

KUCHAŘ, Karol (1937): "A Map of Bohemia of the Time of the Thirty Years' War," *Imago Mundi, 2,* 75-77.

KUCHAŘ, Karol, ed. (1961): *Early Maps of Bohemia, Moravia and Silesia.* Prague, pp. 28-30.

LEMOINE-ISABEAU, Claire (1978): "L'élaboration de la carte de Ferraris," in *La cartographie au xviii* siècle et l'oeuvre du comte de Ferraris (1726-1814).* [Brussels,] Crédit Communal de Belgique, pp.39-52.

SIEGRIST, Werner (1949): "A Map of Allgäu, 1534," *Imago Mundi, 6,* 27-30.

STEGENA, Lajos (1982): *Lazarus Secretarius: The First Hungarian Mapmaker and His Work.* Budapest, Akadémiai Kiadó.

WALLIS, Helen (1975): "Missionary Cartographers to China," *Geographical Magazine, 47,* 751-5.

WALLIS, Helen, and others (1974): *Chinese and Japanese Maps* (an exhibition in the British Library). London, item C9.

6.1220 Lettering

A The process of including in a map words identifying places and things, including auxiliary elements such as titles, explanations, etc.

B It may be assumed that words have appeared on maps since the development of written languages and that maps were drawn on such surfaces as clay tablets, papyrus, cloth and skin, on which characters could be impressed or drawn with quill or brush.

From the latter part of the 15th century the printing processes have dictated how the lettering would be done. The woodcut process (7.1810c) was not sympathetic to lettering because it was difficult to carve and leave standing names small enough to fit in the spaces available. To make matters easier stereotyped (6.1220c) and type-inserted (6.1220d) lettering processes were developed. Lettering was easier in the copper engraving process (7.0920a) but here too stamped lettering (6.1220b) was used very early. Probably most maps made by the lithographic process in the first half of the 19th century were lithographic engravings (7.1210c), the lettering of which was similar to that of copper engravings. After the 1850s photography (7.161) and lithographic transfer techniques (7.1210e) made the reproduction of pen and brush lettering possible. Late in the century pre-printed lettering (6.1220a) was developed.

6.1220A Pre-printed Lettering

A A method of map lettering in which names are pre-printed on paper or stripping film and attached to the manuscript map with wax or some other adhesive. Otherwise known as stick-up lettering.

B While several other methods of combining type with linework in printed maps were used from the 15th century on (for example, type-inserted lettering, stereotyping, and the Eckstein process), the earliest use of true pre-printed lettering seems to have been by one J Rodrigues of Lisbon in 1878, who, after printing the names from movable type on a strip of special paper, placed them in convenient places on the map which he then reduced photographically.

The method did not catch on immediately, however, and around 1920 the United States Geological Survey began experimenting with a method it called stick-up lettering in which the names were pre-printed on a gummed paper stock. Further refinements were made during the 1920s and 1930s using photocomposing machines, rice paper, various adhesives, and transparent stripping film, which, despite the relatively awkward and unsophisticated character of the method, is today the dominant technique for lettering a map to be printed.

C GRANDIDIER, Alfred (1882): ". . . Rapport sur les cartes et les appareils de géographie et de cosmographie, sur les cartes géologiques, et sur les ouvrages de météorologie et de statistique." France, Ministère de l'agriculture et du commerce, *Exposition universelle, Paris, 1878, Rapports de Jury International,* Paris, Imprimerie nationale. Groupe ii. Classe 16.

PEARSON, Karen (1978): "Lithographic Maps in Nineteenth-Century Geographical Journals," Ph.D. dissertation, University of Wisconsin-Madison.

RIDGWAY, John Livesy (1920): *The Preparation of Illustration for Reports of the United States Geological Survey, with Brief Descriptions of Processes of Reproduction.* Washington, DC, US Government Printing Office.

6.1220B Stamped Lettering

A A method of lettering maps in which the raised letters of specially cut punches were driven into the copper plate. The right-reading punches differ from those used in making typographic matrices (which must be wrong-reading). Otherwise known as punched lettering.

B Stamped lettering occurs extremely early in the history of European map printing and is far more prevalent than had been once thought. The method was used on the copper plates for the 1478 Rome edition of Ptolemy's *Geography* (and hence the 1490, 1507 and 1508 editions as well as the *tabulae modernae* in the latter two). Other examples from the 15th century are the 1491(?) map of central Europe purported to be by Nicolas of Cusa and the engraved single-sheet world map of Ptolemy, sometimes misleadingly known as the Taddeo Crivelli world map (*c.* 1490).

More examples abound in the 16th century, including Girolamo Ruscelli's editions of Ptolemy's *Geography* (1561 and later), and Livio Sanuto's atlas of Africa (1588). Much later examples include Richard Horwood's map of London (1792-99), the Ordnance Survey of England and Wales six inch series, and the Admiralty Hydrographic Office charts, for which punches were used until recently.

C ARH [HINKS, Arthur R] (1943): "The Lettering of the Rome Ptolemy of 1478," *Geographical Journal, 101,* 188-90.

6.1220C Stereotyped Lettering

A In the cartographic context, stereotyped lettering was a method of lettering a woodcut map in which small plates cast in type metal were cemented to the wood block. A group of names was set up in type, a matrix obtained in damp paper or some other moulding material, and a plate created from type metal poured on the mould. The individual names were then cut out with metal shears and affixed to the wood block.

B Hard evidence is available for its use by Peter Apian in his heart-shaped world map of 1530, of which the unique copy is preserved in the British Library. The technique was also used by Sebastian Münster in the *Cosmographia* (1544 and later editions), and in his editions of Ptolemy's *Geography;* by Joost Murer's map of the canton of Zurich (1566); and for Philipp Apian's *Bairische Landtaflen* (1568). The wood blocks for the last two works exist. These and a stereotype plate in pristine uncut state (described by Aretin in 1804) confirm the principle and details of the method.

The *Penny Cyclopaedia* (1833-46) also had map names done in this manner.

C ARETIN, J C F (1804): "Die Stereotypen in Baiern in XVI Jahrhundert erfunden," *Beyträge zur Geschichte und Literatur, 2,* 72.

HUPP, Otto (1910): *Philipp Apian's Bayerische Landtaflen und Peter Weiner's Chorographia Bavariae: eine bibliographische Untersuchung.* Frankfurt-am-Main, Heinrich Keller.

STETTER, Gertrud, and FAUSER, Alois (1979): Introduction to a coloured facsimile of the first edition, 1568, of Philipp Apian's *Bairische Landtaflen.* München, Suddeutscher Verlag.

WOODWARD, David (1970): "Some Evidence for the Use of Stereotyping on Peter Apian's World Map of 1530," *Imago Mundi, 24,* 43-8.

WOODWARD, David (1975): "The Woodcut Technique," in Woodward, David, ed., *Five Centuries of Map Printing.* Chicago and London, University of Chicago Press, 25-50.

6.1220D Type-Inserted Lettering

A A system of lettering a map by wedging printers' foundry type in slots cut through the wood block. A variation is found where the type has been overprinted in a second impression. See also Stereotyped Lettering (6.1220c).

B The earliest known use is for the maps in the *Rudimentum novitiorum* (1475). For the setting of cartouches and legends, the technique was used widely in the 15th and 16th centuries. It was revived in the 19th century for the wood-engraved maps, such as those in the *Penny Magazine,* a popular illustrated journal published between 1832 and 1845, using a mechanical slot-cutting machine invented by Edward Cowper.

C WOODWARD, David (1975): "The Woodcut Technique," in Woodward, David, ed., *Five Centuries of Map Printing.* Chicago, University of Chicago Press, 25-50.

6.1310 Map Surface

A Any substance on which a map can be constructed.

B Almost anything that provides a surface which can be permanently marked can be, and has been, used for maps. The range of map materials that have lasted is indeed impressive, including, for example, antlers, birchbark, ceramic ware, clay, cloth, metal, mosaic, paper, skin, stone, tusks, and wooden boards. It is also true, as Lloyd A Brown points out *(The Story of Maps,* p. 6), "No medium or material has yet been discovered which can be used for map making and nothing else." Thus silver and copper, on which some of the finest early maps were made, were too precious to be preserved just for the sake of the maps themselves. Copperplates of maps have survived, however, on account of the use of their surface for oil paintings. The earliest map of London, *c.* 1558, is known from two copperplates with 17th century paintings on their versos, now in the Museum of London, whereas the remaining plates (probably 15 in number) and all the printed impressions have vanished. Manuscript maps have suffered vicissitudes of a different kind. Some have turned up as fragments in the bindings of books. The Cantino world map, 1502, served for many years as a screen at the back of a butcher's shop in Modena, Italy.

The earliest maps were drawn, painted, or inscribed on rock, stone, clay tablets, and cloth. The wall painting recently found at Çatal Hüyük in Anatolia, dated *c.* 6200 BC, is the oldest extant topographical map and town plan (see Topographical map, 1.203; Town plan, 1.1630b). The rock drawings of Val Camonica, Italy, which date from the second and first millennium BC, now rank in second place (see Stone and rock, 6.1310g). A relief plan in terracotta depicting an apsed temple, which was found in Malta, dates from the late Neolithic (see Plan, 1.1630). Carved stones or *kudurrus* of the 2nd millennium BC in ancient Babylon are inscribed with boundary plans (see Plan, 1.1630; Estate map, 2.0210c). A stone statue of *c.* 2200 BC, representing Gudea, ruler of the Sumerian city-state of Lagash and a great builder, portrays him holding on his knee the plan of a temple, together with a measuring rule and a writing tool (Figs.3 and 4).

Babylonian clay tablets dating from the 4th millennium BC provided the first reliable writing materials made artificially (Gaur, p. 42). The earliest known topographical map of a region is the Nuzi clay tablet 2500 to 2300 BC (1.203). The oldest world map is on a tablet dating from the 7th or 6th century BC, found at Sippar, central Iraq (British Museum, Department of Western Asiatic Antiquities, no. 92687; see World maps, ancient, 1.2320a) (Fig.6).

Animal skins, properly treated, were used as written documents which survive from c. 2500 BC in ancient Egypt, but papyrus was for more than four thousand years the chief writing material there and in the Mediterranean world in general. The Egyptian "Fields of the Dead," drawn on papyri c. 1400 BC, depict landscapes and field systems (see Imaginary map, 1.091; Plan, 1.1630). The map of the Nubian gold fields, c. 1200 BC, is preserved on fragments of papyrus (Turin, Museo Egizio; see Mineral map, 3.131). Plans on pottery ostracons (potsherds) include the plan of a temple from western Thebes, c. 1250 BC (see Ceramic, 6.1310b). Ancient Egyptians also used wood. They painted inscriptions on wooden statues and sarcophagi. The maps illustrating the "Book of the Two Ways," a guide to life after death, are preserved on the wooden boards of coffins (see Route map, 1.1810).

Although papyrus must have been a common material for mapmaking in ancient Greece, no papyri survive. The more durable maps, inscribed on stone, include the plan of a mine, 4th century BC, found at Thorikos, east of Attica, on the rock face above the entrance to the mine which it evidently depicts (Dilke, p. 26; see Plan, 1.1630). World maps were engraved on bronze tablets or painted on wooden panels. Aristagoras, tyrant of Miletus, took with him on his tour of mainland Greece, 449-8 BC, "a bronze tablet [pinax] with an engraving of a map of the whole world" (Dilke, p. 23). The Romans used stone or bronze for their surveys. The three stone cadasters of Arausio (Orange), AD 77, and the plan of Rome, Forma Urbis Romae, inscribed on marble tablets fixed to a wall, completed between AD 203 and 208, are the most famous (see Stone and rock, 6.1310g). The mapping registers, which were kept in the Emperor's record office, were on papyrus rolls and later in parchment codices.

Parchment, invented according to tradition by Eumenes II of the Greek city of Pergamum in Asia Minor (197-158 BC), superseded papyrus as the chief writing material in the later years of the Roman empire (see Vellum, 6.1310h). The 4th or 5th century version from which the Peutinger map of the Roman empire was derived was presumably a long narrow papyrus roll wound on wooden cylinders. It was probably a revised version of an earlier original on papyrus (Dilke, p. 114; see Road map, 1.1810d). One of the earliest extant maps on parchment is the Dura Europos shield, comprising a fragment painted with a road map dating from before AD 260 made to cover an infantryman's shield (Bibliothèque Nationale, suppl. gr. 1354-5; Dilke, pp. 112, 120-22; see Road map, 1.1810d).

In ancient China silk was used as a writing material in the 7th or 6th century BC and for the making of maps by the 3rd century BC. The earliest extant Chinese maps, dating from c. 168 BC, are on silk (see Cloth, 6.1310c). With the invention of paper, c. 100 BC, the technique of making impressions from stone began to develop (Gaur, p. 195). The engraving of maps on stone probably started in the 2nd century AD, but the earliest extant are steles of 1137 used for taking rubbings, a form of printing (see

Stone and rock, 6.1310g). For some 1200 years from the 6th or 7th century, printing on paper from the wood block was, however, the principal technique in China and also in Korea and Japan. The oldest complete and dated printed book is a Chinese translation of the Diamond Sutra, AD 868; the oldest printed map is a map of West China, in the *Liu Ching T'u, c.* 1155.

In Europe vellum remained the chief medium for writing from the 4th century to the 15th (see Vellum, 6.1310h). The vellum codex supplanted the parchment roll as the vehicle for the production of books, and monasteries with their scriptoria became important centres. Maps on vellum, notably mappae mundi and celestial maps, were drawn and painted in codices. Large mappae mundi were prepared as separate items on large skins or several skins sewn together. Seamen's charts dating from 1300 onwards were normally on single skins (see Chart, 1.0320).

As in classical times, some of the finest items of medieval map making were on metal (see 6.1310e). The three tablets of Charlemagne, 8th century, now lost, were of silver. The Borgia world map, *c.* 1430, was engraved on copper punched with holes to hold coins from the countries depicted (as suggested by Ulla Ehrensvärd). Arabic celestial globes were normally made of metal. Two were recorded in the public library of Kahira in 1043, one made of brass, allegedly by Ptolemy himself, the other of silver (Stevenson, I, p. 28). One of the oldest extant Arabic globes, made at Valencia by Ibrahim Ibn Said-as-Sahli in the year 473 of the Hegira (AD 1080), is of brass (Istituto di Studi Superiori, Florence; Stevenson, I, pp. 28-9). Such items were mainly presentation pieces.

The material that from 1500 onwards has been the most influential is paper, because it proved to be the best suited for printing. From its beginnings in China, now dated to *c.* 140-87 BC, paper making spread eastward to Korea and Japan in the 7th century and westward to Samarkand and Baghdad in the 8th. Four centuries later, through the Arabs, the knowledge reached Spain and spread through Europe. The earliest European printed maps are the oldest European maps made on paper (see Paper, 6.1310f; Woodcut, 7.1810c). For nearly five centuries the production of printed maps on paper has been widespread. The invention of paper gores for globes likewise made the printed globe, terrestrial and celestial, for at least three centuries, 1530-1830, the most popular instrument in geographical education (see Globe, 1.0710).

The growing interest in geography and maps in the period of the European renaissance also encouraged use of other media. Painted maps on cloth were wall hangings in Henry VIII's palaces in 1547, as the inventory of his belongings at the time of his death reveals. Cloths painted with the earth and the cosmos are recorded as decorations on state occasions, such as ceremonial banquets (see Cloth, 6.1310c). Sixteenth and 17th century tapestries woven in Flanders and England to record historic events included maps. Painted maps of countries and continents decorated the corridors of power in the Vatican and in other princely Italian palaces (see Wall map, 1.231). Maps on stone were inscribed as commemorative pieces. The Burgerzaal of the new city hall of Amsterdam, built in 1648 to 1650, was supplied with a large world map in marble and copper laid on the floor and recording Abel Janszoon Tasman's voyages, 1642-44. When it wore out it was replaced in 1746 in marble by a new world map by Jacob Martensz, although the map was not laid until 1806. Amsterdam could also boast one of the most ephemeral and spectacular of carto-

Fig. 19 – A stone churinga of the aborigines of central Australia with abstract symbols probably representing the travels of mythical ancestors across the landscape. By permission of the Trustees of the British Museum; Museum of Mankind, 1935, 4112.1. (Conventional sign, 5.033; Map surface, 6.1310)

raphic pieces, the fireworks globe. It featured in a display to celebrate the visit of the Russian ambassador to Amsterdam on 29 August 1697. For 42 minutes the globe and the arch of triumph beneath it emitted balls and snakes of fire. I Moucheron and Carel Allard recorded the scene in two prints (British Library, Christoffel Beudeker (compiler), *Germania Inferior*, vol. 7 (1718), ff 63, 64; Maps C.9. d.7.).

As mapmakers extended their activities to serve a wider public from the 18th century onward, many ingenious devices were invented using a variety of surfaces. Inflatable globes, cardboard globes, jigsaw maps, and map games were supplied to the school room (see Map games, 1.131), while maps on fans and handkerchiefs pandered to the demands of fashion. There was even a glove map, the Great Exhibition Glove, patented by George Shove in 1851 (Fig.20). A paper version, "Registered exhibition hand guide to London," was published by J Allen for one penny (Guildhall Library, Print Room; Hyde, p. 47).

Outside Europe and Asia a variety of materials have been used by people of literate and nonliterate cultures. Mayan and Aztec manuscript maps were drawn on deerskin and long sheets of *amatl* "paper" made from the fibre, roots, and inner bark of the wild fig tree (Fig.8). This so-called paper looks like paper and is durable (Gaur, p. 39). The Eskimos carved charts in the form of coastal outlines on pieces of wood. The stick charts of the Marshall Islanders were made from palm fronds representing currents and lines of swell, with shells representing islands (British Museum, Department of Ethnography, no. 2289; Wave directions, 4.231). The Australian aborigines inscribed maps on stone *churingas* to depict the travels of their ancestors (Gaur, pp. 25-6) (Fig.19). Some of the designs on the wampum belts of the Iroquois of North America are interpreted as cartographic in form.

C DILKE, O A W (1985): *Greek and Roman Maps.* London, Thames and Hudson.

GAUR, Albertine (1984): *A History of Writing.* London, The British Library.

HARVEY, Paul D A (1980): *The History of Topographical Maps.* London, Thames and Hudson.

HYDE, Ralph (1986): "Spotlight," *The Map Collector, 35*, p. 47.

STEVENSON, Edward Luther (1921): *Terrestrial and Celestial Globes,* 2 vols. New Haven, Yale University Press for The Hispanic Society of America.

6.1310A Map Surface, Birchbark

A A map on the stripped outer bark of any of several trees of the genus *Betula,* the more or less white-barked tree birches, the inner (cambium) side of which has been used by native peoples in various parts of the northern hemisphere for at least two and a half millennia as a medium on which to engrave, draw or paint records and messages. Of the extant and recorded examples, map-like pictographs constitute a small but significant proportion.

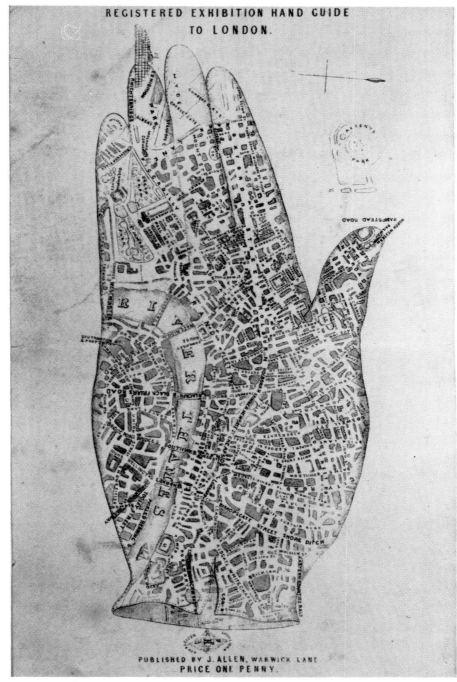

Fig. 20 – Registered exhibition hand guide to London. Published by J Allen, Warwick Lane. Price one penny, [1851]. James Allen's paper version of George Shove's glove map of London, designed as a guide for visitors to the Great Exhibition, 1851. The Crystal Palace appears at the end of the third finger. By permission of the Guildhall Library. (Map surface, 6.1310)

B The taxa of *Betula* together comprise one of the most botanically complicated and geographically extensive of circumpolar tree complexes. A lowland series extends through Eurasia and North America from well north of the Arctic Circle to as far south as northern Burma and the Virginian Piedmont, and a subalpine series occurs as far south as parts of the Himalayas, Caucasus, Alps, Rockies and the Sierra Nevada (Hultén, 1970). The outer bark of several of the taxa exfoliates in layers and has traditionally afforded a raw material for making canoes, covering homes, decorative art, and recording information.

In western Eurasia primitive forms of paper displaced birchbark for recording information at a relatively early date, but in parts of central and eastern Eurasia and of east central North America it remained in use until recently. Alphabetic writings on birchbark dating from as early as the 5th century have survived in the arid environments of Afghanistan, the Tadzhik and Kirghiz Soviet Socialist Republics, and parts of the Indian subcontinent. In 1833 at a fur-trading post in Saskatchewan a few Biblical passages were printed on birchbark in Plains Cree characters, using a mix of indigenous and introduced technology involving ink made from soot and fish oil and type cast in hand-cut wooden moulds with lead from tea-chest linings (Diringer, 1968). Nevertheless, most extant birchbark documents are pictographic.

Nineteenth century exploration in central and eastern Siberia was facilitated by the use of birchbark maps obtained from Tungus, Yukagirs, and other Palaeo-Asiatic peoples (Adler, 1910). Though not necessarily more widely used, more is known about birchbark maps drawn by North American Indians. As used indigenously, they were particularly characteristic of the Algonquian- and Iroquoian-speaking groups of the mixed and deciduous forests around the Great Lakes, where they were part of a pictographic tradition most fully developed in the sacred scrolls of the Ojibway. However, there are occasional references to maps on bark (though not necessarily birch) from well to the south of this region. In 1687 La Salle persuaded Cenis Indians in what is now northern Texas to draw "on bark a map of their country, of that of their neighbors, and of the river Colbert, or Mississippi" (Douay, 1687?), and some years later Lawson (1709) reported of Indians in the southern colonies: "These Maps they will draw in the Ashes of the Fire, and sometimes upon a Mat or Piece of Bark." Baron Lahontan (1703), drawing on his experiences in and to the west of the Great Lakes region between 1683 and 1692, was one of the earliest writers to mention specifically the use of birchbark for this purpose: "These *Chorographical Maps* are drawn upon the Rhind of your Birch Tree; and when the Old Men hold a Council about War or Hunting, they're always sure to consult them." Hunter (1829), having lived with Algonquian- and perhaps Siouan-speaking Indians for some twenty years, described how "they inscribe their correspondence, and such subjects as require to be recorded, on the inner bark of the white birch (*Betula papyracea*) [now *B. papyrifera*], or on skins prepared for the purpose [using] styles [styli] of iron, wood, or stone, and brushes made of hair, feathers, or the fibres of wood . . . to delineate or paint the most prominent objects embraced in their subjects." The medium and the techniques used for marking such documents have recently been described in detail, though with particular reference to sacred scrolls (Dewdney, 1975).

During the 19th century, officials, explorers, and scientists collected maps on birchbark. Relatively few have survived. Extant examples include four of parts of north central Newfoundland, which may have been drawn by Beothucks and hence pre-date 1830 (Newfoundland Museum); one dating from 1841 of part of a route

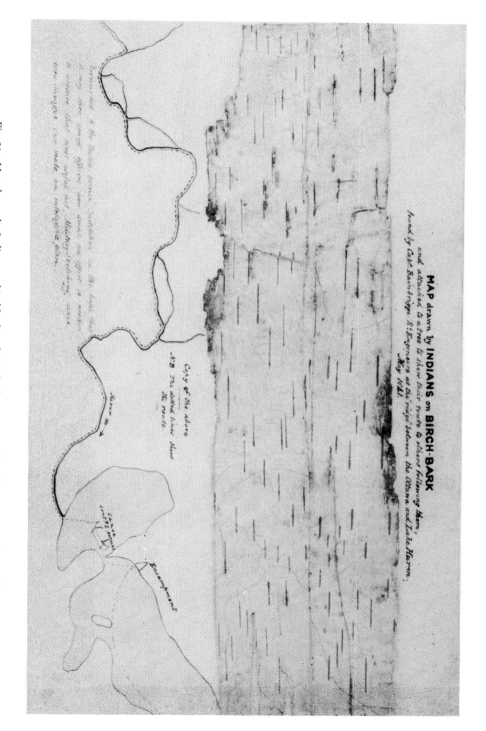

Fig. 21 – Map drawn by Indians on birchbark and attached to a tree to show their route to others following them, found by Capt. Bainbrigge, Royal Engineers, at the "ridge" between the Ottawa River and Lake Huron, May 1841. A copy on paper is pasted below. By permission of the British Library, Map Library, R.U.S.I. Misc. 1. (Map surface, birchbark, 6.1310a; see also Route map, 1.1810; Conventional sign, 5.033)

across the watershed between the Ottawa River and Lake Huron (British Library, Map Library, R.U.S.I. Misc. 1) (Fig.21); four drawn in 1887 by a Passamaquoddy chief of parts of Maine (National Anthropological Archives, Washington, DC); and maps dating from *c.* 1885-1895 on eight pieces of bark of areas in northwestern Ontario (sold by auction in Montreal in October 1978).

C ADLER, Bruno F (1910): "Maps of Primitive Peoples" (in Russian), *Bulletin of the Imperial Society of Students of Natural History, Anthropology and Ethnography, at the Imperial University of Moscow,* vol. 119, *Works of the Geographical Section,* vol. 2. Translated and abridged by H de Hutorowicz in *Bulletin of the American Geographical Society, 43* (1911), 669-79.

DEWDNEY, Selwyn (1975): "Birchbark, a Pictographic Medium," Chap. 2 in *The Sacred Scrolls of the Southern Ojibway.* Toronto and Buffalo, University of Toronto Press, 11-22.

DIRINGER, David (1968): *The Alphabet: A Key to the History of Mankind,* 3rd ed., completely revised with the collaboration of Reinhold Regensburger. London, vol. 1, 132.

DOUAY, Father Anastasius (1687?): "Narrative of La Salle's Attempt to Ascend the Mississippi in 1687," in John Gilmary Shea (1852): *Discovery and Exploration of the Mississippi Valley.* New York, Redfield, 204.

HULTÉN, Eric (1970): "The Circumpolar Plants. II. Dicotyledons," *Svenska Vetenskapsakademiens Handlingar,* 4th Series, 13th Part, No. 1, 340-43, maps 81-2.

HUNTER, John D (1824): *Memoirs of a Captivity among the Indians of North America . . . ,* 3rd ed. London, 185.

LAHONTAN, Louis-Armand de Lom d'Arce, Baron de (1703): *New Voyages to North America.* London, vol. 2, 13-14.

LAWSON, John (1709): *A New Voyage to Carolina.* London, 205.

6.1310B Map Surface, Ceramic

A Map designs on articles made of clay hardened at high temperatures.

B The oldest painted landscape representation on a jar may be that from the Neolithic site of Tepe Gawra (Iraq), dated to the end of the 4th millennium BC (Tobler). Plans on Egyptian pottery survive also from ancient Egypt. A pottery ostracon with the rough plan of a temple from Deir el-Bahari (western Thebes), dated from the New Kingdom *c.* 1250 BC, includes directions for the orientation of the building (British Museum, Department of Egyptian Antiquities, No. 41228).

Many centuries later, in the 18th century, the use of a cartographic motif for the embellishment of pottery became a popular art form in Japan where various kinds of everyday articles were decorated with map designs. Ceramic and porcelain ware were also employed. Early examples of ceramic maps are the so-called Gennai ware plates of the late 18th century which have a world map or a map of Japan drawn on them. The Gennai ware was produced on the model of a southern Chinese pottery by Hiraga Gennai (1726-1779) at Shido in Sanuki province (now Kagawa prefecture). The world map appears on a pair of six-sided plates, one with a map of the Old World and the other with the New World, both of which were based on Matteo Ricci's world map of 1602. The map of Japan, tolerably correct for those days, is designed on a rectangular plate. On both its right and left margins are scales of latitude, and inside, placed north and south of Japan, are two compass roses. Although there are no meridians, parallels, or rhumb lines, the map seems clearly to have been based on a Western-style portolan chart of Japan. Hiraga Gennai was a man of large and varied attainment and had extensive knowledge of Western science. These map plates probably reflect more than anything else Gennai's personal interest in things Western. Apparently only a small number of them were produced, and today they are rare.

By far the greatest number of map-decorated porcelain vessels were produced in the vicinity of Imari in northern Kyushu. There are many Imari plates, bowls, vases, and *sake* cups with maps of Japan and of Kyushu on them, and most bear the date of the Tempo era (1830-44). In addition to Imari ware, there are a few similar plates with a map of Japan, which bear the date of the Bunsei era (1818-1829). In the maps on these plates Kaga province is purposely exaggerated, suggesting that they were made at Kutani in Kaga province (now Ishikawa prefecture). Unlike the Gennai map plates, Imari and Kutani map vessels were unique in that they featured the obsolete Gyoki type of map of Japan accompanied by the fictitious "land of pygmies" and "island of women." Since the appeal of such vessels depended above all on the design, the use of the interesting but highly distorted Gyoki type map testifies to the attempt to be purely decorative rather than providing a map that would be of any practical value.

Maps seem to have been produced on porcelain in Japan until the middle of the 19th century, but after that they disappeared almost entirely. In Europe plates bearing maps of administrative regions in England and France, such as Hertfordshire, Oxfordshire, Somerset, Paris, and Rhône et Loire date from the 19th century.

C AKIOKA Takejiro (1955): *Nihon Chizushi* (History of Japanese Cartography). Tokyo, Kawade Shobo.

GLANVILLE, Stephen R K (1968): *Catalogue of Egyptian Antiquities in the British Museum.* London, British Museum.

NANBA Matsutaro, MUROGA Nabuo, and UNNO Kazutaka (1973): *Old Maps in Japan.* Osaka, Sogensha.

SHIRLEY, Rodney W (1982) "An Early Map of Japan on a Porcelain Plate," *The Map Collector, 20,* 33-4.

TOBLER, Arthur J (1950): *Excavations at Tepe Gawra. Levels 9-20.* Philadelphia, University of Philadelphia Press, 150-1, plate LXXCII b.

6.1310C Map Surface, Cloth

A Maps drawn or painted on cloth, or woven. In China drawn or painted on silk scrolls and screens were common. In Europe linen, cotton, and silk were used. Tapestries were woven in wools and silks. Native Mexican maps were made on agave cloth.

B The reeling of the fibres from the cocoon of the wild, and later the domesticated, silkworm and their use as a weaving material in China dates back to the Shang dynasty (*c.* 1766-1122 BC). But silk only came to be used as a writing material during the Chou dynasty, in the 7th or 6th century BC. It was originally used for important or sacred documents, but by the Han dynasty (206 BC–AD 220) its use for books and correspondence had become widespread. After the invention of paper (*c.* 100 BC), silk was slowly superseded as a popular writing medium, but it continued to be used for special purposes, such as scroll paintings.

The earliest historical reference to a map drawn on silk is for 227 BC, although there are many earlier references to maps in the Chinese sources, some of which must have been on silk. In that year a would-be assassin gained access to the future first emperor of a united China, Shi Huang Ti, on the pretext of presenting him with a silk map, whose case concealed a poisoned dagger. The earliest extant silk maps date from shortly after this, namely the three found in 1972-3 in Han tomb no. 3 and dated *c.* 168 BC, at Ma-wang-tui, near Changsha, in Hunan. The maps represent the area of present-day Changsha and show mountains, rivers, towns and roads. Two of the maps have been restored. One, a relief map, limits itself to topographical features; the other, a garrison map, gives the locations and strengths of military posts, the number of households in each place, the distances between settlements, as well as some topographical information.

The great Chinese cartographers Phei Hsiu (AD 224-271) and Jia Dan (730-805) are known to have used silk for their maps, but apart from the Changsha maps, no others on silk survive from a period earlier than the Ming dynasty (1368-1644). Silk continued to be used for maps during the Ch'ing dynasty (1644-1911) because of its virtually unlimited seamless breadth and also because of its strength and its suitability as a medium for important, precious and finely executed work. It is possible that the Chinese derived the idea of the square grid (4.191), which was in use by 270, from the practice of drawing maps on silk whose warp and weft provide a ready-made system of co-ordinates. The tradition of using cloth continued in the Orient over later centuries. The large Chinese-Korean world map of Ch'uon Chin, "Hun-I Chiang-Li Li-Tai Kuo Tu chi L Thu" (Map of the territories of the One World . . .), dated 1402, was painted on silk (Fuchs).

In Europe maps on cloth were popular features of renaissance cartography. They featured, for example, in the great pageants of state. When Henry VIII banqueted at Greenwich in May 1526, the roof of the specially built hall was designed by the King's astronomer Nicholas Kratzer to show "the whole earth . . . like a very Mappe or Carte." A cloth above displayed the cosmos, and it was described as "a connying thing and a very pleasant syght to behold" (Hall, Edward (1904), *Henry VIII, I,* 219-20).

Tapestries with cartographic motifs provided a more permanent legacy. A set of panoramas by Jan Cornelis Vermeyen of Beverwijck, near Haarlem (1509-1559), depicts the conquest of Tunis in 1535 and includes a portolan chart of the western part of the Mediterranean (Destombes, 67). Two famous sets of tapestry maps were made in England in the late 16th to 17th centuries. William Sheldon, who was reputed to have introduced tapestry weaving into England, initiated before his death in 1570 the project of making maps of the English counties. The maps were woven by the family's weavers at Weston and Barcheston in Warwickshire and Beoley in Worcestershire, and date from the later years of the 16th century to the Civil War, 1644. They are preserved in the Victoria and Albert Museum, the Bodleian Library (16th century) and the Yorkshire Philosophical Society (17th century); while others are in private hands. The tapestries are in coloured wools and silks on wool warps. The antiquary Richard Gough described them in 1783 as "this earliest memorial to two arts among us, Tapestry and Map-making" (Nichols). The second set comprised ten great tapestry maps to celebrate the defeat of the Spanish Armada in 1588. They were commissioned in or after 1590 by Queen Elizabeth I's Lord High Admiral, Lord Howard of Effingham (Charles Howard). The designs were made by the Dutchman Hendrik Cornelis Vroom, founder of the Dutch school of marine painters, and the weaving was done by another Dutchman, Francis Spiring of Haarlem. The maps were based on the charts drawn by Robert Adams and engraved by Augustine Ryther for his *Expeditionis Hispanorum in Angliam vera Descriptio* [1590]. They hung in the House of Lords from 1616 until 1834, when they perished in the great fire. They are now only known from John Pine's engravings, 1739.

Sixteenth century topographical and administrative maps were also sometimes drawn and painted on cloth. The large map of the Franc de Bruges, 3.35 metres high and 6.20 metres wide (Archaeological Museum, Bruges), was painted on canvas by Pierre Pourbus, painter of Bruges, who completed the work in 1571. It exemplifies an important body of topographical maps which were not only drawn but painted like pictures (Smet). Another map on canvas made for official purposes depicts plantations in Leix and Offaly, Ireland, *c.* 1565 (British Library, Cotton MS Augustus I.ii.40.).

The durability and ease of folding of fabric maps, especially those made in silk, proved advantageous in military campaigns. The "Quartermaster's Map" of England and Wales, engraved by Wenceslaus Hollar, and originally published by Thomas Jenner, London, 1644, for use by the armies in the Civil War, was available for purchase on cloth from John Garrett, Jenner's successor, *c.* 1675. Examples were backed on linen or silk, not painted on the cloth, as copies in the British Library show: an edition of about 1676 is on cloth, Maps C.7.b.15, and one of *c.* 1690 on silk, Maps C.7.b.14. Both have the six sections of the map bound in volume form. Military maps of German territories printed on silk in the late 18th century are preserved in the Duke of Cumberland's Collection in the Royal Library at Windsor Castle and in King George III's Topographical Collection, the British Library, London. The Ordnance Survey of Great Britain produced monochrome silk versions of one inch to a mile sheets of Ambleside and Keswick in the Lake District in 1890-94. A War Office Committee of 1892, asked to make proposals for a military map of the United Kingdom, referred to the advantage of linen over paper as a material on which maps could be "printed for use in the field." The most extensive production of silk, tissue and rayon maps in Great Britain occurred in the Second World War, 1939-45, when hundreds of sheets were issued to combatants and secret agents to aid escape and evasion. Many ended up after the war as items of ladies' apparel.

Examples of cloth maps made as memorabilia of a more modest kind than tapestries are those embroidered on samplers in the 18th and 19th centuries, often as beginners' exercises in embroidery. London mapsellers and commercial entrepreneurs traded on the growing popular interest in geography. In 1690 John Seller offered a new map of Ireland on paper or silk. About 1700 Philip Lea advertised his maps as follows: "Any of these . . . may be printed on Silk for a Sarsh Window, or to carry in the Pocket in a little Room, as an Handkercheife" (Tyacke). The ingenious John Spilsbury (1739-69), inventor of the jigsaw puzzle, as a sideline to his trade as an engraver and mapmaker sold silk kerchiefs printed with a map as souvenirs. An example, "A New and Most Accurate Map of the Roads in England and Wales; with the Distances by the Milestones," with the imprint, "Printed for Spilsbury, Engraver, Russel Court," c. 1760, is preserved in the British Museum, Department of Prints and Drawings (Hannas).

With the diversification in globe manufacture in the 19th century, globes on cloth came into the market. William Stokes, "Teacher of Memory," printed on cloth and published in 1868 "Stoke's Capital Mnemonical Globe" which displayed a human face superimposed on the countries and oceans of the world (Hill). "John Betts's Portable Terrestrial Globe," London, c. 1860, was printed on linen and could be opened to form a globe by means of an umbrella-like mechanism (British Library, Maps C.3.bb.6.). A Japanese globe printed on cotton, with a similar mechanism, "Chikyu-gi, Shintei" [A terrestrial globe, revised], was published in Tokyo by Mombusho, c. 1880 (British Library, Maps C.3.bb.3.).

One other cultural area notable for its maps on cloth was ancient Mexico. Hernando Cortés, conqueror of Mexico, recorded that some of the Indians "drew on a cloth a figure of the whole land, whereby I calculated that I could very well go over the great part of it." Bernal Díaz del Castillo in his *Historia verdadura de la conquista de la Nueva-España* (1632) stated that the maps were made on agave cloth. The "Map of Metlatoyuca," so named after the place where it was found, is painted in colours on cloth and measures 102 x 175 cm (British Museum, Department of Ethnography).

C BOND, Barbara (1983): "Maps Printed on Silk," *The Map Collector, 22,* 10-13.

BOND, B (1984): "Silk Maps: The Story of MI9's Excursion into the World of Cartography 1939-1945," *Cartographical Journal, 21,* 141-4.

CHANG, Kuei-sheng (1979): "The Han Maps: New Light on Cartography in Classical China," *Imago Mundi, 31,* 9-17.

DESTOMBES, Marcel (1959): "A Panorama of the Sack of Rome by Pieter Bruegel the Elder," *Imago Mundi, 14,* 64-73.

FUCHS, Walter (1953): "Was South Africa Already Known in the 13th Century?" *Imago Mundi, 10,* 50.

GUZMÁN, Eulalia (1939): "The Art of Map-making among the Ancient Mexicans," *Imago Mundi, 3,* 1-6.

HANNAS, Linda (1972): *The English Jigsaw Puzzle 1760-1890*. London, Wayland Publishers, 19.

HILL, Gillian (1978): *Cartographical Curiosities*. London, 18-19.

HSU, Mei-Ling (1978): "The Han Maps and Early Chinese Cartography," *Annals of the Association of American Geographers, 68,* 45-60.

NEEDHAM, Joseph (1959): *Science and Civilisation in China*. Cambridge, At the University Press, *III,* sect. 22, 534-5 and 541.

NICHOLS, John (1812): *Literary Anecdotes of the Eighteenth Century*. London, *Vol. 6, 327.*

PINE, John (1739): *The Tapestry Hangings of the House of Lords*. London, J Pine.

SCHRIRE, D (1963): "Adams' and Pine's Maps of the Spanish Armada," *Map Collectors' Series, 4.*

SMET, Antoine de (1947): "A Note on the Cartographic Work of Pierre Pourbus Painter of Bruges," *Imago Mundi, 4,* 33-36.

STUDY GROUP FOR HAN SILK MANUSCRIPTS FROM MA-WANG-TUI (1975): "The Reconstruction of the Maps Excavated from the Third Han Tomb at Ma-wang-tui, Changsha" (in Chinese), *Wen Wu, 225* (No. 2), 35-43.

TAN Chi-hsiang (1975): "A Map More Than Two Thousand One Hundred Years Old" (in Chinese), *Wen Wu, 225* (No. 2), 43-48.

THOMPSON, Henry Yates (1919): *Lord Howard of Effingham and the Spanish Armada*. London, Roxburghe Club.

TSIEN, Tsuen-hsuin (1962): *Written on Bamboo and Silk*. Chicago, University of Chicago Press, 114-30.

TYACKE, Sarah (1973): "Map-sellers and the London Map Trade," in Wallis, Helen, and Tyacke, Sarah, *My Head Is a Map*. London, Francis Edwards and Carta Press, p.65.

VICTORIA AND ALBERT MUSEUM PORTFOLIOS (1915): *Tapestries, Part III. Tapestry Maps. English; 16th and 17th Cent*. London, HMSO.

6.1310D Map Surface, Hard Animal Tissues

A Bone, tusk, antler, horn and shell have all been used in various parts of the world as surfaces on which to inscribe, among other things, maps and topographical drawings.

B Hard animal tissues, although portable and fairly durable, are not particularly suitable materials for maps because they do not afford sufficiently extensive flat surfaces. Certain cultures, however, have had long traditions of decoratively engraving artifacts made from such tissues. Powder horns were frequently used for this purpose. They had been carried in Europe since gunpowder came into general use in the early 15th century. Those designed for coarse powder, which included most of the larger horns, were made from cow or ox, and those for fine powder from calf. European horns were also made from stag and tortoise shell. An Italian bovine horn dating from 1560, minutely engraved with scenes on land and sea, and another with the arms of the Duke of Parma, 1773, are preserved in the Metropolitan Museum, New York. Powder horns with maps were most commonly used in North America, where they became part of the standard equipment of the huntsman. In the period of the French and Indian Wars, 1754 to 1763, horns engraved with maps showing the routes between New York and Canada were used even by high ranking officers instead of printed maps which were in short supply. When farmers and townsmen were called to military service in the American War of Independence they took their powder horns with them. About one hundred American powder horns engraved with maps and dating from the 18th century are known today, made either in North America or imported during the colonial period. Some were homemade and others the work of professional engravers or gunsmiths produced for sale. Most show areas within the Atlantic Colonies, with the Hudson River–Lake Champlain and the Hudson–Mohawk river routes particularly popular subjects. One *c.* 1770, displaying the Mohawk and Hudson rivers, with a view of New York and the royal arms, is in the Armouries at the Tower of London, Armouries Inventory Number XIII. 126 (reproduced in British Library (1975): *The American War of Independence 1775-83: A Commemorative Exhibition.* London, p.79, no. 92). Another shows a map of the Cherokee villages on the Little Tennessee River (Rijksmuseum voor Volkenkunde, Leiden, The Netherlands, series 924, no. 82). After contact was established with the Indians the engraved powder horn became a commercial art form. It would appear that because of their mnemonic appeal for tourists, powder horns were engraved by Europeans and Indians with a view to barter or sale.

Eskimo examples of maps engraved on ivory are relatively common from *c.* 1890 onwards, especially from the northwest coast of Alaska, where the Alaska Commercial Company fostered the souvenir trade and supplied whole walrus tusks. Engravings of maps on whole tusks are included in the collections of the Yale University Map Library, the Lowie Museum of Anthropology at Berkeley, California, and the Alaska State Museum. Ray (1977) gives an account of this art form and reproduces a photograph of a letter opener carved in ivory with a map engraved on one side of the blade. Both the County Museum, Liverpool, and the University of Alaska Museum possess cribbage boards carved in ivory with a map engraved on the back. Almost all the Eskimo examples display coastlines and include English or Anglicized toponymies. Although the material's elongated shape led to geographical distortion, the images appear to have been derived from the printed maps of the period. This tradition may have been a development of the late 18th and 19th century tradition of scrimshaw on whalebone and whale tooth in New England and other maritime regions, one early example of which shows a chart of Boston harbour.

Maps have also occasionally been made on antlers. A moose antler, supposedly brought back by the expedition of Meriwether Lewis and William Clark to the Pacific Ocean (1804-6), has a series of holes close to the edge which, according to tradition, were made by a Shoshoni squaw to represent the expedition's route. However, since the information is rudimentary and the geometry of the dots is determined by the shape of the antler, it is probably only a record of camp sites or days' travel.

In eastern North America Indians formerly used beads (wampum) made from black and white shell to construct belts incorporating mnenomic patterns, some of which were topological maps. The structure of wampum belts imposed geometrical rather than realistic designs. Even before the replacement of shell by glass beads, some belts incorporated designs which were intended to represent spatial relationships in the real world. The original confederation of Iroquois villages and the later Iroquois Confederacy of Five Nations were usually represented by diamondshaped symbols, with connecting lines indicating trails.

In the 18th century, Europeans began to present glass bead belts to the Indians. One accepted from the French by the Indians of western Pennsylvania in a treaty identifies four French forts with English territory by representations of the forts woven into it. In 1758 an English agent, while attempting to persuade Indians on the Ohio River to go to Philadelphia, presented a white bead wampum belt with the figure of a man at each end (one representing the Indians and the other the colonists) and connecting streaks of black (indicating the road between the Ohio and Philadelphia).

Maps were also carved on Japanese netsuke, toggles usually made of ivory and attached at one end of the cord worn round the waist to hold keys or purses. A 19th century map of Japan giving a table of distances carved on a Manju-type netsuke, so called because it is round and flat like a cake called *Manju,* is preserved in the British Museum (OA 1945-10-17-552) (Wallis et al. (1974): *Chinese & Japanese Maps,* item J25).

C CARPENTER, Charles H, Jr (1972): "Early Dated Scrimshaw," *Magazine Antiques, 102,* 414-9.

DU MONT, John S (1985): "Engraved Maps on Powder Horns," *Map Collector,* No. 33, 2-7.

GRANCSAY, Stephen V (1945): *American Engraved Powder Horns* . . . New York, Metropolitan Museum of Art, 11-14, 77-80; pl. i-vi, xxiii, xxiv, xxix, xxxi-xxxix.

HOLMES, William H (1883): "Art in Shell of the Ancient Americans," *Second Annual Report of the Bureau of Ethnology to the Smithsonian Institution, 1880-81,* by J W Powell. 47th Congress, 1st Session, House of Representatives, Misc. Doc. No. 61, Washington, DC, pp. 243 and 250.

RAY, Dorothy J (1977): *Eskimo Art: Tradition and Innovation in North Alaska.* Seattle and London, University of Washington Press, 24-8, 43-5, 160 and 234-9.

SPECK, Frank G (1935): *Naskapi: The Savage Hunters of the Labrador Peninsula.* Norman, Oklahoma, University of Oklahoma Press, Chapter 6.

6.1310E Map Surface, Metal

A A map, usually monumental or decorative, designed as a metal artifact.

B Metal, usually gold, silver, bronze, or copper, has been used as a medium for maps at least since the time of Charlemagne (8th century), when reference is made to three silver tablets in his possession engraved with maps of Rome, Constantinople, and the world. A coin from the Greek colony of Zancle (now Messina) in Sicily dates from about 525 to 494 BC. Zancle meant sickle and referred to the curving sandbank forming the harbour. The coin represents a map of the harbour, showing a sickle blade enclosing a dolphin and the town's name. (Harvey, p.53). An Ionian Greek coin of the 4th century BC is also thought to bear a relief map of the Ephesus district on its reverse (Johnston, 1967), although some doubt is now cast on this interpretation (see Topographical maps, 1.203). An even earlier gold disk of questionable interpretation may show a schematic image of the Middle East and thirty-two triangular surrounding islands (Kish). Other examples include a bronze medal (22 BC) honouring the Roman Emperor Marcus Cocceius Nerva (Miller, III, 129-31); the Borgia world map (c. 1430) engraved on copper and filled with black pigment in the *niello* manner (Almagià); the Laurana medal with a circular world map (c. 1460) (Destombes, p. 241); and a reliquary with a tripartite world, offered to Louix XI by the city of Dijon (c. 1470) (Destombes, p. 5). The silver map of the world (1589), a rare medallion commemorating Sir Francis Drake's circumnavigation, comprises two carefully engraved hemispherical maps on its two faces (Christy; Wallis, 1979 and 1984).

Globes (1.0710), as decorative map forms and symbols of royal power, were frequently made of the finest precious metals with intricate engraving. Medieval Arabic and Persian globes were made of brass. Many European examples exist from the 15th to 17th centuries, such as the so-called globe of Laon in red copper (c. 1493), the gilt copper globe in the Jagiellonian University Museum (c. 1510), the Lenox globe in the New York Public Library (c. 1510), and those by Robertus de Bailly (1530), Abraham Gessner (c. 1600), Johann Hauer (1620), and others illustrated in Stevenson and in Schramm. (See Celestial globe, 1.0710a.)

C ALMAGIÀ, Roberto (1944): *Monumenta Cartographica Vaticana.* Vatican City, Biblioteca Apostolica Vaticana, vol. 1, 27-29.

CHRISTY, Robert Miller (1900): *The Silver Map of the World.* London.

DESTOMBES, Marcel, ed. (1964): *Mappemondes, AD 1200-1500,* vol. 1. of *Monumenta Cartographica Vetustioris Aevi.* Amsterdam, N. Israel.

HARVEY, P D A (1980): *The History of Topographical Maps.* London, Thames and Hudson.

JOHNSTON, A E M (1967): "The Earliest Preserved Greek Map," *Journal of Hellenic Studies, 87,* 86-94.

JOHNSTON, A E M (1971): "Maps on Greek Coins of the 4th Century BC," *Imago Mundi, 25,* 75-6.

KISH, George (1980): *La Carte.* Paris, Editions du Seuil, 90.

MILLER, Konrad (1895-1898): *Mappaemundi; Die altesten Weltkarten,* 6 vols. Stuttgart, Jos. Roth'sche Verlagshandlung.

MURIS, Oswald, and SAARMANN, Gert (1961): *Der Globus im Wandel der Zeiten.* Berlin & Beutelsbach bei Stuttgart, Columbus Verlag, Paul Oestergaard.

SCHRAMM, Percy Ernst (1958): *Sphaira, Globus, Reichsapfel.* Stuttgart, Anton Hiersemann.

STEVENSON, Edward Luther (1921): *Terrestrial and Celestial Globes.* New Haven, Yale University Press for the Hispanic Society of America.

WALLIS, Helen (1979): *The Voyage of Sir Francis Drake Mapped in Silver and Gold.* Berkeley, Friends of the Bancroft Library, University of California.

WALLIS, Helen (1984): "The Cartography of Drake's Voyage," in N J W Thrower, *Sir Francis Drake and the Famous Voyage, 1577-1580.* Berkeley, Los Angeles, London, University of California Press, pp. 121-63.

6.1310F Map Surface, Paper

A Paper is a thinly moulded, compactly interlaced sheet or leaf formed by the aqueous deposition of macerated plant fibre, this being obtained either directly from the plant, as in the case of mulberry paper, or indirectly, as in the case of rag (i.e., linen, cotton, etc.) papers.

B The technique of paper making had its origins in China but like many "inventions" its development was gradual. The oldest examples discovered thus far were excavated from an Early Han (*c.* 206 BC–AD 8) site in Sian in Shensi Province. These have been scientifically dated to *c.* 140–87 BC, showing that the technique is of greater antiquity than has heretofore been thought.

From China the technology moved first eastward to Korea and then on to Japan where paper was made early in the 7th century. Travelling westward the art reached Samarkand in 751 and Baghdad in 793. The Arabs are generally credited with the invention of rag paper, but strictly speaking this is incorrect, although they were the first to produce paper solely from rags. It is to them that Europe owes its knowledge of the art. The technique taught by the Arabs to the Spanish and Italians in the 13th century was almost exactly that which they had learned from the Chinese some four centuries earlier. From Spain the art moved slowly throughout Europe, paper being made in France at Herault in 1320 and in Germany at either Köln in 1320 or Nuremburg in 1390. The first mill in England was set up by John Tate in 1494, and the first in Holland approximately a century later. Paper making was introduced into the New World by William Rittenhouse in the mid-17th century.

Paper was traditionally hand moulded by scooping the "stock," or pulp-water mix, onto a fine wooden or wire screen held in a wooden frame; but in France in 1798, Nicolas-Louis Robert constructed the first continuous screen moulding machine. Such machines became a practical reality when two English engineers, familiar with Robert's ideas, built an improved version for the Fourdriner brothers in 1807. Other innovations include the "hollander" or macerater of the early 18th century and M F Illig's discovery in 1800 of the fact that paper could be sized in vats with rosin and alum rather than only by impregnation with animal glue or vegetable gums. Sizing affords a superior printing surface.

When paper was first used for maps is not clear, but in China it almost certainly dates from the 3rd century AD when its quality and durability were sufficient to challenge silk as a formal stationery (see Map surface, cloth, 6.1310c). In Japan paper production increased during the 8th century, when it became widely used for official documents and Buddhist writings. The cadastral maps of those days were mostly drawn on hemp cloth, but a few of those surviving are drawn on paper.

Toward the close of the 8th century the Japanese developed a special technique for making paper using the bark of *Broussonetia kozinoki, Edgeworthia papyrifera,* and others to which a glutinous vegetable substance was added, creating high quality, strong, and durable paper. After that, maps in Japan were made almost exclusively on paper. Some are in the form of scrolls or screens, but most of them were drawn or printed on a single large sheet of paper, which was folded into a much smaller size for preservation. Since Japanese paper is both soft and strong, it can be easily folded and folding does not damage it.

The Meiji government launched a long-range project of surveying the whole of Japan, and for this purpose in 1887 special map paper was produced in a government-managed factory. It was handmade using traditional materials, despite the fact that Western machine-made paper had been produced since 1872 for ordinary use. It was not until 1908 that paper for the exclusive use of maps began to be mass-produced by machine in Japan.

The strongest single stimulus to the adoption of paper as a medium on which to reproduce maps in both East and West was the development of printing. The earliest printed maps (see Woodcut, 7.1810c) also happen to be some of the oldest maps extant committed to paper.

B HICKMAN, Brian (1977): "Japanese Handmade Papers: Materials and Techniques," *Bulletin of the Japan Society* (London), *81,* 9-16.

HUART, C, and GROHMANN, A (1974): "Kaghad," *The Encyclopaedia of Islam,* 2nd ed., *IV* (fasc. 67-68), 419-20.

HUNTER, Dard (1932): *Old Papermaking in China and Japan.* Chillicothe, Ohio.

HUNTER, Dard (1957): *Papermaking: The History and Technique of an Ancient Craft,* 2nd ed. London.

HUNTER, Dard (1971): *Papermaking through Eighteen Centuries.* New York (reprint of 1930 edition).

JUGAKU Bunsho (1959): *Paper-making by Hand in Japan.* Tokyo, Meiji Shobo.

JUGAKU Bunsho (1967): *Nihon no Kami* (Japanese Paper). Tokyo, Yoshikawa Kobunkan.

KUME Yasuo (1976): *Washi no bunka-shi.* Tokyo.

SOKURYO CHIZU HYAKUNENSHI HENSHU IINKAI (1970): *Sokuryo Chizu Hyakunenshi* (A Hundred Years of Surveying and Map-making). Tokyo, Nihon Sokuryo Kyokai.

6.1310G Map Surface, Stone and Rock

A   Maps or plans incised in solid slabs of stone or on rock surfaces.

B   The tradition of inscribing or painting messages, prayers and records on rock surfaces, cliff faces, and natural or cut stones dates back in Europe to the Upper Paleolithic (*c.* 30,000 years ago). Much of this rock art is comprised of naturalistic representations of animal and anthropomorphic figures and other recognizable objects, but there remains a large body of abstract and symbolic signs. Some of these, dating particularly from the late Neolithic period, would seem to be topographic signs showing outlines of huts, fields and orchards, streams and paths as seen from above, e.g. in plan. The majority of the topographical compositions so far known come from the Monte Bego and Val Camonica regions of the Alps. The functions of maps in the prehistoric period is unclear, but it is unlikely that they served the same recording or directional purposes as maps in the historic period.

The earliest surviving inscribed stones in China are ten stone drums, currently thought to have been cut in 770 BC. In the course of the Chou, Ch'in and Han dynasties (1122 BC–AD 220), stone came to replace bronze as the medium preferred for the preservation of important records, culminating in the inscription on stone tablets of the whole Confucian canon in AD 175-183. No maps inscribed on stone survive from this period, but given the wealth of references to maps in the historical sources, there is little doubt that some were so engraved.

The earliest extant maps on stone in China are a pair which were inscribed within a few months of each other in AD 1137 and now are in the Bei Lin (Forest of Steles) Museum at Sian. The first of these, the "Hua I Thu" (Map of China and the Barbarian Countries), contains textual material no later than 1043 and is almost certainly based on the map, now lost, made in 801 by the Tang cartographer Jia Dan. The second, the "Yü Chi Thu" (Map of the Tracks of Yu), is remarkably modern in appearance, having a square grid (see 4.191) and depicting both the coastline of China and the courses of the major rivers with great accuracy; its purpose, however, seems to have been to instruct students in the most ancient historical geography of China. A further set of three maps, inscribed in 1247, representing a planisphere, the Chinese empire, and the city of Suchow, survive in the Confucian temple at Suchow.

A number of plans incised in marble survive from the Roman period, notably the substantial fragments of the "Forma Urbis Romae," a marble plan of the City of Rome. The plan originally measured 18.3 m x 13.03 m and is on the approximate scale of 1:300, but the scale varies considerably in different parts of the plan. It is thought to date from AD *c.* 203-208 and may be derived from an earlier map of the late 1st century compiled at the instigation of Vespasian and Titus, who are known to have ordered the measurement of the city in 74. A reconstructed copy is to be seen in a courtyard of the Capitoline Museum.

Other surviving Roman plans on marble are two surveys recording the extent of burial plots and surrounding lands, which are preserved at Perugia and Urbino respectively. The first is a plan of the mausoleum of a freedwoman of Octavia, wife of

Nero, with measurements of each room in feet; the second, found at the side of the Via Labicana, shows a burial monument, a private road and a public highway (each with given measurements of length), two areas of reed-beds and a ditch.

The principal Roman maps on stone other than marble are the cadastral maps of the Roman colony of Arausio (Orange). A small Roman plan on stone is described in the *Corpus Inscriptionum Latinarum* (1876), fasc. 1261, as a diagram of some waterworks in or near Rome. One entry may be translated "C. Iulius Hymetus, Aufidian estate, 2 water channels, from 2nd to 6th hour."

Globes made of marble include the earliest extant celestial globe, which is part of the Roman statue known as the Atlante Farnesiano. The globe, second century or later, is borne on the shoulders of the mythical figure of Atlas. The statue is now in the Naples Museum (Stevenson). A marble terrestrial globe of the second quarter of the 16th century is preserved in the Schlossmuseum, Gotha (Horn).

C CARETTONI, G, et al. (1960): *La pianta marmorea di Roma antica*, 2 vols. Rome, S.P.Q.R.

CHAVANNES, Edouard (1903): "Les deux plus anciens specimens de la cartographie chinoise," *Bulletin de l'École Française de l'Extrême Orient, 3*, 214.

Corpus Inscriptionum Latinarum (1876, 1894, 1887). Berlin, G Reimer, *VI*, pt I, 274; *VI*, pt 4.1, 2897, no. 29846-7, 29847a; *XIV*, 409.

DELANO SMITH, Catherine (1982): "The Emergence of 'Maps' in European Rock Art: A Prehistoric Preoccupation with Place," *Imago Mundi, 34*, 9-25.

HORN, Werner (1976): *Die alten Globen der Forshungsbibliothek und des Schloss museum Gotha*. Gotha, pp. 13-18.

HUELSEN, C (1890): "Piante iconografiche incise in marmo," *Mitteilungen des kaiserlich deutschen archäologischen Instituts, roemische Abtheilung, 5*, 46-63 and Taf. III.

NEEDHAM, Joseph (1959): *Science and Civilisation in China*. Cambridge, At the University Press, vol. 3, sect. 22, 543-51.

STEVENSON, E L (1921): *Terrestrial and Celestial Globes*. New Haven, Hispanic Society of America. *Vol.1*, 15-17, fig. 8.

TSIEN, Tsuen-hsuin (1962): *Written on Bamboo and Silk*. Chicago, University of Chicago Press, 64-83.

6.1310H Map Surface, Vellum

A Vellum is an animal skin from calves, cows, sheep or goats that has been treated with lime and stretched and scraped rather than tanned. The skins are sometimes split, one layer of which is finished on both sides for use as writing materials. Such split skins, most commonly sheepskins, are called parchments, although sometimes parchment and vellum are used interchangeably (Middleton, p.41). Vellum or parchment is thought to have been invented in the 3rd or 2nd century BC. The Greeks and Romans considered it an inferior medium to the papyrus roll which they used for literary works. When Constantine the Great established Christianity as the religion of the Roman Empire, parchment in the form of the codex (the paged book) became the accepted medium for written texts in Europe. Thus the Codex Palatinus in the Vatican Library, of the 4th or 5th century, comprising the almost complete text of Virgil's works, is on parchment. The Greek Bible, the only work with a fuller early manuscript tradition than Virgil, is known in three complete or nearly complete manuscripts of the 4th and 5th centuries, all on parchment (Williams and Pattie, 74-5).

B It is reasonable to assume that maps were drawn on vellum at the same time that texts were written in codices. The maps derived from the famous Roman map, the "Orbis terrarum," or "Survey of the World," executed by M Vipsanus Agrippa and completed after Agrippa's death in 12 BC, were probably on vellum. For about a thousand years vellum remained the chief medium for European mapmaking and the scriptoria in the monasteries were the main centres for production. The *Phaenomena* of Aratus of Soli, late 10th century, Latin, on vellum (British Library, Harley MS 2506), was the basic text for medieval astronomers (Backhouse, and others, no. 43); the drawings of the constellations were made by an artist who had spent some time at Fleury in France where the philosopher and mathematician Abbo of Fleury was a scholar of renown (see Astronomical map, 3.011). The "Anglo-Saxon" world map of about the year 1000 is preserved in a volume comprising a collection of computational, astronomical and geographical texts, one of the most lavishly illustrated non-liturgical books made in the Middle Ages (British Library, Cotton MS Tiberius B.V. part 1) (see also Fig.7). The map is attached to a copy of Priscian's *Periegesis*.

The large mappae mundi were made up of single or multiple sheets of vellum, according to size. The Ebstorf world map, *c.* 1240 (3.58 x 3.56 m, destroyed in 1943), comprised 30 sheets of different dimensions, on average 740 x 800 mm. The world map by Richard of Haldingham in Hereford Cathedral, *c.* 1290, is circular and painted on a single skin of vellum, 1.63 x 1.35 m, with the neck at the top. The culminating work in this genre was the world map of 1459 made by Fra Mauro, a monk of Murano near Venice. It is circular on parchment, 2 metres in diameter (Biblioteca Marciana, Venice). The earliest extant terrestrial globe, by Martin Behaim of Nuremberg, 1492, is inscribed in manuscript on vellum.

Sea charts, also known as portolan charts, comprised another major class of medieval map. The charts were most commonly centred on the Mediterranean Sea and were the work of Italian and Catalan chartmakers. They were usually drawn on single skins of vellum which kept their natural outline and were arranged with the neck to the left or right; they varied in size from 90 x 45 cm to 140 x 75 cm. Some were bound in atlas form. The navigator could prick his course on the vellum using a pair of dividers.

The revival of Ptolemy's Geography was a major event in the geographical renaissance in the 15th century. The manuscript Ptolemys were normally drawn on vellum and prepared as presentation pieces. Examples in the British Library are the Latin codices executed by Donnus Nicolaus Germanus, a Benedictine monk of Reichenbach, *c.* 1468, for Borso d'Este, Duke of Modena (Harley MS 7182), and by Jacobus Angelus, dedicated to Pope Alexander V, *c.* 1410 (Harley MS 6195), and the

Greek Ptolemy of the later years of the 15th century (Burney MS 111) which formerly belonged to Charles Maurice, Prince de Talleyrand. From 1477, when the first printed Ptolemy with maps was published, it became common for Ptolemys to be issued on paper, but illuminated copies on vellum for presentation would be prepared in the same workshop (Skelton, p.IX). The woodcut Ptolemys, such as the *Cosmographia* published at Ulm, 1482, rather than the copperplate, were printed on vellum. As Cornelis Koeman points out, letter press from a raised surface had been used for printing on thin flexible parchment since Gutenberg's time, but intaglio printing on parchment or vellum was not practiced before the late 16th century. Vellum was stiff and resisted the application of printing ink under the pressure of a copperplate (Koeman, p.53). It is significant that only one copy printed on vellum of Christopher Saxton's Atlas of England and Wales, 1579, is known, namely that in the Lessing J Rosenwald's collection of printed works on vellum, now in the Library of Congress, Washington, DC (Note of an exhibition of printing on vellum at the Library of Congress, *The Map Collector (1978), No. 3*, 35).

By the end of the 15th century the best illuminators were more often working in the service of princes and the aristocracy than in the monasteries. Beautiful atlases of charts on vellum made for royal or noble presentation include such works as Jean Rotz's "Boke of Idrography," 1542, presented to Henry VIII (British Library, Royal MS 20.E.IX) and Diego Homem's atlas made for Queen Mary, 1558 (British Library, Add. MS 5415.A.). The durability of skin explains the continued production of manuscript charts on vellum over a period of six centuries. The importance of the medium can be judged from the fact that the Département des Cartes et Plans of the Bibliothèque Nationale had in its collection in 1946 a total of 132 charts from the 13th to the end of the 18th centuries, representing all schools of chartmaking, Pisan, Genoese, Venetian, Majorcan, Portuguese, Spanish, English, Dutch, Norman, Marseillais, and Basque. A further 32 charts on vellum transferred from the Archives du Service de la Marine date from the 16th to the mid-18th century, and 18 from the Société de Géographie from the mid-16th to the end of the 18th centuries.

Towards the end of the 16th century the advantages of vellum for use at sea led to the development of a new technique, copperplate engraving on vellum. Although no engraved portolan chart on vellum is known before 1580, printers in Amsterdam seem to have known the secret at that time (Koeman, p.53). Lucas Janszoon Waghenaer of Enkhuizen and Cornelis Doetsz (or Doedes) of Edam were the first Dutch chartmakers to print portolan charts on vellum. It then became a common practice to print charts for atlases on paper and to print charts for use at sea on vellum. An example is the chart of Western Europe by Willem Jansz. Blaeu [1606]. Other well-known charts on vellum are W J Blaeu's "Generale Pascaerte vande gheheele oostersche westersche . . . zeevaert," 1605, and his "West Indische Pas-kaert," 1630, together with Jacob Aertszoon Colom's copy, *c.* 1640, and Pieter Goos's chart, "Oost Indien," Amsterdam, *c.* 1681. The Bibliothèque Nationale lists nearly 40 charts printed on vellum. Land maps were also printed on vellum in the Nether-lands, and it became a tradition to print dedicatory copies of maps and charts on vellum.

Some English and French maps and charts were also printed on vellum in the 17th and 18th centuries. John Man's second map of Jamaica, 1671, appears in two versions in *The Blathwayt Atlas,* a volume of maps of America collected by William Blathwayt for the use of the committee of the Lords of Trade and Plantations,

1675-83. One (no. 35 in the atlas) is on paper, the other is a special copy on vellum, revised in manuscript with some colour added, and may be the one recorded as presented to the governor Sir Thomas Modyford (Black). The deletions of printed features illustrates the fact that vellum as a medium for the printed map permitted this type of revision of individual copies. Another example is the Ulm Ptolemy in the New York Public Library, in which the name of Johann Schnitzer, the engraver of the world map (the earliest printed map to bear an engraver's signature), has been scraped off.

C BACKHOUSE, Janet, TURNER, D H, and WEBSTER, Leslie (1984): *The Golden Age of Anglo Saxon Art.* London, British Museum and British Library Board.

BLACK, Jeannette D (1975): *The Blathwayt Atlas. Volume II. Commentary.* Providence, Brown University Press, pp. 188-92.

CRONE, Gerald R (1978): *Maps and Their Makers.* London, Dawson.

FONCIN, Myriem, DESTOMBES, Marcel, and LA RONCIÈRE, Monique de (1963): *Catalogue des cartes nautiques sur vélin.* Paris, Bibliothèque Nationale.

GAUR, Albertine (1984): *A History of Writing.* London, The British Library.

KOEMAN, Cornelis (1980): "The Chart Trade in Europe from Its Origin to Modern Times," *Terrae Incognitae, 12,* 49-64.

MIDDLETON, Bernard C (1972): *The Restoration of Leather Bindings.* Chicago, American Library Association.

REED, Ronald (1972): *Ancient Skins, Parchments and Leathers.* London and New York, Seminar Press.

REED, R (1975): *The Nature and Making of Parchment.* Leeds, Elmete Press.

SKELTON, R A (1963): *Claudius Ptolemaeus Cosmographia Ulm 1482.* Amsterdam, Meridian Publishing Co.

WILLIAMS, R D, and PATTIE, T S (1982): *Virgil . . . His Poetry through the Ages.* London, The British Library.

6.132 Mosaic Map

A Maps consisting of small coloured stone cubes (tesserae) set in a matrix.

B Mosaic maps are a rare form of cartographic expression. As a graphic art, the mosaic flourished in Byzantine times, and it is from then that the best known examples originated. Since each graphic element in the map (point, line, area and lettering) is composed of discrete units in the form of 1.0–1.5 cm squares, the graphic resolution of an inlaid mosaic map is naturally low, giving rise to discontinuities of both form and colour. Mosaic maps need to be viewed from a certain distance, and therefore their scales, together with overall sizes, must be relatively large. As a consequence, mosaic maps were used chiefly as floor and wall decorations.

Although countless decorative mosaics – both horizontal and vertical – have survived from antiquity, only a few of these show maps. One, a representation of Egypt, was found in the temple of Fortuna Primigenia, Praeneste (Palestrina), southeast of Rome. A multi-coloured mosaic plan of baths in Rome, found in the praetorian camp and now in the Capitoline Museum, contains measurements of rooms in Roman feet (*Corpus Inscriptionum Latinarum,* 1894). A mosaic world map at Nicopolis in Epiras dates from the 6th century AD.

The best known example is the Madaba Map, found in 1884 in the ruins of a Byzantine church in the village of Madaba, Jordan, 30 km south of Amman. The map, constituting the floor and originally measuring some 22 m by 7 m, was partly destroyed during reconstruction of the church in 1896. It shows the Holy Land and until recently was the oldest known map of that area. It was based chiefly on the *Onomastikon* by Eusebius, Bishop of Caesarea (*c.* 260-340), compiled as a list of Biblical place names. Judging by certain buildings shown in the map, the dates of construction of which are known, the map itself can be dated between 560 and 565.

The Madaba Map is truly oriented, i.e., east is east. While neither a map projection nor a proper scale factor was used, general direction and relative location, though not distances, were kept to a marked degree. Features are represented with the aid of colours and include towns (partly shown by graded symbols), rivers and mountain ranges (as elongated clusters of brown, yellow and red). With its vignettes of buildings, animals, plants and ships, the Madaba Map can be regarded as a forerunner of medieval and modern pictorial maps. It carries many inscriptions in Greek, including place names, historical and geographical notes, and allusions to passages in the Old Testament. The most detailed representation is that of Jerusalem; present archaeological excavations verify its authenticity and adherence to detail.

AVI-YONAH, Michael (1954): *The Madaba Mosaic Map.* Jerusalem, Israel Exploration Society.

BAGATTI, B (1957): "Il significato dei mosaici della scuola di Madaba (Transgiorda-nia)," *Rivista di Archeologia Cristiana, 23,* 139-60.

BAGROW, Leo (1964): *History of Cartography.* Revised and enlarged by R A Skelton. London, C A Watts and Co.; Cambridge, Mass., Harvard University Press, 20-21, 217 and Plate XIV.

Corpus Inscriptionum Latinarum (1894): *VI,* pt 4.1, 2987, no. 29845.

DALTON, O M (1911): *Byzantine Art and Archaeology.* Oxford, Clarendon Press, 422, figs. 248-9.

LEVI, Annalina, and LEVI, M (1967): *Itineraria Picta: Contributo allo studio della Tabula Peutingeriana.* Rome, L'Erma di Bretschneider, 51-5.

MILLER, Konrad (1898): *Mappaemundi; Die ältesten Weltkarten.* Stuttgart, Jos. Roth'sche Verlagshandlung, vi. 148-54.

O'CALLAGHAN, R T (1953): "Madeba (carte de)," in Pirot, Louis, and Robert, André: *Dictionnaire de la Bible.* Paris, supplement, fasicule 26.

PALMER, P, and GUTHE, H (1906): *Die Mosaikkarte von Madeba.* Leipzig.

SCHULTEN, A (1900): "Die Mosaikkarte von Madeba," *Abh. Ges. d. Wiss. Göttingen,* N.F. IV, 2, Berlin.

6.231 Watermark

A The translucent image formed in paper as a result of an emblem or design affixed to the mesh of a paper mould, resulting from a reduction of thickness and fibre density. Watermarks in handmade papers were produced by the design being shaped in wire which was sewn to the mesh. A smaller secondary mark, often a single letter but well separated from the main mark, is called a countermark. Watermarks are usually found in handmade Western papers made on wire moulds but not on their Eastern counterparts, in which bamboo was characteristically used for the mould.

B The first use of watermarks appears to have been in Fabriano, Italy, around 1282, and by the time of the invention of printing around 1440, they had become a common practice with hundreds of designs in use. The functions of these marks are not properly understood but include, among others, the trademark of the mouldmaker or the papermaker, the arms of a patron, a special design for a large custommade batch, and, later, the size of paper (e.g., foolscap, elephant, etc.).

Watermarks are of great value in dating textual and graphical material on paper, but the prime date is that of the manufacture of the sheet, not the creation of the map or document produced on it, and it is not unknown for old paper to be re-used for honest or spurious purposes. The date of the mark can be derived either historically from the records of the papermill or from the dates borne by text, etc., or similar paper. Recent advances in watermark photography and radiography have demonstrated that minute differences in the state of the paper mould over its life can be detected and chronicled; an example of this analysis is found in Stevenson.

When combined with other physical evidence and a proper understanding of the historical context of papermaking, watermarks can be a valuable aid in identification and dating. Of the numerous manuals of early watermarks, that of Briquet is the best known.

C BAYNES-COPE, A D (1981): "The Investigation of a Group of Globes," *Imago Mundi, 33,* 9-19.

BRIQUET, Charles Moise (1968): *Les Filigranes,* 4 vols. Amsterdam, The Paper Publications Society. (Facsimile of 1907 edition.)

HUNTER, Dard (1957): *Papermaking: The History and Technique of an Ancient Craft,* 2nd ed. London.

Monumenta Chartae Papyraceae (1950–). Amsterdam, The Paper Publications Society. (14 volumes to date covering the history of paper and watermarks in various countries and periods.)

SCHOONOVER, David (1982): *Techniques of Reproducing Watermarks.* New York, Modern Language Association.

STEVENSON, Allan Henry (1967): *The Problem of the Missale Speciale.* London, Bibliographical Society.

GROUP 7

Methods of Duplication

7.011 Anaglyptography

A The art of engraving representations made by an anaglyptograph of objects with raised surfaces (coins, medals, relief models, etc.); also called medal-engraving. The result gives a three-dimensional, "plastic relief" effect. The anaglyptograph is an elaborate machine which has a pointer that is taken across the three-dimensional object along successive, closely spaced, equidistant lines, i.e., "scanned." A pencil attached to the device makes a drawing which is the guide for the engraver.

B The anaglyptograph was patented by John Bate in the first half of the 19th century. The later work of Alfred Robert Freebairn (1794-1846) was entirely confined to engraving representations made by an anaglyptograph. Among the most notable examples of his work were the maps made from Colonel William Siborne's relief model of the area where the battle of Waterloo occurred. Colonel Siborne was ordered to make this model by the army, and while preparing it he collected so much information about the battle that he wrote a book. This was illustrated by anaglyptograph portraits of the generals and accompanied by anaglyptograph maps made from the model (Fig. 9). Although there are references to anaglyptography in the 1870s, the early deaths of Siborne and Freebairn combined with the invention of photography seem to have caused a decline in the use of the system. Even the later three editions of Siborne's books were not accompanied by anaglyptographic maps but instead contained maps showing relief by hachures, in which Siborne was greatly interested.

C HARRIS, Elizabeth M (1968): "Experimental Graphic Processes in England 1800-1859," *Journal of the Printing Historical Society* (London), *4*, 74-86.

HARRIS, Elizabeth M (1975): "Miscellaneous Map Printing Processes in the Nineteenth Century," in Woodward, David, ed., *Five Centuries of Map Printing.* Chicago and London, University of Chicago Press, 129-30.

SIBORNE, William (1844): *History of the War in France and Belgium, in 1815, Containing Minute Details of the Battles of Quatre-Bras, Ligny, Wavre, and Waterloo*, 2 vols. and atlas. London.

7.012 Anastatic Process

A A means of reproducing either drawn or printed images employing a method of transfer to a zinc plate and subsequent planographic printing.

B As originally conceived, the process was concerned with transferring from paper a page of letterpress, an illustration, or a map. After proper preparation and inking, multiple reproductions could be printed. Before photography was invented, anastatic printing was regarded as a technique for reproducing facsimiles of works previously printed from type or from engraved plates. Subsequently it was discovered that original drawings made on properly prepared paper with appropriate ink could also be transferred to zinc plates and duplicated.

The technique, which only later became known as the anastatic process, was invented about 1840 by Charles Frederick Christopher Baldamus of Erfurt, Germany. It was promoted in England by a civil engineer, Joseph Woods, who in 1844 took out an English patent on the process, for which he proposed the name "anastatic printing." The process achieved its greatest fame in England between 1845 and 1855. Several maps were reproduced by means of it at Cowell's Anastatic Press, London, in the early 1850s, including Peter Brutt's "Plan . . . for Harwich Docks," 1852, and R A Shafto Adair's "Military Sketch-Map of England and Wales . . . ," 1855. The process was introduced to the United States in 1846 by John Jay Smith, Librarian of the Library Company of Philadelphia, and his son Robert Pearsall Smith. The Smiths published (in 1846) one map by means of the process, a facsimile edition of Thomas Holme's "Map of the Province of Pennsylvania," 1681.

By 1862 the process had fallen into abeyance. Despite the high hopes it had raised, the results had been disappointing and unremunerative. Nevertheless, it had contributed to the advancement of graphic reproduction of materials, including maps, in the transition from engraving to lithography. The technique may also have stimulated the development of the transfer process and the use of zinc plates for lithographic reproduction.

C COWELL, S H (1852): *A Brief Description of the Art of Anastatic Printing.* Ipswich, England, Anastatic Printing Office.

DE LA MOTTE, P (1849): *On the Various Applications of Anastatic Printing and Papyrography.* London.

RISTOW, Walter W (1972): "The Anastatic Process in Map Reproduction," *Cartographic Journal, 9*, 37-42.

RISTOW, Walter W (1975): "Lithography and Maps, 1796-1850," in Woodward, David, ed., *Five Centuries of Map Printing.* Chicago and London, University of Chicago Press, 101-2, 111-12.

WAKEMAN, Geoffrey (1970): *Aspects of Victorian Lithography: Anastatic Printing and Photozincography.* Wymondham (Leics.), Brewhouse Press.

7.021 Ben Day

A An apparatus for transferring line, stipple, or other area patterns to a lithographic or a relief printing surface. A flexible gelatine film with a raised pattern on one side was inked and suspended in an adjustable frame just above the surface to receive the image. A pointed instrument was used to press down the film and transfer the image. Areas not intended to receive the image were masked. The workman could see through the film to guide his work. The film could be shifted and patterns superimposed in subsequent transfers to form a graduated series of flat tones.

Described as Printing-Film in the original patents, the method also came to be known after its inventor as the Ben Day stippling machine, the Ben Day tint, or the Ben Day shading medium.

B Patented by Benjamin Day of Hoboken, New Jersey, in 1878 and 1881 (U.S. Patents 214,493, and 250,211), by 1883 the Ben Day apparatus was being marketed in England and on the continent where it rapidly gained popularity in the next decades. In lithography, the Ben Day process and improved European versions of it supplanted the older method of transferring machine-ruled area patterns from master plates. Ben Day tints found similar application in the preparation of relief-etched metal plates. The Ben Day process remained an important nonphotographic method of creating the area image on maps well into the 20th century, long after photomechanical reproduction of the line image had taken over in lithographic and relief printing.

C ANONYMOUS (1883): "Day's Lithographic Shading Mediums," *Printing Times and Lithographer, 9,* 171-72.

ANONYMOUS (1897): "Day's Shading Mediums," *Process Year Book (Penrose's Annual), 3,* 122.

FRITZ, Georg (1901): *Handbuch der Lithographie und des Steindruckes,* Bd. 1, *Handbuch der Lithographie.* Halle a. S., Wilhelm Knapp.

7.031 Colour Printing

A The colouring of maps by some form of mechanical impression, as contrasted to hand or stencil colouring with watercolour washes. Different colours were sometimes applied in a single impression from a selectively inked printing form but more frequently by the separate printing of colour plates in register with the base plate. Sometimes the base plate was executed in a technique best suited to fine line work and lettering, as copper engraving, while colours were added by another method more suitable for the area image, as woodcut or lithography.

B In the Western world, although the colour printing of maps was tried early in the 16th century, the problems of expense and especially of registering precisely different colours kept it from becoming a common practise until the 19th century. The scattered examples of earlier colour printed maps generally reflect contemporary trends in other areas of colour printing. For example, the Venice Ptolemy of 1511, with classes of geographical names differentiated in red and black print, follows the 15th century practice of printing book headings and initials a different colour from the body of the text. The map of Lotharingia in the 1513 Strassburg edition of Ptolemy was printed by Johann Schott from three woodblocks just at the time that similarly executed chiaroscuro woodcuts in imitation of wash drawings were enjoying brief popularity in Germany and Italy.

During the 18th century, a general rising interest in intaglio colour printing is reflected by Jacques Nicolas Bellin's "Carte Reduite des Isles Philippines", 1752, which has a network of rhumb lines in a second colour, and by Thomas Jefferys' *A General Topography of North America and the West Indies* (1768) which includes a map printed in colour from engraved metal plates.

In 1818 Aloys Senefelder, the inventor of lithography (1798), suggested a method of producing coloured lithographs from separately inked stones, but it was not until the later 1820s that innovations in the preparation and printing of colour plates made the colour-printed map generally feasible. In 1823 Firmin Didot of the French printing house, Paris, patented a multicolour map printing process. Colour lithography was introduced in 1826 by the Militärgeographisches Institut in Vienna, and was employed effectively in the production of various maps of city environs, notably the land-use map in 112 sheets "Umgebungen von Wien", made at a scale of 1:14,000, begun in 1830, and believed to be the earliest chromolithographic map. Ernest von Nischer von Falkenhof (1925) credits this innovation in lithographic techniques to Franz Ritter von Hauslab. In 1841 the "Tableau d'Assemblage des Six Feuilles de la Carte Géologique de la France" was printed in twenty-three colours. In 1843 Victor Raulin's "Carte Géognostique du Plateau Tertiaire Parisien," produced by the Kaeppelin lithographic printing firm of Paris, included eleven colours.

In the latter half of the century many new ideas were advanced. For example, colour printing by a three-colour process, making use of three stones whereon parallel line screens of uniform density had been drawn mechanically, was invented about 1860 by Charles Eckstein, director of the Topographical Department of the Ministry of Defence in the Netherlands. Lithography, introduced at the beginning of the 19th century, helped to facilitate this change, but new relief and intaglio printing methods also contributed substantially to the adoption of colour printing in cartography.

In the Eastern world colour printing was invented in the 16th century when Chinese woodblock printing reached its golden age. In an historical work entitled *Yü-shih Yüo-shu* by Wang Kuang-lu, published in 1638, the place names on the map of China are printed in red and black, and in the 1767 China-centred world map by Huang Ts'ien-jen the grid lines are printed in brown. But on the whole colour printing scarcely seems to have been applied to maps. In the early 19th century, however, polychrome printed maps began to appear in increasing numbers. Among them were several derivatives of Huang Ts'ien-jen's world map which, starting with the 1812 edition, were printed in several colours.

Japanese woodblock printing also has a long tradition. The art reached a new level in the late 17th century, when the popular genre of woodblock prints known as *ukiyo-e* made their first appearance, and printing techniques made important strides forward. Around the mid-18th century artists began to practise a delicate technique using a number of carved blocks applied one after another to achieve fine polychrome printing. Maps which until then had been hand-coloured began to be printed in colour with such methods. The oldest example of these is said to be a 1764 map of Nagasaki in which most colour was applied by hand, but the seas and rivers were printed in blue. Soon afterwards, in 1774, a fully printed multicoloured map of Kyoto appeared, and in 1785, one of Edo (Tokyo). From then on, maps printed in colour appeared in great numbers and variety. Printed multicolour maps, however, tended to cater to popular tastes for decoration even at the cost of accuracy, whereas maps of Japan and the world made by competent scholars continued to be hand coloured.

Copperplate printing of maps by Dutch techniques started in Japan toward the close of the 18th century, and by the middle of the 19th century, it had become the dominant method of map printing. From that time onward, colour printed woodcut maps faded into oblivion.

C BOSSE, Heinz (1951): "Kartentechnik I," *Petermanns Geographische Mitteilungen*, Ergänzungsheft Nr. 243.

BOSSE, Heinz (1951): "Kartentechnik II," *Petermanns Geographische Mitteilungen*, Ergänzungsheft Nr. 245.

NISCHER [-FALKENHOF], Ernst von (1925): *Österreichische Kartographen, Ihr Leben, Lehren und Wirken*, Vienna.

ODA, Takeo, MUROGA, Nabuo, and UNNO, Kazutaka (1972): *Nihon Kochizu Taisei* (Monumenta Cartographica Japonica) [in Japanese]. Tokyo, Kodansha.

ODA, Takeo, MUROGA, Nabuo, and UNNO, Kazutaka (1975): *Nihon Kochizu Taisei, Sekaizuhen* (The World in Japanese Maps until the Mid-19th Century). Tokyo, Kodansha.

PEARSON, Karen S (1980): "The Nineteenth-Century Colour Revolution: Maps in Geographical Journals," *Imago Mundi, 32,* 9-20.

WOODWARD, David, ed. (1975): *Five Centuries of Map Printing.* Chicago and London, University of Chicago Press.

7.041 Dry Impression

A The printing of lithographs on dry rather than damp paper.

B Registration of colour plates for copper and lithographic engraved maps was complicated by the need to dampen the paper and thus render it soft enough to pick up the ink from the lines on the printing surface. The printer had to compensate intuitively for the stretching of the damp paper in the press by letting it dry a little between each colour impression. The solution for lithography was to print on dry paper, sized to resist the dampness of the lithographic stone.

Godefroi Engelmann introduced the practice of printing lithographs on dry paper in Paris during the 1830s in conjunction with his important invention, the registration frame. The colour plates used to produce the 1841 "Tableau d'Assemblage des Six Feuilles de la Carte Géologique de la France," one of the earliest geologic maps reproduced by colour lithography, were printed on dry paper at the Imprimerie nationale using Engelmann's device. However, the lithographic image printed in this fashion had to be one drawn on the surface of the stone with crayon or ink since the slightly incised lines of a lithographic engraving, the most popular method of executing lithographic maps during the mid-19th century, would not print well on dry paper. The problems occasioned by printing on damp paper were finally eliminated in the 1860s when the printing of transfers from lithographic engraved master stones was adopted. Although the transferred image was somewhat lower in quality than the original, the master image was preserved from wear, and the ability to print on dry paper was a great advantage in colour registration.

C ANONYMOUS (1875): "Godefroi et Jean Engelmann. Introduction de la lithographie en France. Invention de la chromolithographie et la diaphanie," *L'Imprimerie, 2,* 719-20.

ANONYMOUS (1894): "Camaïeu Typographique, Lithographique et Taille-Douce," *L'Imprimerie, 31,* 114-15.

LEMERCIER, Alfred (1899): *La Lithographie française de 1796 à 1896 et les arts qui s'y rattachent: manuel pratique s'adressant aux artistes et aux imprimeurs.* Paris, C. Lorilleux et Co.

VERNEUIL, Charles (1880-1883): "Des Procédés de la gravure de la nouvelle carte de France," *L'Imprimerie, 5,* 451-52 and 472-74.

7.051 Electrotyping

A The preparation of, or the duplication of, a printing surface or object by electrolysis. An electrotype is made by depositing a thin layer of metal (shell) either on a matrix for a relief plate (7.1810) or on an engraved intaglio plate (7.0920), the electrotype of the latter becoming the mould for then duplicating the original. The shell is backed with metal for strength.

B Following Michael Faraday's discovery of electrolysis in 1833, and Warren De La Rue's observation of a separable deposit in 1836, the first public announcement of the utility of electrolysis in the reproduction of line engravings was by M H von Jacobi before the Academy of Science in St. Petersburg, October 5, 1838, soon followed by claims in England of invention by C J Jordan and Thomas Spencer. In

America Joseph A Adams had employed the process in 1839 to produce duplicate plates from wood engravings (7.1810d). Elsewhere, William Dalgleish of the Ordnance Survey of Ireland elaborated the process in relation to large engraved map plates between 1837 and 1840, and it was in regular use in Dublin from 1840. Electrotyping became a normal part of map production for the Ordnance Survey of Great Britain at Southampton from 1847. The electrotype remained an essential feature in many practical and experimental reproduction processes.

The search for a satisfactory source of electricity to replace chemical batteries and thermopiles for electrolytic processes was a fundamental cause of the development of the electric generator. However, electric power provided by Smee's cells and later Clamond's thermopiles continued to be used after the development of electric generators because of the overriding need for a steady as well as a powerful electrical current over lengthy periods of deposition to make the electrolysis uniform.

Simultaneous developments in a number of centres, particularly in the later 1830s, led initially to two main types of electrotyping in map production. Rapid expansion of letterpress printing by power presses required a quicker and cheaper method for the preparation and duplication of relief blocks. The established process of stereotyping was inadequate for, or unfitted to, maps. The solution applied by Sidney Edwards Morse in America (1839) led to the rapid development of cerography or wax engraving (7.1810a) to provide an intaglio matrix for electrotype casting of line illustrations in relief suitable for letterpress reproduction.

In the case of copperplate engravings, the application of the principle to maps was developed most notably by William Dalgleish of the Ordnance Survey Office, Dublin (1837-40). From the engraved plate a relief mould was cast. From this master mould a subsequent electrotype produced an intaglio plate identical with the original engraving, which could be used either for printing or for the addition of new detail – e.g., contours – thus preserving the original engraving from the normally heavy wear and tear in the press. The mould also offered the facility for correcting an engraved plate without the undesirable results of scraping away unwanted detail on the original engraving followed by "hammering up" which might introduce imperfections in the surrounding detail. Detail scraped away on the relief mould left a flat surface which after reduplication was suitable for adding information by engraving.

The electrotype was normally cast in copper which had properties not identical with commercial copper. This fact and the need for longer press runs in high pressure powered presses plus the effects of corrosion of copper by new and especially red-coloured inks led to the application of superficial electrolytic deposits of steel (in actual fact iron), nickel, chromium, etc., as a preservative to the engraving or relief plate.

C ANDREWS, John Harwood (1975): *A Paper Landscape. The Ordnance Survey in Nineteenth-Century Ireland.* Oxford, Clarendon Press.

SMITH, Cyril Stanley (1974): "Reflections on Technology and the Decorative Arts in the Nineteenth Century," in Quimby, Ian M G, and Earl, Polly Ann, eds., *Technological Innovation and the Decorative Arts*, Winterthur Conference Report, 1973. Charlottesville, Virginia, University Press of Virginia, 1-64.

VOLKMER, Ottomar (1884): "Die Verwertung der Elektrolyse in der graphischen

Künsten," *Mitteilungen der k.k. militär-geographischen Institut* (Wien), *4*, 65-88.

WOODWARD, David (1977): *The All-American Map, Wax Engraving and Its Influence on Cartography.* Chicago and London, University of Chicago Press.

7.081 Halftone

A A continuous tonal image which has been translated photomechanically into a printable pattern of very small marks whose varying sizes give the impression of continuous tones. During the 19th century many methods of breaking up the tonal image for this purpose were tried, including exposure of the tonal image onto a rough surface; exposure of a light-sensitive gelatine surface (Collotype printing, 7.1210b); use of an open ground as a mask whose physical contact with the tonal image protected portions of it from chemical etching; and, finally, the halftone photographic screen. The openings in the halftone screen function as tiny pinhole cameras which permit varying amounts of light to pass through (in direct proportion to the reflectance of the tonal image) and strike the light-sensitive photographic emulsion, creating dots of varying size.

B Experimenters seeking to produce photomechanical halftones tried assorted methods with varying success from the early 1850s through the 1860s. Cartographic applications were limited, but such early processes were used occasionally to reproduce terrain shading and relief models. Lithographic halftones were achieved in 1852 by Lemercier, Barreswil, Davanne and Lerebours of Paris, who used a grained (roughened) lithographic stone to break up the tonal image. Paul Pretsch of Vienna produced halftone relief and intaglio (gravure) plates *c.* 1854 by making electrotypes from swelled gelatine plates. The idea of the modern photographic screen had been suggested by William Henry Fox Talbot of England in a patent of 1852.

The first workable halftone screen was patented in the United States by Friederich W von Egloffstein in 1865 and used by him to print shaded relief maps of the western United States from intaglio plates. By the late 1860s, the potential of the photographic screen had been generally recognized, and efforts were concentrated on perfecting its use. Georg Meisenbach of Germany achieved the first commerical success with his relief Autotypie process (patented in England in 1882), which employed a single line screen turned during exposure. The crossline halftone screen as it is known today was perfected independently during the 1880s by Frederick E Ives and by Louis and Max Levy in Philadelphia. General application in intaglio and relief printing (and somewhat later in lithography) followed.

C KAINEN, Jacob (1951): "The Development of the Halftone Screen," *Smithsonian Report,* 409-25.

KOEMAN, C (1975): "The Application of Photography to Map Printing and the Transition to Offset Lithography," in Woodward, David, ed., *Five Centuries of Map Printing.* Chicago and London, University of Chicago Press, 137-55.

7.091 Ink

A A fluid used for writing, printing and drawing, composed of some dye or pigment in a vehicle. The identification of anomalous or anachronistic inks may be helpful in dating a suspect map or globe.

B Writing ink was known to early Chinese and Egyptian civilizations, and its use dates from the 3rd millennium BC. The basis of ancient ink was the carbon of lampblack, which was baked with a gum and made into blocks or sticks. Writing fluid was produced by adding water. Greek and Roman writing ink, as used for lettering on papyrus and skin, was usually made from pitch-pine soot mixed with gum or vinegar. Lampblack could also be obtained from the combustion of a variety of materials, such as rice straw, pinewood, vine twigs, and sesame oil. Today it is a by-product of tar, turpentine, oil and natural gas. Carbon ink is the most durable and resistant to atmospheric exposure; its defect is that it clogs and needs constant stirring. The Romans also made a sepia ink from cuttle fish. For coloured inks the ancients used vegetable dyes and minerals such as red ochre, yellow ochre, and vermilion.

In early Christian times an alternative formula was evolved. Ferrous sulphate (Fe SO_4) was mixed with an infusion of nut galls, which contain tannic acid. Small quantities of gum and acid (e.g. vinegar) were added to render the ink stable. It is still the basis of many writing inks, but when the steel pen superseded the quill, anticorrosive modifications were introduced.

Printing ink was manufactured by using varnish or boiled linseed or similar oil of the correct thickness and grinding it with lampblack. Joseph Moxon in *Mechanick Exercises* (2nd ed., 1683, London) especially praises Dutch ink-makers and describes their methods. In the 1770s a patent was issued in England for making coloured inks, and later various drying agents were added.

C LAMB, Cecil Mortimer, ed. (1968): *The Calligrapher's Handbook,* 2d ed. London. Faber and Faber.

MITCHELL, Charles Ainsworth (1937): *Inks: Their Composition and Manufacture,* 4th ed. London.

WHALLEY, Joyce Irene (1975): *Writing Implements and Accessories from the Roman Stylus to the Typewriter.* Newton Abbot, David and Charles.

7.0920 Intaglio Technique

A A method of duplicating achieved by creating a recessed image consisting of grooves or depressions in a smooth material, filling them with ink,, wiping the rest of the surface clean and transferring the ink held in the depressions onto another material.

B The technique was used to make prints in the mid-15th century, and was first employed for maps as copper engraving (7.0920a) in 1477. It was the preferred technique for printing separate maps (maps not printed along with letterpress) until lithography began to supplant it in the 19th century. Although other materials can be used (Steel engraving, 7.0920d), copper has been by far the most favoured.

Fine lines and intricate work could be obtained in copper engraving, but it took much skill and was expensive, both because of the cost of the copper and the fact that it was slow. One advantage of copper engraving was that corrections or changes on the printing plate could be made with relative ease. One drawback of the intaglio technique for map work was that extensive areas of solid printing could not be obtained with the technique; instead large areas of uniformity had to be lined, cross hatched, or stippled. Toward the end of the 18th century such patterns could be produced more or less mechanically on the copper plate. Other intaglio methods of obtaining tones are by etching (7.0920b) and gravure (7.0920c).

7.0920A Copper Engraving

A A method of engraving in which copper plates are incised with a graver or stamping tool. The plates are inked, wiped clean to remove all ink other than in the engraved lines, and printed on dampened paper in a rolling press under great pressure. Re-inking is necessary after each impression. While other soft metals, such as pewter, gold, silver, and zinc, were occasionally used, copper was the preferred medium for map engraving.

B While the roots of the copper-engraving process lie in antiquity in the methods used by goldsmiths to record the progress of engraved decoration on their work, the oldest dated print from copper is one of a series in the *Passion of Christ* printed in 1446 in Berlin. The earliest confirmed use for maps is in the twenty-six maps in the 1477 edition of Ptolemy's *Cosmographia* printed in Bologna. There is a possibility that some of the plates for the 1478 Rome edition of the same work were engraved earlier, and it may be that the undated single-sheet world map of Ptolemy also predates 1477, but this requires further study of the watermark and paper evidence. Other important examples of the technique from the 15th century are found in the 1482 edition of Ptolemy edited by Francesco Berlinghieri, the map of central Europe purported to be by Nicolas of Cusa and bearing the questionable date 1491, and various separate maps engraved by Francesco Rosselli. By the middle of the 16th century, the technique was preferred for all maps except those printed with text by letterpress, and was to dominate map printing for the next three centuries until supplanted by lithography in its various forms.

C DESTOMBES, Marcel (1952): *Catalogue des cartes gravées au XVe siècle.* International Geographical Union.

VERNER, Coolie (1975): "Copperplate Printing," in Woodward, David, ed., *Five Centuries of Map Printing.* Chicago and London, University of Chicago Press, 51-75.

7.0920B Etching

A A method for producing an intaglio image in a metal printing plate, usually copper, by means of "biting" with acid rather than by manual incising. The plate is coated with an acid-resistant substance, called a ground, such as wax, bitumen, etc. Removal of the ground by scratching or scraping, much as the scribe-coat is removed in modern scribing practice, exposes the metal which is then etched by acid to produce intaglio ink-holding grooves and pits.

There are two varieties of etching. One uses a ground that covers the plate completely. The other, called aquatint, employs a discontinuous, granulated ground. When bitten the aquatint plate will print various tonal values depending on the nature of the ground and the exposure of the plate to acid baths of varying strengths and for different periods.

B Etching was introduced in the early 16th century, and since then it has been primarily a medium for artistic expression. It was normally used only for the ornamentation and pictorial components of maps, but on occasions etching was applied to cartography notably by Wencelaus Hollar (1607-77) in England. It never became a popular method of map production and was utilised only sparingly thereafter. Many more and better impressions could be obtained from a manually engraved plate.

Mechanical ruling machines for use in copperplate engravings could also be used to produce the closely spaced grooves in a wax ground for subsequent etching, and various kinds of rough-surfaced rockers, rollers, etc., produced patterns of small pits. By the 1820s these techniques were used occasionally to obtain flat tones and areas of pattern on maps. Large areas of solid colour could not be printed from an incised plate, and when colour printing became more common, after the mid-19th century, patterns and lining were used to obtain tints.

The aquatint process was occasionally used to obtain the effect of continuous tone (smooth gradations of tone) as well as flat tone. One of the earliest, if not the earliest, attempts to employ aquatint to obtain continuous tone was by J Carl Ausfeld, the engraver for Carl Ritter's atlas *Sechs Karten von Europa* (1806). He attempted, not very successfully, to produce a copperplate to portray in black ink Ritter's continuous tone, colour wash drawings for a landform map of Europe, entitled (in translation) "The Surface of Europe Represented as a Bas-Relief." Early precipitation maps (3.161) by Oluf Nikolay Olsen (1839), Heinrich Berghaus (1841), and A K Johnston (1848) employed shading by aquatint to show relative amounts where data were insufficient to employ isarithms. Varying densities of other distributions were also so delineated. An early and almost unique attempt to use the aquatint process to obtain flat tones was by J Gardner, the engraver for Henry Drury Harness's population density map in the 1838 *Atlas to Accompany the Second Report of the Railway Commissioners, Ireland.*

Aquatint was never widely employed, and even simple etching for maps was superseded by lithographic and photo engraving techniques after the mid-19th century.

C ENGELMANN, Gerhard (1966): "Carl Ritters 'Sechs Karten von Europa,' mit einer Abbildung," *Erdkunde, 20* (part 2), 104-10.

HARRIS, Elizabeth M (1975): "Miscellaneous Map Printing Processes in the Nineteenth Century," in Woodward, David, ed., *Five Centuries of Map Printing.* Chicago and London, University of Chicago Press, 113-36.

VERNER, Coolie (1975): "Copperplate Printing," in Woodward, David, ed., *Five Centuries of Map Printing.* Chicago and London, University of Chicago Press, 51-75.

7.0920C Gravure

A Gravure is a term with both a general and a specific denotation. In the general sense it refers to the preparation of an intaglio printing plate in which the ink is held in recesses below the surface. In the specific sense, treated here, gravure is a process in which an intaglio printing plate covered with a pattern of closely spaced, tiny, square depressions is used to reproduce continuous tone images. In its early form in the 1860s and 1870s the pattern of depressions (dots) was irregular and was obtained by using resinous powders, but after the development of the halftone screen (7.081) in the 1880s a regular pattern came to be used.

A gravure printing plate differs from an ordinary halftone in the chemical and technical procedures used and in the fact that one is intaglio and the other is not. Of the resulting contrasts in stylistic effect, most noticeable is the appearance of the dots, which in gravure are all the same size and shape, whereas in halftone they vary. In a gravure plate each square dot is a tiny well filled with ink; the deeper the well, the more ink is transferred to the paper, and the darker is the resulting image. The pattern of dots in a halftone reproduction is more noticeable than in gravure, because in gravure the lines separating the dots are thin and uniform and the ink tends to spread, making for smoother transitions.

B There were various forerunners of the gravure process among the numerous experimental printing techniques developed during the later part of the 19th century. Because gravure requires that everything be screened, i.e., broken into a pattern of dots, the process was not particularly suitable for maps with much line work. One occasional mapmaker, Friederich W von Egloffstein, did use a gravure-type process as early as the 1860s, but gravure did not become a practical process until near the end of the century, mainly through the work of Karl Klič of Bohemia. Few maps were printed by gravure.

C HARRIS, Elizabeth M (1975): "Miscellaneous Map Printing Processes in the Nineteenth Century," in Woodward, David, ed., *Five Centuries of Map Printing.* Chicago and London, University of Chicago Press, 113-36.

7.0920D Steel Engraving

A A method of intaglio engraving using steel plates.

B The hardness of steel offered a far longer printing run than the traditional copper, but the engraving was more difficult with the traditional copper engraver's tools. In a variation of the process for small plates, known as siderography, a soft steel plate was engraved and then hardened. It is thought to have been introduced around 1820, but since there is no positive method of distinguishing steel engravings from copper engravings except by designation on the plate, it is difficult to provide a precise date of introduction. It was used fairly widely in both Europe and America for the engraving of school and general atlases, for which long runs and multiple editions were expected.

C HARRIS, Elizabeth M (1975): "Miscellaneous Map Printing Processes in the Nineteenth Century," in Woodward, David, ed., *Five Centuries of Map Printing.* Chicago and London, University of Chicago Press, 114.

7.1210 Lithographic (Planar) Technique

A A method of reproduction utilizing the phenomenon that an oleate ink image on a flat surface can be reproduced by moistening the surface, inking it with an oily ink, and then pressing the surface against a sheet of paper to transfer the image. Since oil and water will not mix, the oleate image accepts the ink but not the water and the moist surface accepts the water but not the ink.

B The principle of lithography was discovered by Aloys Senefelder in Munich in 1798. It was quickly adopted as an easy, fast way to make duplicate copies of a drawing by autographic methods (7.1210a). Because the flat surface used exclusively until the mid-19th century (and often until into the 20th century) was a smoothed limestone slab, it was first called stone printing *(Steindruck)* in German. In French it was called *lithographie,* which name persisted even after slightly roughened or "grained" zinc plates were used in place of the limestone slabs.

Lithography was quickly adapted to map reproduction, but the early crude autographic transfer methods (7.1210a) soon gave way to lithographic engraving (7.1210c) as the preferred technique. Lithographic engraving rivalled copper engraving as a technique for map reproduction until superseded by photographic techniques after the 1860s.

Because the printing surface is essentially planar, lithography differs fundamentally from those duplicating processes which depend upon a raised surface (Relief technique, 7.1810) or an incised surface (Intaglio technique, 7.0920).

7.1210A Autography

A As its name suggests, autography or self-writing permitted the draftsman to create an acceptable rightreading image on paper and see it transferred to the lithographic surface and printed just as he had drawn it. Untreated paper could be employed, but special transfer paper coated with any of several mixtures of thickening, adhesive, hygroscopic, colouring and sensitizing agents permitted more perfect transfer of the image. Drawing a high quality image with lithographic ink or crayon on transfer paper was more difficult than ordinary drafting, as the coarse and rather irregular quality of autographic linework frequently attests.

B The application of the transfer process of lithography to maps dates from 1808 when autography was employed in England for quick reproduction of battle plans and other military maps during the Napoleonic wars. In August 1807 Lieut. Colonel John Brown, Assistant Quarter Master General, bought the secret of Senefelder's invention from the German lithographer J G Vollweiler. With the help of D J Redman, a former employee of Vollweiler, Lieutenant G Pawley printed on 7 May 1808 the earliest extant lithographic map, a plan of Bantry Bay. With the appointment of Sir Willoughby Gordon as Quarter Master General in 1811 lithography was increasingly used for maps, plans and circulars. A collection of some 115 or more maps executed in the Quarter Master Generals Office in the Horse Guards, Whitehall, between 1808 and 1828 are in the British Library (Maps C.18.l.1.; Maps C.18.m.1.) (Fig. 22).

The relatively low quality of the transferred image restricted the widespread use of autography in map reproduction until after the mid-century. During the 1860s lithographic transfer techniques improved, and autography briefly replaced the more expensive technique of lithographic engraving for many purposes until autography, in its turn, gave way to photolithography.

C MUMFORD, Ian (1972): "Lithography, Photography and Photozincography in English Map Production before 1870," *Cartographic Journal, 9,* 30-36.

PEARSON, Karen (1978): "Lithographic Maps in Nineteenth-Century Geographical Journals," Ph.D. dissertation, University of Wisconsin-Madison.

RICHMOND, W D (1878): *The Grammar of Lithography.* London, Wyman & Sons.

TWYMAN, Michael (1970): *Lithography 1800-1850.* London, Oxford University Press.

7.1210B Collotype

A A printing process which simulates continuous tone without a halftone screen. A plate, formerly glass and now often aluminium, is coated with a colloidal bichromated, light-sensitive gelatine solution. The coating is exposed through a continuous tone photographic negative and, after chemical treatment, becomes differentially hardened according to the quantity of light passed by the negative. The coating, when soaked in glycerin, swells more in the lighter image areas and less in the darker. Under controlled humidity conditions the coating will accept less ink in the lighter image areas. The almost invisible grain results in good tonal gradations of near photographic quality.

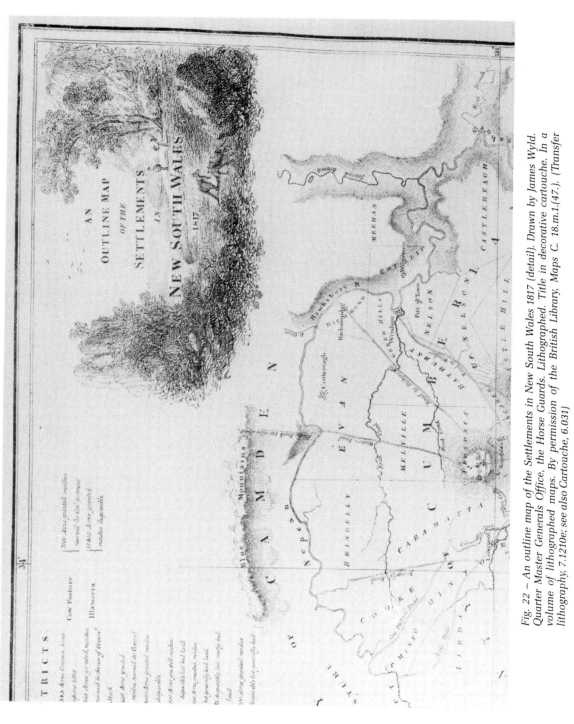

Fig. 22 – An outline map of the Settlements in New South Wales 1817 (detail). Drawn by James Wyld. Quarter Master Generals Office, the Horse Guards. Lithographed. Title in decorative cartouche. In a volume of lithographed maps. By permission of the British Library. Maps C. 18.m.1.(47.). (Transfer lithography; 7.1210e; see also Cartouche, 6.031)

B The collotype concept was probably discovered by Louis Poitevin in 1855. It was patented that year and called photocollography. It was reported by Waterhouse as little known in 1868 when Joseph Albert produced the first power collotype press, and in 1878 as not much used for reproduction of maps in government establishments. The process depended on individual professional manipulation and for long was shrouded in commercial secrecy. The development of the halftone screen combined with photolithography to take much of the interest away from this photographic process in normal map production when long runs were required.

The capability of great fidelity and beauty in copying coloured originals led to its application to the reproduction of old maps (see Facsimile atlas, 8.0110b). Difficulties of colour control and registration may result in a lack of uniformity over the complete print run, which is often difficult to evaluate in individual prints.

C HÖDLMOSER, K (1882): "Über ältere und neuere Reproduktions-Verfahren und deren Verwertung für die Kartographie," *Mitteilungen der k. k. militär-geographischen Institut* (Wien), *2*, 41.

WATERHOUSE, James (1879): "Extract from Report No. $\frac{433}{P}$, dated 14 December 1878, by Captain J. Waterhouse, S. C., on recent progress in Europe of Photography as applied to the Reproduction of Maps, . . . ," in [First] *General Report on the Operations of the Survey of India . . . during 1877-78.* Calcutta, Office of the Superintendent of Government Printing, 147-55.

7.1210C Lithographic Engraving

A Although closer in quality to copper engraving than any other lithographic technique, lithographic engraving clearly stemmed from intaglio etching on metal. Both employed the scribing concept, in which portions of a soft, superficial ground were removed to expose the underlying surface for chemical treatment. The lithographic engraver scratched lines in a coloured gum arabic ground with a steel needle or a diamond point, either handheld or fixed in a ruling machine. Although the aim was simply to expose portions of the stone for subsequent oiling and inking, the slight incisions in the soft stone made a lithographic engraving harder to print than other lithographic methods.

B Lithographic engraving was discovered by Aloys Senefelder, the inventor of lithography, about 1800. As lithography emerged from its initial experimental period during the second decade of the 19th century, lithographic engraving became the preferred technique for executing high quality lithographic maps. Around 1812, for instance, the desire to improve image quality at the Bavarian Cadastral Survey, Munich (where Senefelder was employed from 1809 to 1827), led to a shift from pen and ink to lithographic engraving. Lithographic engraving usurped much of the copper engraving's former role in map reproduction, especially in Germany and France, until itself challenged by transfer and photolithography from the 1860s onward. Lithographic engraving, which had entered the scene as a cheap competitor to copper engraving, by the end of the century was similarly declining in favour owing to the high cost of production.

C AMANN, Joseph (1908): *Die bayerische Landesvermessung in ihrer geschicht-lichen Entwicklung. Erster Teil: Die Aufstellung des Landesvermessung-swerkes, 1808-1871.* München, Verlag des K.B. Katasterbureau.

DESPORTES, Jules (1838): "Gravure sur pierre," *Le Lithographe* (Paris), *1,* 158-67 and 183-96.

WATERHOUSE, James (1870): *Report on the Cartographic Applications of Photography as Used in the Topographical Departments of the Principal States in Central Europe, with Notes on the European and Indian Surveys.* Calcutta, Office of the Superintendent of Government Printing.

7.1210D Photolithography

A Any of a variety of techniques by which the processes of photography (7.161) and lithography (7.1210) are combined to produce a printing surface, usually by transfer methods (7.1210e).

B Experiments in photolithography were initiated around 1860, most notably by John W Osborne, Chief of the Australian Department of Lands and Surveys, and Colonel Sir Henry James, Director of the Ordnance Survey in Great Britain and Ireland. Before photolithography was developed, the Australian surveyors in the field would send in sketches which were redrawn for lithographic or copperplate reproduction. After photolithography became practical, the then more carefully drawn surveyors' maps were themselves photographed. Later, Osborne went to the United States and established the American Photo-Lithographic Company in New York City, which specialized in plans and maps.

Photolithography was widely used after the 1860s.

C HARRIS, Elizabeth M (1975): "Miscellaneous Map Printing Processes in the Nineteenth Century," in Woodward, David, ed., *Five Centuries of Map Printing.* Chicago and London, University of Chicago Press, 113-36.

KOEMAN, C (1975): "The Application of Photography to Map Printing and the Transition to Offset Lithography," in Woodward, David, ed., *Five Centuries of Map Printing.* Chicago and London, University of Chicago Press, 137-55.

OSBORNE, J W (1860): "Specimen of Photo-lithography Copied from the Ordnance Map of the County of Cork by J W Osborne." Melbourne, Deparment of Lands and Surveys.

SMITHSONIAN INSTITUTION, Washington, DC: J W Osborne papers.

VICTORIA PARLIAMENT (1860-61): "Photolithography. Report of the Board appointed by the Hon. the President of the Board of Lands and Works," *Victoria Parliamentary Papers,* 1860-61, vol. 3, No. 11, 1-75.

7.1210E Transfer Lithography

A All of the varied lithographic transfer methods employed in the 19th century utilized the idea that a printable image could be transferred from another surface to the lithographic surface. The transfer image could be drawn on special paper with lithographic ink or crayon, a method known as autography (7.1210a). A transfer could also be made from an existing map plate (most often copper or lithographic engraved). Such a transfer could also form a base image to which special information was added, either on the same plate or by printed colour. In anastatic printing (7.012), another type of transfer lithography, either an existing map print or one drawn expressly for anastatic transfer, was treated chemically to render the image more readily transferable. The transfer of type lettering was sometimes used as well to speed the execution of lithographic maps. Finally, early photolithography generally involved the photographic production of transfers.

B Soon after 1798, the date generally accepted for Aloys Senefelder's invention of lithography, lithographic presses were established in Munich, London and Paris, with output mainly artwork. Nearly all failed after a short time. When the first map was printed by the lithographic process is not known, but it was probably a plan of a small area or a city and done before 1808. Senefelder himself became involved in extensive map printing only in late 1809 when he was appointed superintendent of lithographic printing at the Bavarian Cadastral Survey in Munich, which had decided to reproduce all its maps by lithography. The system used there after 1812 was lithographic engraving (7.1210c).

The first lithographic press set up outside Germany was in London, which Senefelder visited in 1800 to 1801, taking out a patent on 20th June 1801. In August his press passed with the patent to his London associate Philipp Andre. It was licensed in 1806 to J G Vollweiler, who had been assistant to Andre's eldest brother Johann in Offenbach, Hesse. Before returning to Germany in 1807, Vollweiler sold to the Quarter Master Generals Office in the Horse Guards, Whitehall, the press and the secrets of the process, which he called *polyautography*. Employing the system of autographic transfer, the Horse Guards now became the centre for the rapid production of military maps, with thousands of copies issued for official purposes between 1808 and 1828 (see Autography, 7.1210a) (Fig.22). The great advantage of the process was speed of production. For example, a "Sketch of the Action between the British and French Forces at Vimiera in Portugal, August 21 1808" was printed on 5 September 1808 in the Horse Guards.

By the 1820s lithographic presses were well established in cities throughout Europe, but the maps produced were generally crude and not widely popular. In France Godefroi Engelmann and Charles Philibert de Lasteyrie claimed a major role in the development of lithography in the second decade of the 19th century, but few lithographic maps were published before 1820. The first colour maps produced by the lithographic process were of the environs of Vienna, 1830 (see Colour printing, 7.031). Attempts to apply lithographic transfer in more formal production (i.e. transfer of copperplates for lithographic printing, transfer of crayon shading, and typolithographic transfer) met with limited success during the first half of the 19th century.

The low quality of the transferred image restricted its widespread adoption in map reproduction until the 1860s by which time transfer techniques had improved. Common applications in map lithography from that time on included transfer to save the master plate from wear, transfer of copper engraved plates for cheaper lithographic reproduction, autography, transfer of the crayon-shaded and the machine-ruled area image, anastatic printing and typolithographic transfer. These forms of transfer were soon joined by photolithography, first introduced separately for map reproduction in 1859 by Colonel Sir Henry James of the Ordnance Survey in Great Britain and Ireland and by John W Osborne in Australia. Photolithographic transfer, then frequently more practical than direct exposure on the printing surface, replaced many of the non-photographic methods of transfer during the last third of the 19th century.

C FRITZ, Georg (1901): *Handbuch der Lithographie und des Steindruckes.* Bd 1, *Handbuch der Lithographie.* Halle a. S., Wilhelm Knapp.

LEMERCIER, Alfred (1899): *La Lithographie française de 1796 à 1896 et les arts qui s'y rattachent: manuel pratique s'adressant aux artistes et aux imprimeurs.* C. Lorilleux et Co.

MUMFORD, Ian (1972): "Lithography, Photography and Photozincography in English Map Production before 1870," *Cartographic Journal, 9,* 30-36.

RISTOW, Walter W (1972): The Anastatic Process in Map Reproduction," *Cartographic Journal, 9,* 37-42.

WAKEMAN, Geoffrey (1970): *Aspects of Victorian Lithography: Anastatic Printing and Photozincography.* Wymondham (Leics.), Brewhouse Press.

7.151 Offset Lithography

A The basic principle of offset printing is the transfer (offsetting) of the printing image from the printing form to another surface from which the final impression is made. The image on the printing form is right reading, is reversed when offset, and then printed again right reading. When the printing image is first offset onto a rubber cylinder it makes better contact during impression than in direct rotary printing and reduces the wear on the printing form by eliminating contact with the hard-surfaced printing paper.

B Starting with Paul's rotary relief offset press of 1796 (British Patent 2084), persistent attempts to incorporate the offset principle in intaglio, relief and lithographic printing were made during the course of the 19th century. Early offset lithographic presses were flatbed ones with offset cylinders of varying composition and were used mainly for printing on rough surfaces. The idea of offsetting onto paper and thence onto canvas for the imitation of coloured oil paintings was suggested by Aloys Senefelder in the earliest comprehensive lithographic textbook (1819). Flatbed offset presses were being employed for tin printing by the 1880s, a period when direct rotary presses were also in use. Commercial success in printing on paper came when the offset idea was combined with the rotary principle in a more efficient press design.

The first rotary offset press capable of printing on paper was invented by Auguste Hippolyte Marinoni and Jules Michaud of Paris and patented in England by Clark in 1884. During the first decade of the 20th century advances by inventors in the United States and Europe led to the general adoption of rotary offset presses in lithography. Acceptance for map printing followed soon in the United States but more slowly in Europe where map editions were smaller and investments in existing equipment high. The Austrian Military Topographical Survey installed an offset press as early as 1910, but general use of offset lithography, especially in commerical cartography, was delayed in Europe until after 1945.

ANONYMOUS (1924): "Who Invented the Offset Machine?" *Modern Lithographer & Offset Printer,* 290.

ANONYMOUS (1925): "The Genesis of Offset," *Modern Lithographer & Offset Printer,* 307-8 and 356.

ANONYMOUS (1925): "Who Invented Offset Printing?" *Modern Lithographer & Offset Printer,* 48.

EVANS, A B (1925): "A Short History of Offset Printing," *Modern Lithographer & Offset Printer,* 55-7.

KOEMAN, C (1975): "The Application of Photography to Map Printing and the Transition to Offset Lithography," in Woodward, David, ed., *Five Centuries of Map Printing.* Chicago and London, University of Chicago Press, 137-55.

SENEFELDER, Aloys (1819): *A Complete Course of Lithography.* London, R. Ackermann. (Reprinted New York, Da Capo Press, 1968.)

7.161 Photography

A A process in which a negative or positive image on a sensitized surface is produced directly or indirectly as a result of exposure to radiant energy. Variables are the surface, the emulsion carrying the photo-chemical substance, the radiant energy, the manner of processing, and if used, the camera lens. The ultimate objective is to produce a useful image. Since the variety of surfaces, emulsions, light sensitive substances, processes and lenses is large, numerous combinations have been found useful for mapmaking, even before the advent of the air photograph.

B Practical developments in photography began in the early decades of the 19th century, and by 1850 most of the basic photochemical reactions in use today, involving light-sensitive bitumin, chromium and silver compounds, were known. However, none seems to have been adapted to mapmaking.

The first use of photography in map production occurred in the decade 1850-60 in Britain, Germany, Australia and the United States. Initial applications were confined mostly (a) to making photographs of maps as an alternative to making copies some other way, and (b) to the changing of size, thereby making it unnecessary to do all construction work at reproduction scale. After the 1850s experimentation with a variety of gummy emulsions led to the production of an image on a stone or zinc plate as a guide for subsequent ink drafting or engraving. Another development at that time was the preparation of an image which could then be transferred to stone or zinc thereby producing a printing plate (see Photolithography, 7.1210d). The use of the image as a guide for drafting or engraving was adapted in the early 1860s by Charles Eckstein, director of the Dutch Topographical Department, to flat colour map reproduction using red, yellow, blue, and black printing plates and even pin registry. The photographic production of a printing plate, when used with metal, was known as photozincography and was employed by the Ordnance Survey of Great Britain in the early 1860s to make facsimiles of maps.

Photography was used in various other techniques of map reproduction involving photochemical etching as well as electroplating. Probably best known was the heliogravure-electroplating production of intaglio printing plates for the Austrian 1:75,000 during the period 1872-85, by reproduction from 1:60,000 drawings. A less elaborate but widely used application of photography was the production of relief printing "line blocks" after the mid-1870s and "halftone blocks" after the mid-1880s (see Metal engraving, 7.1810b). Both involved exposing a negative to a zinc plate which had a sensitized coating. After removing the unexposed soluble portions, the plate was etched leaving the printing surface, protected by the exposed hardened coating, standing in relief.

Two characteristics of photography and mapmaking before 1900 are worthy of mention. One is that by the end of the 19th century, in one way or another, photography was involved in most mapmaking and reproduction. The other is that only after the 1880s did electric carbon arc lights begin to be used for photographic illumination, and only gradually did they supersede direct sunlight and daylight.

C ECKSTEIN, Charles (1876): *New Methods for Reproducing Maps and Drawings.* Topographical Department, War Office (Netherlands), and The Hague, Giunta d'Albani Bros.

HARRIS, Elizabeth M (1975): "Miscellaneous Map Printing Processes in the Nineteenth Century," in Woodward, David, ed., *Five Centuries of Map Printing.* Chicago and London, University of Chicago Press, 113-36.

KOEMAN, C (1975): "The Application of Photography to Map Printing and the Transition to Offset Lithography," in Woodward, David, ed., *Five Centuries of Map Printing.* Chicago and London, University of Chicago Press, 137-55.

MUMFORD, Ian (1972): "Lithography, Photography and Photozincography in English Map Production before 1870," *Cartographic Journal, 9,* 30-36.

7.1810 Relief Technique

A Any of several printing processes in which the non-printing portions of a block are removed leaving the printing image surface in "relief." The image portion is successively inked and pressed on the receiving surface, usually paper. An advantage of the relief technique is that the printing block can be locked in the form with type and then both are printed simultaneously.

B The earliest of the relief techniques to be used for maps is the woodcut (7.1810c), in which the image was produced on a plank cut block, i.e., cut with the grain. It was first employed for maps in China in the 12th century and in Europe in the 15th. No change in the basic process took place until the first part of the 19th century when the image was produced on a cross cut block, i.e., on the end grain. That technique is known as wood engraving (7.1810d), and it should not be confused with the intaglio technique (7.0920). In wood engraving the image is in relief, not recessed as in copper engraving (7.0920a).

In the 19th century several other relief-technique processes were developed and widely employed. The first, known by several names and introduced *c.* 1840, is generally called wax engraving or cerography (7.1810a). It enjoyed considerable popularity into the present century. In the 1870s the technique known as metal engraving (7.1810b) came into use, combining the technique of photography to fix an image on a zinc plate and the process of acid etching (with manual routing) to remove metal from the non-printing portions.

Until well after 1900 the great majority of the maps that appeared in printed books were prepared by one of the relief techniques.

7.1810A Cerography

A A method of producing a mould from which a printing plate could be cast. The wax engraver applied a thin layer of wax to a plate, usually of copper. The lines and symbols were engraved through the wax to the metal beneath, using special tools. After engraving, most of the wax left between the lines was built up further to give depth to the subsequent printing plate. The printing plate was cast from the wax by the process of electrotyping (see 7.051) and then backed with metal.

Cerography is also known as wax engraving, wax process, glyphography, electrographic printing, typographic etching.

B The technique was invented simultaneously by a number of individuals. Sidney Edwards Morse published a map of Connecticut produced by cerography in the *New York Observer* of June 29, 1839. His *Cerographic Atlas* was published in 1842. The process was pioneered in Britain by Edward Palmer under the name of glyphography, for which a patent was taken out in 1842. The technique of cerography was widely used by commercial map publishers in the United States from 1870 to 1940. The use began to decline after the introduction of offset lithography in 1904 and was completely superseded by 1950 by the improvements in photomechanical methods made during and immediately following World War II.

C HARRIS, E M (1969): "Experimental Graphic Processes in England, 1800-1859, Part 3," *Journal of the Printing Historical Society* (London), *5,* 63-80.

WOODWARD, David (1976): *The All-American Map. Wax-engraving and Its Influence on Cartography.* Chicago, University of Chicago Press.

7.1810B Metal Engraving

A A term usually referring to the relief technique of direct engraving in metal. It includes metalcut, the photographic line block (line cut) and the halftone engraving, all of which could be printed on a common printing press. In the metalcut, analogous to woodcut (7.1810c), a relatively soft metal block was manually engraved with gravers or punches. The line block was produced by exposing a negative to a sensitized zinc plate, and etching the unprotected portions. The relief halftone block used a similar photomechanical principle, but employed halftone screens (7.081) to break up continous tones on the original.

B The metalcut is known to have been used for book illustrations in the 15th and 16th centuries, but no maps engraved by the process have apparently been documented. On the other hand, the photographic line block and halftone block were frequently used at the end of the 19th century for small maps in books, periodicals, and newspapers, although rarely for large maps requiring accurate registration of colours. Charles Gillot is usually credited with the commercial application of the line block in 1872, and by the end of the 1880s, line blocks were appearing in many publications side by side with wood engravings and wax engravings. All three were commonly used for maps. The photographic halftone block was commercially available in the mid-1880s, following the patents of Frederick E Ives of Philadelphia (1881) and Georg Meisenbach of Munich (1882), but the quality was significantly improved by the screen developed by Louis and Max Levy of Philadelphia, patented in 1888. Halftone blocks were used for maps routinely by the 1890s, especially for rendering relief shading, for which they were printed in registration with line blocks or wax engravings.

C WOOD, H Trueman (1890): *Modern Methods of Illustrating Books.* London, Elliot Stock.

7.1810C Woodcut

A One of the oldest methods of map printing, in which a smooth, plank surface of wood (cut with the grain) is carved in relief and ink is transferred from the raised surface to paper using vertical pressure in a hand-printing press. The non-printing areas are carved away with various knives and chisels.

B The oldest known printed map in any culture is a woodcut map of Chinese origin, comprising a map of West China, "Ti Li Chih Thu", in *Liu Ching Thu* (Illustrations of Objects Mentioned in the Six Classics), an encyclopaedia edited by Yang Chia in about 1155 (copy in Peking National Library).

The *Etymologiae* of St Isidore, Augsburg, 1472, contains the first printed map, a world map of T-O style by Günther Zainer (see Ethnographic map, 2.052). The second printed map is the more elaborate world map in the anonymous *Rudimentum novitiorum*, Lubeck, 1475. Because woodcuts and text could be passed through the press together, the technique continued to be employed for maps in books until the second half of the 19th century. In the mid-16th century, woodcut was superseded in Europe by copperplate engraving for maps, especially large maps and maps requiring more precise delineation. Refinements to the woodcut technique included the use of letterpress type for type-inserted lettering (6.1220d; earliest identifiable use: *Rudimentum novitiorum,* 1475) and stereotyping (6.1220c), found in confirmed use on Peter Apian's world map, 1530.

Woodcut is not the same as wood engraving (7.1810d), which produces a relief printing surface by the use of a burin or graver on the polished cross-sections of a close-grained wood, such as boxwood. Wood engraving of maps during the 19th century was used mainly for small or simple maps to be printed with text.

C HIND, Arthur M (1935): *An Introduction to the History of Woodcut,* 2 vols. London, Constable. (Reprinted, New York, Dover Publications, 1963.)

NEEDHAM, Joseph (1959): *Science and Civilisation in China, vol.3.* Cambridge, At the University Press, pl. LXXII, 549.

WOODWARD, David (1975): "The Woodcut Technique," in Woodward, David, ed., *Five Centuries of Map Printing.* Chicago and London, University of Chicago Press, 25-50.

7.1810D Wood Engraving

A A technique in which a burin or graver is applied to the polished cross-sections of a close-grained wood, such as boxwood. It was used during the 19th century for small or simple maps requiring a relief printing process.

B Wood engraving allows finer lines and hence a greater variety of renderings than woodcut, since the graver can be moved in any direction on the end grain. It was used in the usual "black lines on a white ground" mode for thousands of small and usually substantively insignificant maps in textbooks, newspapers, and encyclopaedias. Few maps conceived as "white lines on a black ground" were made. The technique was in full swing by the 1830s, but was quickly superseded by competing methods of producing relief printing surfaces, such as wax-engraving, by the 1870s.

C WOODWARD, David (1975): "The Woodcut Technique," in Woodward, David, ed., *Five Centuries of Map Printing.* Chicago and London, University of Chicago Press, 25-50.

GROUP 8

Atlases

8.0110 Atlas

A A particular collection of maps, usually bound together.

B In the later Middle Ages groups of maps or portolan charts of the known world were often brought together by their authors in the form of a codex or bound volume. Among the earliest examples are the so-called "Atlas of Islam," dating from the 10th century (see Nautical chart, 1.0320c) and, in European cartography, the atlas of Pietro Vesconte, 1313 (Paris, Bibliothèque Nationale, Cartes et Plans, Rés. Ge. DD 687).

As a combination of map and book, the atlas or book of maps is relatively modern in the history of cartography, if the early editions of Ptolemy's Geography (printed with maps from 1477) are considered to be illustrated books. The earliest atlases date from the 15th century and consist of maps of island areas (see Island atlas, 8.0110c). In the 1530s and 1540s some of the hydrographers and cosmographers of Dieppe produced world maritime atlases in manuscript for presentation to royal or noble persons. Jean Rotz claimed to be the first with his "Boke of Idrography," presented to Henry VIII of England in 1542 (British Library, Royal MS.20.E.IX). In this he converted a large chart of the world into a book containing "all hydrography or marine science," explaining that such a book would be easier and more convenient to handle. In Italy an important centre of production for world maritime atlases (in manuscript) was in the workshop of Battista Agnese at Venice, 1536 to 1564.

From 1477 Ptolemy's *Geography* and its later editions published from 1540 by Sebastian Münster, together with Münster's popular *Cosmographia,* published at Basle from 1544 to 1628, undoubtedly served as world atlases and inspired publishers with the idea of making new atlases. Italian mapsellers created the prototype of the modern atlas with their sets of maps, worldwide in compass, taken from the mapseller's stock (Skelton, p. 43). The earliest appears to date from 1568-69, and the concept probably originated in Venice (Tooley, p. 14). Antonio Lafreri of Rome was a leading practitioner, hence the name suggested by A E Nordenskiöld, "Lafreri atlas." An alternative term suggested is IATO — "Italian atlases assembled to order." Some collections have title pages engraved between 1570 and 1572 and worded "Geografia: Tavole moderne di Geografia de la maggior parte del mondo di diversi autori raccolte et messe secondo l'ordine di Tolomeo . . . stampate in rame . . . in Roma." The designed title page normally displayed a picture of Atlas holding the world (not the heavens) on his shoulders. The maps in these atlases were arranged in the standard order of Ptolemy, as the title indicated.

Abraham Ortelius of Antwerp was the first to compile and publish a modern atlas in the form of a systematic, bound collection of maps of various parts of the world in consistent format, compiled on principles laid down by the editor. His *Theatrum Orbis Terrarum,* Antwerp, 1570, was an immediate success and ran to many editions. Ortelius was essentially a cartographic editor using maps of other authorities, whose names he listed in his "Catalogus auctorum." As a companion to Ortelius's *Theatrum* Georg Braun and Frans Hogenberg completed their great city atlas of the world, the *Civitates Orbis Terrarum,* Cologne, produced in six volumes from 1572-1618.

Gerard Mercator of Louvain, later Duisburg, described by his friend Ortelius as "the Ptolemy of our day," achieved the next innovation with the preparation and publication of an atlas comprising new maps from the editor's own hand. The work was issued in parts from 1585 and then published as a bound collection by Mercator's son Rumold in 1595, a year after his father's death. It carried the title chosen by the elder Mercator, namely *Atlas sive cosmographicae meditationes de fabrica mundi et fabricati figura* (Atlas, or cosmographical reflections concerning the structure of the universe and the nature of the universe as created). Henceforward the term "atlas" was generally accepted. The splendour of Dutch atlases inspired the poet Vondel to coin the phrase "a banquet of maps."

The greatest Dutch atlases of the 17th century were produced in competition by the Amsterdam firms of Hondius, later in partnership with, and succeeded by, Johannes Janssonius, on the one hand, and Willem Janz., Joan and Cornelis Blaeu on the other. They established the multi-volume atlas as a popular genre. Father Vincenzo Coronelli continued the tradition from his convent in Venice, emulating the Dutch with his *Atlante Veneto,* Venice, 1690-1701, which ran to 13 volumes and included a variety of atlas forms, for example, *Isolario, Teatro dell Città,* and the innovative *Libro de' Globi,* 1697 (see Globe, 1.0710) (Fig.23). Coronelli's *Theatro della Guerra* in 20 volumes, Venice, 1706-8, ranks as a notable military atlas, though traditional in style.

Specialized atlases having to do with more limited areas also began to be published in the 16th century (Regional atlas, 8.0110f). After the third quarter of the 16th century all kinds of atlas publications burgeoned in a variety of sizes and shapes, from those that would fit in pockets (8.0110e) to some of monumental size and weight that were ornamental as well as informative. The three atlases made in Amsterdam, probably under the direction of Prince Maurice of Nassau, *c.* 1660, rank as the largest

Fig. 23 – Vincenzo Coronelli, surrounded by examples of his cartographic works, presents to Pope Clement XI volume I of his Biblioteca Universale, *the first modern encyclopedia. On the shelves are displayed the volumes of his atlas series,* Atlante Veneto, *Venice, 1691-1698. In* Biblioteca Universale, *Venice, 1701,* I, *frontispiece. (Atlas, 8.0110)*

313

early atlases in the world: namely, the atlas of Charles II, presented by Johan Klencke, 1660 (London, British Library), that made for the Great Elector (East Berlin, Deutsche Staatsbibliothek), and that made for the Duke of Mecklenburg (Rostock University Library). Other Dutch made-up atlases augmented from the works of Blaeu and expanded into many volumes are the Van der Hem atlas (Vienna, National Bibliothek) and Christoffel Beudeker's atlas, *Germania Inferior*, in 24, originally 27, volumes (London, British Library, Maps C.9.d.e.).

As time went on atlases became more and more specialized both as to area and subject matter; the first non-general atlas, a mineralogical atlas, was published in 1780: J-É Guettard and A-G Monnet, *Atlas et description minéralogiques de la France*, Paris (see Mineral map, 3.131). Johan Matthias Korabinsky's pocket atlas of Hungary, *Atlas Regni Hungariae Portalis*, Vienna, 1786 (?), was a forerunner of the economic atlas (see Atlas, pocket, 8.0110e). The mid-19th century saw the publication of major atlases of thematic maps (1.202) and many school atlases, as well as the county atlas (8.0110a) in the United States. The 19th century also saw the publication of atlases of facsimiles of older maps (8.0110b). National and state atlases (8.0110d), regional in style, date from the 16th century, but the national atlas in modern terms, containing a variety of economic and cultural data about a particular country or subdivision, is largely a 20th century development. The first such atlas, the *Atlas de Finlande*, appeared in 1899.

The later history of maritime atlases parallels that of the general atlas. Gerard Mercator's plan to publish his chart of 1569 in atlas form was abortive, although one version in atlas format is now known (Rotterdam, Maritiem Museum, Prins Hendrik). Lucas Janszoon Waghenaer's *Spieghel der Zeevaerdt* (Leyden, 1584-85) was the first printed sea-atlas (if we discount Piero Coppo's little printed *Portolano*, with woodcut charts, published at Venice in 1528). The pirated English version of Waghenaer, *The Mariner's Mirror*, appeared in London in 1588. Robert Dudley, an English emigré in Florence, incorporated various innovative features in his *Dell' Arcano del Mare*, Florence, 1646-47, and this was the first sea-atlas with all the charts drawn on Mercator's projection, but its influence on contemporary and later hydrographic charting was limited. Dutch maritime atlases were prominent in the 17th and early 18th centuries, with the firm of Van Keulen outstanding as the successor to that of the Blaeus. The *Zee-Fakkel*, published at Amsterdam by Johannes van Keulen and his successors, 1681 to 1753, in four language editions, achieved the highest development of the atlas in Dutch pilot-book style. His *Zee-atlas*, 1680–c. 1710, was, in its 1695 edition, rightly claimed by Van Keulen as a masterpiece of its kind. Joseph Frederick Wallet Des Barres in the monumental *Atlantic Neptune*, London, 1777, introduced new standards for coastal and hydrographic depiction. In the 19th century the *Justus Perthes See-atlas*, Gotha, 1894, illustrates the introduction of thematic mapping to the oceans of the world.

C CORTESÃO, Armando, and TEIXEIRA DA MOTA, Avelino (1960): *Portugaliae Monumenta Cartographica*. Lisboa, Comemorações do V Centenário da morte do Infante D. Henrique.

KOEMAN, Cornelis, ed. (1967-71, 1985): *Atlantes Neerlandici*, 6 vols. Amsterdam, Theatrum Orbis Terrarum.

SIMONI, Anna E C (1985): "Terra Incognita: the Beudeker Collection in the Map Library of the British Library," *The British Library Journal, 11*, 143-75.

SKELTON, R A (1952): *Decorative Printed Maps of the 15th to 18th Centuries.* London, Spring Books.

TOOLEY, R V (1939): "Maps in Italian Atlases of the Sixteenth Century," *Imago Mundi, 3,* 12-47.

WALLIS, Helen (1981): *The Maps and Text of the Boke of Idrography Presented by Jean Rotz to Henry VIII.* Oxford, Roxburghe Club.

WAWRIK, Franz (1982): *Berühmte Atlanten.* Harenberg, Die bibliophilen Taschen-bücher.

8.0110A County Atlas (N. Am.)

A A bound collection of maps, often at a uniform scale, of the administrative units (townships, towns) of a particular county of the United States or Canada. Such maps or plats, characteristically, each occupy one page and show property boundaries, size, and ownership; roads; railroads; and drainage. These township maps are usually preceded by a general county map, and sometimes by a state, national and even a world map. In many instances these privately produced county atlases contain local, state, and regional geographical descriptions as well as local history and biography. In addition, some have perspective landscape views and portraits of leading citizens and their families (see also, Plat, 1.1630a).

B County atlases first appeared in the early 1860s as an outgrowth of county landownership wall maps, and increased in number as the 19th century progressed. Some publishers extended the illustrated county atlas format to cover whole states in modified form in the 1870s. By about the time of World War I the county atlas gave way to the simpler and less expensive plat book. The first county maps and atlases were produced in the middle and north Atlantic states, especially Pennsylvania and New York, and depicted counties in those states (Figs.24 and 25). However, it was in the Middle West that the private county atlas business had its greatest impact and success. Large areas of the United States lack coverage by county atlases, notably the Old South and the sparsely populated West. Many eastern Canadian counties were also covered. Some counties of the Middle West were the subjects of more than ten separate editions of one or more county atlases. Publishers normally produced the county atlas on a subscription basis. Some county-atlas publishers in the 1870s developed separate city real-estate atlases, which highlighted individual property ownership.

C HARRINGTON, Bates (1879): *How 'tis Done: A Thorough Ventilation of the Numerous Schemes Conducted by Wandering Canvassers.* Chicago.

LE GEAR, Clara Egli (1950, 1953): *United States Atlases: A List of National, State, County, City, and Regional Atlases in the Library of Congress.* 2 Vols. U.S. Library of Congress, Map Division, Washington, DC.

RISTOW, Walter W (1985): *American Maps and Mapmakers. Commercial Cartography in the Nineteenth Century.* Detroit, Wayne State University Press.

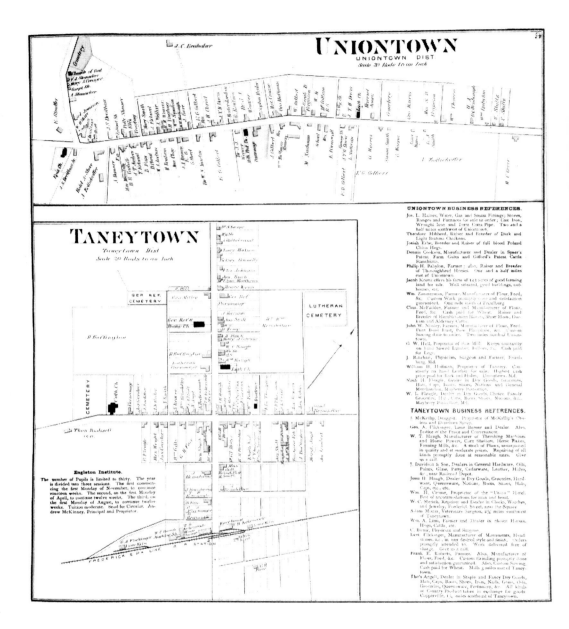

Fig. 24 – Plans of Uniontown and Taneytown. From An Illustrated Atlas of Carroll County, Maryland, *compiled, drawn, and published by Lake, Griffing and Stevenson, Philadelphia, 1877, pl. 9. By permission of the British Library. (County Atlas, (N.Am.), 8.0110a; see also Property map, 2.0210e)*

Fig. 25 – Illustrations of farmers' properties and public buildings in Carroll County, Maryland. Atlas, pl. 21 (see Fig. 24).

THROWER, Norman J W (1961): "The County Atlas of the United States," *Surveying and Mapping, 21,* 365-73.

THROWER, Norman J W (1972): "Cadastral Survey and County Atlases of the United States," *Cartographic Journal, 9,* 43-51.

8.0110B Facsimile Atlas (Map)

A (MDTT 817.7) "A printed reproduction of an Old Map (815.1) identical with the original." Preparation or reproduction of hand-drawn copies of originals are also called facsimiles. A facsimile atlas is either a collection of facsimile maps or the reproduction of an entire older atlas. A E Nordenskiöld gave his collection of maps published in 1889 the title *Facsimile-Atlas,* and the term rapidly gained general acceptance.

B The earliest cartographic facsimile appears to be the engraving of the Peutinger map, of which two sections were published by Marcus Welser at Antwerp in 1591, and eight sections were redrawn by Abraham Ortelius for the *Parergon* (the historical maps) of the 1598 edition of his *Theatrum Orbis terrarum.* The facsimile of this Roman map was reprinted in subsequent issues of the *Theatrum* to 1612, and the *Parergon* to 1624. The twelve sections were specially engraved at Nuremberg in 1682 for an edition of Welser's complete works *Opera Historica et Philologica.*

During the 17th and 18th centuries a limited number of facsimile maps were published, among them a reproduction by Richard Gough in 1780 of a manuscript map of Great Britain originally drawn in *c.* 1360. This interesting map has subsequently been known as the "Gough Map."

Reproduction from engraved copper plates was not conducive to the widespread production of facsimile maps. One of the first significant developments in facsimile publishing, therefore, occurred following the application of lithography to map printing and reproduction in the first half of the 19th century. Between 1842 and 1849, Manuel Francisco de Barros e Sousa, Viscount de Santarem, a Portuguese historian living in exile in Paris, published three editions of his *Atlas composé de mappemondes, de portulans et de cartes hydrographiques et historiques depuis le VIᵉ jusqu'au XVIIIᵉ siècle.* In its final, most complete form the atlas included some 80 sheets on which were reproduced 146 old maps and charts. Approximately 10 percent of the sheets were printed from engraved plates, and the remainder were lithographically reproduced. All were based on hand-drawn copies of originals in various European archives and libraries.

Publication of Santarem's facsimile atlas greatly exercised Edme François Jomard, who was at the time director of the map department in France's Bibliothèque Royale. Jomard challenged Santarem's claim to priority in planning and producing a collection of facsimiles of historical maps. He asserted that by 1842 he was well along on the task of having reproductions made of selected ancient maps. Jomard proceeded to publish in 1854 (or according to others, in 1842), the first of his facsimiles. The entire series of maps was not issued until 1862, after Jomard's death.

Entitled *Les Monuments de la géographie, ou recueil d'anciennes cartes européenes et orientales*, the atlas includes twenty-one maps on 81 sheets. All were lithographically reproduced from hand-drawn copies. The third compiler and producer of a facsimile atlas during this early period was Joachim Lelewel of the Polish government in exile in Brussels who drew and engraved the plates for his *Géographie du moyen-âge ... Accompagné d'atlas et de cartes dans chaque volume*, 5 vols., 1852-57.

A notable but never-reproduced collection of facsimiles was assembled at this time by Johann Georg Kohl, a native of Germany. By 1854 he had copied more than 450 16th to 19th century maps illustrating American discovery and exploration. He brought these to the United States in 1856, and they are now in the Library of Congress. The first application of photography to the production of map facsimiles was made by Ferdinando Ongania in his *Raccolta*, 1875.

The next major advance in cartographic facsimile publishing occurred during the last two decades of the 19th century, following the introduction of large cameras and the successful marriage of photography and lithography. This made possible for the first time the easy reproduction of map facsimiles with true fidelity to the originals. Two facsimile atlases, reproducing old maps and charts preserved in the Bibliothèque Nationale, were published in the 1880s. The first, *Choix de documents géographiques conservés à la Bibliothèque Nationale*, includes eleven maps and four illustrations. The *Choix de documents ...* was prepared under the direction of France's Minister of Public Instruction for exhibit at the Third International Geographical Congress in Venice, 1881. The second, a group of portolan charts selected by Gabriel Marcel from the collections of the Bibliothèque Nationale, was reproduced in heliogravure and published in 1886 under the title *Recueil de portulans*.

Cartographic history took a great stride forward in 1889 with publication of A E Nordenskiöld's *Facsimile-Atlas*, with full-size photolithographic reproductions of 50 historical maps and a number of reduced facsimiles. Less than a decade later, in 1897, Nordenskiöld's *Periplus: An Essay on the Early History of Charts and Sailing-Directions* was published in Stockholm. The utility of Nordenskiöld's work is apparent from the fact that facsimile editions of the 1889 volume, i.e., facsimiles of a facsimile, have been published in recent years.

In the 20th century advances in techniques of reproduction, such as collotype, and a greatly increased interest in "old maps" have led to a great proliferation of facsimile reproductions.

C FAIRCLOUGH, Roger H (1972): "Original or Facsimile," *New Library World* (London), *73*, 291-94.

HEAWOOD, Edward (1930): "Reproductions of Notable Early Maps," *Geographical Journal, 76*, 240-48.

HODGKISS, Alan G (1978): "Facsimiles of Early Maps," *Bulletin, Society of University Cartographers, 12*, No 2, 1-12.

KOEMAN, Cornelis (1964): "An Increase in Facsimile Reprints," *Imago Mundi, 18*, 87-8.

MARGARY, Harry (1973): "The Facsimile Reproduction of Early Engraved Maps," *Bulletin, Society of University Cartographers, 7*, No. 2, 1-7.

MASON, K T (1976): "Reproductions of Historical Maps and Atlases: A List of British Publishers," *Bulletin,* Special Libraries Association, Geography and Map Division, *106,* 3-10.

NOE, Barbara R (1980): *Facsimiles of Maps and Atlases: A List of Reproductions for Sale by Various Publishers and Distributors,* 4th ed. Washington, DC, Library of Congress.

ONGANIA, F (1875-81): *Raccolta di mappamondi e carte nautiche del XIII al XVI secolo.* Venice.

RISTOW, Walter W (1967): "Recent Facsimile Maps and Atlases," *The Quarterly Journal of the Library of Congress, 24,* 213-29. Reprinted as "New Maps from Old, Trends in Cartographic Facsimile Publishing," *Canadian Cartographer, 5,* 1-17.

SKELTON, Raleigh A (1972): *Maps: A Historical Survey of Their Study and Collecting.* Chicago, University of Chicago Press, 70-85.

WALLIS, Helen, and SIJMONS, A H (1985): *Atlas de Santarem: Facsimile of the Final Edition 1849: Explanatory Notes.* Amsterdam, Rudolf Muller.

WOLTER, John A (1981): "Johann Georg Kohl and America," *Map Collector, 17,* 10-14.

WOOD, Fergus J (1976): "J. G. Kohl and the 'Lost Maps' of the American Coast," *American Cartographer, 3,* 107-15.

YONGE, Ena L (1963): "Facsimile Atlases and Related Material: A Summary Survey," *Geographical Review, 53,* 440-46.

8.0110C Island Atlas

An island atlas is a specially designed collection of graphic materials relating to islands, mainly comprising maps, town plans and views, usually with accompanying text. Since the term "atlas" for a collection of maps dates only from the time of the publication of Mercator's atlas (1595), other titles were used by their authors: *islario* (Spanish), *insulaire* (French), *isolario* (Italian), *Liber insularum* (Latin). Modern translations are *island atlas* (English) and *Inselbuch* (German). In principle, the island atlas is a sub-category of the regional nautical atlas, but by extension certain island atlases have been conceived as universal atlases (Alonso de Santa Cruz, *c.* 1560; André Thevet, 1587). There are two categories of island atlases: regional (general class) and universal (exceptional class).

By its nature the island atlas is therefore selective, in that islands are given priority over other regional elements; and it is normally devoted to those regions where islands are numerous. The original objective clearly was to show the islands at a larger scale than on regional nautical maps, thus reflecting the special interest in islands. The island atlas was related to the *portolano,* or pilot book; it originated in the Mediterranean and was intended as an illustrated guide for travellers.

B The *isolario* developed in Italy during the 15th and 16th centuries. The earliest known is that of Cristoforo Buondelmonte, entitled *Liber insularum Archipelagi* and written at Florence in 1420. Buondelmonte was an ecclesiastic with antiquarian interests who travelled in the Aegean and eastern Mediterranean between 1414 and 1420. His book comprises a description of islands in the Aegean and Ionian Seas, illustrated by 78 maps. It has been described as a traveller's guide intermediate in form between a *portolano* and an historical-descriptive chorography (Skelton, bibliographical note to Bordone, p. v). The work exists in numerous manuscript copies of the 15th century and served as the prototype for many later isolarii.

The first printed *isolario,* which was copied partly from Boundelmonte, was that by Bartolommeo dalli Sonetti (Bartolommeo "of the Sonnets"), published at Venice in or after 1485. The 49 woodcut maps are without names, descriptions in verse are printed on the facing pages, and many examples are coloured and have names added by hand. At least one manuscript version is known, *c.* 1485 (National Maritime Museum, Greenwich, 38 MS.9920). The book was reprinted in 1532. Benedetto Bordone (1460-1539) published at Venice in 1528 the *Libro . . . de tutte l'Isole del mondo,* which is based on the *Isolario* of 1485, enlarged to 64 maps (Aegean, Adriatic, islands of the Atlantic and Indian Oceans). The subsequent editions of 1534, 1537 and 1547 are similar. Bordone's *Libro* has been described as "the earliest complete work of its kind to have been produced by the printing-press in Italy or anywhere else" (Almagià, 1937, in Skelton, bibliographical note to Bordone, p. v).

Much more accurate and complete is the *Kitab-i-Bahriye* (Atlas of the Sea), a nautical manual in the Turkish language by the Turkish admiral Piri Re'is (*c.* 1470-1554). This extensive work, with two versions dated 1521 and 1525, comprises 215 detailed maps of islands and roadsteads of the Mediterranean. Some thirty manuscripts survive. In Western Europe Alonso de Santa Cruz (1500-1572), cosmographer of Charles V of Spain, prepared *c.* 1560 the *Islario general de todos las islas del mundo* consisting of 120 maps accompanied by an explanatory text. The original was identified in 1909 as the work of Santa Cruz rather than that of Andrés Garciá de Céspedes as the title indicates. In addition to 80 maps of Europe (of which 50 are of the Greek archipelago), the work extends to Asia, Africa, and America.

In Italy Antonio Lafreri (1512-1577) included in his various atlases assembled to order from 1553 to 1577 some 20 maps of islands. Similar collections put together by Venetian editors also have about 20 maps of islands. Other examples of collections of island maps are the small Venetian atlases illustrating the war between Venice and the Turks: Simon Pinargenti (1573), *Isole che son da Venetia nella Dalmatia et per tutto l'arcipelago, fino a Constantinopoli* (53 maps); Gian Francesco Camocio (1574), *Isole famose porti fortezze e terre maritime sottosposte alla . . . Venetia . . . et al Signor Turco* (83 maps); and the small collection of Thomaso Porcacchi (1530-1585), engraved by Girolamo Porro, entitled *L'Isole piu famose del mondo,* 1572 (30 maps), 1576, 1590, 1605 (*c.* 47 maps).

In France, André Thevet (1504-1592), royal cosmographer from 1558 to 1589, undertook a new comprehensive work, namely, "Grand Insulaire et Pilotage . . . Dans lequel sont contenus plusieurs plants d'isles habitées et deshabitées." Abraham Ortelius was informed in 1590 that the work had been started in 1585, and Thevet wrote to him in 1588 that he had engraved 250 plates of plans of the islands of the whole world. He planned to include "cinq cents Isles par moy recouvertes en divers endroits des quatres parties de l'univers"; thus the work, when completed, would have ranked as one of the most imposing atlases of the 16th century (Lestringant, p. 481). The manuscript, consisting of 640 folio sheets is entirely devoted to pilotage, to navigation, and to the description of islands and ports (Bibliothèque Nationale, Paris, cotes fr. 15452 and 15453). The text has more value than the maps which are uneven because they have been borrowed from the small collections of Camocio and Porcacchi. Thevet also appears to have used a copy of the *Bahriye* of Piri Re'is.

At the end of the 16th century a small island atlas entitled *Insularium: Orbis aliquot insularum,* Cologne, 1601, with 60 woodcut maps (Mediterranean, Azores, Cape Verde, Madagascar, Socotra, etc.) on 36 sheets, was published as a supplement to *Speculum Orbis Terrae,* a posthumous work of Johannes Metellus Sequanus. In the 17th century regional maritime mapping developed rapidly and doomed the island atlas. Porcacchi was reprinted in 1606, 1620 and 1686. Marco Boschini (1613-1678), an eminent painter and engraver, produced in 1658 a handsome revision of the Isolario of Sonetti under the pompous title: *L'Arcipelago con tutte le isole, scogli secche e bassi*; this work owes its success mainly to the beauty and originality of its engraving.

The Venetian cosmographer Father Vincenzo Coronelli (1650-1718), aspiring to embrace all the cartographic art forms in his monumental atlas series, the 13-volume *Atlante Veneto* (1690-1701), included as the second and third volumes the *Isolario,* part 1, 1696-1698, covering the islands of the Mediterranean, and part 2, 1697, for the other islands of the globe. The work was traditional in style with some plates, for example, taken directly from Thevet. Appropriately, Venice was the site of the final flowering of the island atlas. Olfert Dapper's *Description exacte des isles de l'Archipel . . . ,* Amsterdam, 1703, was derivative of Venetian works and especially of Coronelli's *Isolario.*

C ALPAGOT, Haydar, and KURTOGLU, Fevzi (introduction by) (1935): *Piri Reis Kitabi Bahriye.* Facsimile edition of Aya-Sofya MS 2612 (in Turkish). Istanbul, Devlet Basimevi.

BAGROW, Leo (1951): *Die Geschichte der Kartographie.* Berlin, Safari Verlag.

BAGROW, Leo (1964): *History of Cartography.* Revised and enlarged by R A Skelton. London, C A Watts and Co.; Cambridge, Mass., Harvard University Press.

BORDONE, Benedetto (1528): *Libro . . . de tutte l'Isole del mondo . . .* Venice. Facsimile edition, Theatrum Orbis Terrarum, Third Series, Vol. 1, Amsterdam, 1966. Bibliographical note by R A Skelton.

DESTOMBES, Marcel (1952): *Catalogue des cartes gravées au XVe siècle.* International Geographical Union.

DESTOMBES, Marcel (1972): "A. Thevet (1504-1592)," *Proceedings, Royal Society of Edinburgh, 72B,* 127-29, 2 plates.

KAHLE, Paul, ed. (1926): *Piri Re'is Bahrije: Das türkische Segelhandbuch für das Mitteländische Meer.* Berlin and Leipzig, Walter de Gruyter, 2 vols.

LEGRAND, E (1897): *Description des îles de l'Archipel par C. Buondelmonti.* Paris, Leroux.

LESTRINGANT, Frank (1984): "Thevet, André," Chapter XXIX in Pastoureau, Mireille, *Les Atlas français XVIᵉ-XVIIᵉ siècles: répertoire bibliographique et étude.* Paris, Bibliothèque Nationale, pp. 481-95.

NORDENSKIÖLD, A E (1889): *Facsimile-Atlas to the Early History of Cartography.* Stockholm. (Kraus reprint, New York, 1961; Dover Publications, 1973.)

8.0110D National (State) Atlas

A A national (state) atlas is now defined as a publication, in one or several parts, of a series of maps with text, figures, and diagrams concerning a country, its nature, population, and economic and cultural circumstances. As such it is a late 19th century innovation. There is, however, an earlier type of national atlas based on provincial or county maps.

B The first national atlas in the world was the *Suomen Kartasto (Atlas de Finlande)* which appeared in 1899 published by the Geographical Society of Finland. It contained 32 plates with a wide range of topics. Although very different in style and content from the earlier type of national, or provincial, atlas, an evolution towards the Atlas of Finland can be traced.

Claimants to be the first European national atlases of the earlier type are Johan Stumpf's atlas of Switzerland, entitled *Landtaflen . . . ,* Zurich, 1552; followed by Wolfgang Lazius's atlas of Austria, *Typi chorographici Prouin: Austriae,* Vienna, 1561 (see Regional atlas, 8.0110f). As these works have been little known until recent years, it has been more usual to identify as the first "national" atlas Christopher Saxton's Atlas of England and Wales, published in 1579. This comprises a general map and 34 maps of counties or groups of counties, and it gained for Saxton the title "optimus Chorographus" (William Camden, *Britannia,* 1607). John Speed's *Theatre of the Empire of Great Britain,* London, 1611-12, which followed, covered the whole of the British Isles. Both Saxton's and Speed's atlases were revised (notably with the addition of roads) and reissued over a period of two centuries. In France the first national atlas was Maurice Bouguereau's *Le Théatre Françoys,* Paris, 1594, comprising 14 to 16 maps. Jean Le Clerc's *Théatre géographique du Royaume de France,* Paris, 1619, used some of Bouguereau's maps and expanded the work (Drapeyron, Dainville.)

Until the end of the 18th century national atlases remained regional in style, despite some attempts to be more comprehensive or more thematic, foreshadowing the national atlas of today. The French Field Marshal Sebastien Le Prestre de Vauban (1633-1707) had planned to compile an atlas of France, which would show "everything in the Kingdom which is significant and worthy of note." In Russia Ivan K Kirilov (1689-1737), an outstanding cartographer and geographer, tried to publish a large "Atlas of the Russian Empire" in which he hoped to include historical and economic maps in addition to general geographical maps. Neither De Vauban nor Kirilov ever completed his atlas. In the later years of the 18th century and early 19th century more information was given both in textual form and cartographically. Thus the *Atlas national et général de la France,* Paris, Chez Desnos, 1790, showed the new *départements,* with text on each side, but it was in the traditional form, despite its title. Augustin Codazzi in his *Atlas Físico y Politico de la Republica de Venezuela,* Caracas, 1840, presented historical and political maps of Venezuela and adjoining states, but the result would not qualify the work as a national atlas in the modern sense. In contrast, the *Atlas nacional de España* by A H Dufour, printed in Paris, 1838-39, may be regarded as intermediate between the county or provincial atlas and the national atlas in the modern sense.

The change in conception from locational and topographic maps to a statistical map followed the development of thematic maps in the 19th century. As the first national atlas, the Atlas of Finland (1899) remained unique for a long time. Later, revised editions of the Atlas of Finland appeared in 1910-11, 1925-28, and 1960. Other national atlases only appeared in the 20th century.

C DAINVILLE, François de (1961): "Le premier atlas de France. Le Théatre Françoys de M Bouguereau 1594," *Actes du Quatre-Vingt-cinquième Congress national des Sociétés Savantes,* Séction de Géographie (Extract). Paris.

DRAPEYRON, Ludovic (1890): "L'Évolution de notre premier atlas national sous Louis XIII," *Acta Geographica, 20,* 1975, 162-86.

DRECKA, Jolanta, and TUSZYŃSKA-RĘKAWEK, Halina (1964): *National and Regional Atlases: Sources, Bibliography, Articles.* Warsaw, Polish Academy of Sciences, Institute of Geography, Dokumentacja Geograficzna No. 1.

HODSON, Donald (1984): *County Atlases of the British Isles, Published after 1703,* Vol.1. Tewin, Welwyn, Hertfordshire, Tewin Press.

PASTOUREAU, Mireille (1984): *Les Atlas français XVIᵉ-XVIIᵉ siècles.* Paris, Bibliothèque Nationale.

SALICHTCHEV, Konstantin A (1960): *Atlas nationaux: Histoire, analyse, voies de perfectionnement et d'unification.* Moscow and Leningrad, Édition de l'Académie des Sciences de l'URSS. Translated into English (1972): in *Cartographica,* Monograph No. 4, Toronto, University of Toronto Press.

SKELTON, R A (1970): *County Atlases of the British Isles 1579-1850. A Bibliography.* London, Carta Press.

YONGE, Ena L (1957): "National Atlases: A Summary," *Geographical Review, 47,* 570-80.

8.0110E Pocket Atlas

A An atlas of octavo size or smaller, usually a reduced version of one that has already been published in larger format.

B In 1548 Giacomo Gastaldi produced *La geografia di Claudio Ptolemeo,* which measured 5 inches by 6¾ inches and was intended to be carried "nella manica" (in the sleeve). It consisted of reduced versions of Ptolemy's maps with additions and corrections by Sebastian Münster.

Thenceforth most major atlases were published in both small and large versions.

The rapid rise to popularity of the pocket format is illustrated by Abraham Ortelius's *Theatrum Orbis Terrarum* (1570). A small "epitome" of the *Theatrum* was published by Ortelius himself in 1577 with a Dutch text that was later translated into French, Latin, Italian and English. A rival pocket edition was published in 1601 by Michel Coignet. New editions of both *Epitomes* appeared in London in 1603; these were the first world atlases published in England and the first world atlases with English text.

For most travellers a miniaturized "national" atlas was probably more useful than a world atlas. An early example was Pieter van den Keere's atlas of the British Isles, prepared from the late 1590s onwards and finally published by Willem Janszoon Blaeu in 1617. A later example notable for its economic information is Johan Matthias Korabinsky's *Atlas Regni Hungariae Portalis. Neue und vollständige Darstellung des Königreichs Ungarn, auf 60 Tafeln in Taschenformat*, Vienna, 1786(?).

Portable atlases were also essential for military use. In the American War of Independence the London publishers Robert Sayer and John Bennett printed in 1776 *The American Military Pocket Atlas*. Popularly known as the "Holster Atlas," it was issued to British officers in the field, and comprised six folding maps of the theatre of war.

C SKELTON, R A (1968): "Bibliographical Note," to Abraham Ortelius, *The Theatre of the Whole World*, 1606. Amsterdam, Theatrum Orbis Terrarum, v-xvii.

WALLIS, Helen (1972): "Preface," *Atlas of the British Isles by Pieter van den Keere c. 1605*. Lympne Castle, Harry Margary, i-iv.

8.0110F Regional Atlas

A A collection of maps depicting an area either smaller or larger than that of a state or country. Modern examples of the smaller kind are the *Economic Atlas of Ontario*, the *Atlas du Languedoc-Roussillon*, the *Atlas of the Trans-Baikal Area* (in Russian), and of the larger, the *Atlas der Donauländer*, the *Oxford Economic Atlas of Western Europe*, and the *Atlas international de l'Ouest Africain*. The first of the two interpretations, which is prevalent today, covers maps indicating a region's administrative, historical, and, sometimes, natural geographical divisions. In this sense, regional atlases are subordinate to national atlases. The second interpretation defines regional atlases as sets of maps depicting parts of continents. In renaissance Europe the term "chorography," derived from Ptolemy, was applied to such atlases.

B The first European regional atlases appeared in the 16th century. The centuries-old interest in the Holy Land may explain the prime place of the set of seven maps of Palestine and adjoining countries which, together with Scandinavia, illustrate Jakob Ziegler's *Quae intus continentur. Syria . . . Palestina . . . Arabia . . . Aegyptus . . . Schondia . . . Holmiae . . . historia. Regionum superiorum, singulae tabulae Geographicae*, published at Strassburg in 1532. The earliest regional atlas purporting to be one for Europe itself, by Johan Stumpf, was in fact an atlas of Switzerland, and might therefore claim to rank as the first national atlas. The work was entitled *Landtaflen, Hierinn findst du lieber Läser schöner recht und wolge-*

machter Landtaflen XII, namlich ein Allgmeine Europæ . . . and issued at Zurich by Christoffel Froschauer in 1552. Stumpf's earlier *Chronik* of 1548, which contained 23 maps, was a forerunner but does not qualify as an atlas of Switzerland.

Wolfgang Lazius of Vienna was next in the field with his twelve-sheet atlas of Austria, entitled *Typi chorographici Prouin: Austriae . . . ,* published at Vienna by Michaël Zimerman in 1561. As the maps cover the hereditary lands of the Austrian Crown, this might also be considered a national atlas. Both Stumpf's and Lazius's atlases thus predate Christopher Saxton's of 1579 (see National atlas, 8.0110d). Later examples of regional atlases are *Newe Landesbeschreibung der zweij Hertzogthümer Schleswich und Holstein,* by the mathematician Johannes Mejer, 1652; *Atlas nuevo de la extrema Asia . . . ,* J Blaeu, 1659; *Semyon u. Remezov's Book of Maps of Siberia* (in Russian), 1701; *Atlas of Silesia . . . ,* Homann's Heirs, 1750. The firm of Johann Baptist Homann of Nuremberg (1664-1724), established in 1702 and continuing as Homännische Erben (1730-1813), was one of the leading continental publishers of regional atlases.

For the Americas the first published atlas was *Descriptionis Ptolemaicae Augmentum sive Occidentis notitia brevis commentario,* by Cornelis van Wytfliet, published at Louvain in 1597. As the title indicates, it was designed by the author as a supplement to Ptolemy's Geography. The *American Military Pocket Atlas,* London, R Sayer and J Bennet, 1776, and William Faden's *North American Atlas,* London, 1777, followed in the 18th century.

More modern, special subject atlases, providing detailed information about the climate, geology, resources, population, economy, and culture, represent a high point in the development of the regional atlases. There are countries whose regional atlases give a complete representation of their administrative or other subdivisions. Some regional atlases are of a multi-purpose nature and could be used for scientific research, planning, etc., while others serve a special purpose. Conversely, the *Atlas national et général de la France,* Paris, Chez Desnos, 1790, despite its title, is not strictly speaking a national atlas. Each of its maps shows one or more of the new *départements,* with text on each side.

In China there was a long tradition of regional cartography, with local topographies dating from the 3rd and 4th centuries. An 18th century example is a manuscript Atlas of Kiangsi Province (London, British Library, OMPB Add. 16356). Each opening represents one of 13 prefectures (*fu*), and illustrates the distribution of centres of government down to the level of the *hsien* or county capital.

C PHILLIPS, Philip Lee (1909-1920): *A List of Geographical Atlases in the Library of Congress,* Vols. 1-4; LE GEAR, Clara (1958-1974): *Idem,* Vols. 5-8. Washington, DC, Library of Congress.

SALICHTCHEV, Konstantin A, ed. (1964): *Regional Atlases: Tendencies of Development, Subject Matters of the Maps on Natural Conditions and Resources.* Moscow and Leningrad, Publishing House "Science."

SALICHTCHEV, Konstantin A, ed. (1976): *Complex Regional Atlases.* Moscow University Press.

WALLIS, Helen, and others (1974): *Chinese and Japanese Maps.* London, British Museum Publications for the British Library Board.

List of Contributors and Correspondents

A John H Andrews
Bruce Atkinson

B Peter M Barber
A D Baynes-Cope
R P Beckinsale
Roy C Boud
Anna-Dorothee von den Brincken
Elspeth Buxton

C Eila M J Campbell
Tony Campbell
Gregory Chu
Peter K Clark
Elizabeth Clutton
John Conroy
Michael P Conzen
Brian F Cook
Karen S Cook (Pearson)

D Peter F Dale
Lisette Danckaert
Catherine Delano Smith
Roger Desreumaux
Louis De Vorsey
*Marcel Destombes
O A W Dilke

E Ralph E Ehrenberg
Ulla Ehrensvärd
Joan Eyles

F Akio Funakoshi

G *Richard A Gardiner
Wilma George
Jean Gottmann

H J B Harley
Paul D A Harvey
Francis Herbert
B Hickman
Gillian Hill
B P Hindle
Alan Hodgkiss
Yolande A Hodson (O'Donoghue)
A A Horner
Melvyn Howe
Derek Howse
Mei-Ling Hsu
Ralph Hyde

J T G H James

K Naftali Kadmon
Roger J P Kain
George Kish
Cornelis Koeman
Jerzy Kostrowicki

L Lucie Lagarde
Paul Laxton
Malcolm Lewis
Mary Alice Lowenthal

M Alan L Mackay
George F McCleary, Jr.
Malcolm D McLeod
Douglas Marshall
Leena Miekkavaara
Karl-Heinz Meine
Terence C Mitchell
Andrew M Modelski
Ian Mumford
*Nobuo Muroga

N Howard Nelson
T R Nicholson

O *A S Osley

P Mireille Pastoureau
Monique Pelletier
Dorothy Prescott

Q David B Quinn

R William L D Ravenhill
Walter W Ristow
Arthur H Robinson
William F Ryan
Wislaw Rybotycki

S Konstantin Alekseevich Salichtchev
Günter Schilder
Edward F Schnayder
Albert H Sijmons
Anna E C Simoni
Antoine de Smet

T Norman J W Thrower
Waldo Tobler
Sarah Tyacke

V Marie-Antoinette Vannereau
*Francisco Vazquez Maure
*Coolie Verner

W Christopher Walker
Helen Wallis
Deborah Warner
James Welu
Joseph W Wiedel
H R Wilkinson
David Woodward
John A Wolter

Y James A Yelling

Z Lothar Zögner

* = deceased

General Index
(Main entry in boldface)

Index to Bibliographies, Section C

Chen, C.S. (1978), 4.141
Chorley, R.J., et al (1964, 1973), 3.021
Christy, R.M. (1900), 6.1310e
Chu, G. H-Y. (1974), 4.191
Clemente, G. (1968), 2.0210d
Cline, H.F. (1972), 2.0210e
Close, C. (1926), 1.203, 5.0910a
Clüver, P. (1629), 2.081
Cocchiara, G. (1963), 2.192
Coe, R.T. (1976), 1.1810
Colles, C. (1961), 1.1810d
Cook, A. (1981), 1.0320b
Coronelli, V. (1697), 2.202
Cortazzi, H. (1983), 4.122
Cortesão, A. (1969-71), 1.0320d
Cortesão, A., and Teixeira da Mota, A.
    (1960), 1.0320a, 1.0320b, 3.031, 4.121, 4.122,
    4.131, 4.133, 5.162, 8.0110
Cowell, S.H. (1852), 7.012
Craig, G.Y., et al (1978), 3.021
Crome, A.F.W. (1782), 2.163
Crone, E. (1966), 1.191
Crone, G.R. (1954), 1.231, 1.2320b
    (1961), 4.151
    (1978), 1.0320b, 1.2320, 1.2320b, 4.031, 4.121,
    4.122, 6.1310h
Cuénin, R. (1972), 5.081
Cumming, W.P. (1972), 1.0320a
Curtis, J.B. (1928), 1.201
Czoernig, K. von (1855), 2.052
Dahlberg, R.E. (1962), 4.134
Dainville, F. de (1956), 2.181
    (1958), 1.203, 3.081, 5.0910a, 5.0910d, 5.192,
    5.193
    (1961), 1.0320a, 1,1810a, 8.0110
    (1964), 1.033, 3.022, 3.132, 5.033, 5.081, 5.122,
    5.1610a, 5.1610c, 6.031, 6.121
    (1968), 2.1310
    (1970), 1.1630, 5.122
Dale, P.F. (1976), 2.0210b
Dalrymple, A. (1774-75), 1.0320b
Dalton, O.M. (1911), 6.132
Darby, H.C. (1935), 4.122
    (1970), 2.121
Davenport, W. (1960), 1.0320c, 4.231
Davey, M. (1866), 3.231
Dawson, R.K. (1831-32), 2.0210a
Day, A. (1967), 4.032
De Boer, G. and Skelton, R.A. (1969), 5.192
De La Motte, P. (1849), 7.012
Delano Smith, C. (1982), 3.011, 5.033, 5.1610c,
    6.1310g
    (1984), 6.032, 6.121
    (1985), 5.033, 5.122, 6.121
Denaix, M.A. (1829), 2.052
Denucé, J, and Gernez, G. (1936), 5.192
Desportes, J. (1838), 7.1210c
Destombes, M. (1952), 1.0320d, 7.0920a, 8.0110c
    (1955), 4.191
    (1959), 1.161, 6.1310c

(1964), 1.162, 1.2320, 1.2320b, 6.1310e
(1968), 5.192
(1972), 8.0110c
DeVorsey, L. (1966), 1.1810
    (1976), 1.0320c, 3.151
Dewdney, D. (1968), 6.1310a
Dicks, D.R, ed (1960), 4.121, 4.122, 4.134
Diderot, D, and d'Alembert, J.L. (1753), 1.033
Diemer-Willroda, E. (1939), 2.1310a
Dieterici, C.F.W. (1859), 1.041
Digges, L., and T. (1591), 1.164
Dilke, O.A.W. (1967), 5.032
    (1971), 2.0210, 2.0210b. 2.0210c, 2.0102e,
    2.1310, 4.121, 4.141
    (1973), 2.0210b
    (1974), 4.141
    (1985), 1.0710a, 1.1630b, 1.231, 1.2320,
    1.2320a, 1.2320b, 4.121, 4.134, 4.141, 4.151,
    5.033, 5.122, 6.1310
Dilke, O.A.W., and Dilke, M.S. (1975), 1.192
Dinse, P., et al (1897), 4.134
Douay, A. (1687?), 6.1310a
Downing, A.M.W. (1900), 4.041
Drapeyron, L. (1890), 8.0110d
Drecka, J., and Tuszyńska-Rękawek, H. (1964),
    8.0110d
Du Bus, C. (1931), 2.161, 5.041
Du Carla, M. (1782), 5.0910a
Dudich, E., ed (1984), 3.021, 3.131
Dujardin-Troadec, L. (1966), 3.201
Du Mont, J.S. (1985), 6.1310d
Dunlop, R. (1905), 2.0210a
Dupain-Triel, J.L. (1798-99), 5.121
Dupin, C. (1827), 1.034
Durand, D.B. (1952), 4.121
East, W.G. (1966), 2.081
Easton, W.W. (1977), 2.1310a
Eckert, M. (1921, 1925), 1.202, 3.021, 3.022,
    3.132, 3.231, 3,262, 4.134, 5.0910c, 5.121
Eckstein, C. (1876), 7.161
Eden, R. (1561), 3.262
Edgerton, S.Y. (in press), 4.141
Ehrensvärd, U. (1977), 3.131
Ehrensvärd, U, and Pearson, K.S. (in press),
    5.032, 5.121
Elias, W. (1981), 1.1810d
    (1982), 1.043
Engelmann, G. (1964), 5.0910g
    (1966), 7.0920b
Evans, A.B. (1925), 7.151
Evans, E.J. (1976), 2.202
Eyles, V.A. (1972), 3.131
Fairclough, R.H. (1972), 8.0110b
Fauser, A. (1967), 1.0710, 1.0710a
Febvre, L.P. (1925), 2.0210d
Filteau, C. (1980), 1.091
Fiorini, M. (1899), 1.0710
Fischer, J., and Wieser, F. von, eds (1903), 1.231
Fischer, K. (1967), 1.0710b
Fischer, T. (1886), 1.0320d

Hunter, J.D. (1824), 6.1310a
Hupp, P. (1910), 6.1220c
Hurst, H. (1899), 1.161
Hyde, C.C. (1933), 2.0210
Hyde, R. (1976), 1.1630b
(1979), 1.043
(1986), 6.1310
Imhof, E. (1965, 1982), 3.132, 5.081
Ireland, H.A. (1943), 3.071
Jacunski, W.A. (1955), 2.081
Jarcho, S. (1970), 2.041, 2.191
(1973), 2.161, 5.041
(1979), 1.044
(1983), 2.191
Jervis, J. L. (1985), 3.011
Johnston, A.E.M. (1967, 1971), 6.1310e
Johnston, A.K. (1848), 3.161
Jones (O'Donoghue), Y. (1974), 5.081
Jordan, E.K. (1958), 4.071
Jouguet, P., ed (1928), 2.0210c
Jugaku Bunsho, (1959, 1967), 6.1310f
Jusatz, H.J. (1939, 1969), 2.191
Kahle, P., ed (1926), 8.0110c
Kahn, C.H. (1960), 1.2320a, 3.011
Kain, R.J.P., (1974, 1975), 2.201
Kain, R.J.P., and Prince, H.C. (1985), 2.202
Kainen, J (1951), 7.081
Kamal, Y. (1926-53), 1.0320d, 1.1810, 4.121
Kamela, C.Z. (1955), 4.071
Kämtz, L.F. (1831-36), 5.0910c
Kant, E. (1970), 1.044, 2.161
Keates, J.S. (1973), 5.123
Keill, J. (1698), 4.051
Kelley, J.E., Jr. (1979), 1.0320d
Keuning, J. (1952), 1.1810a
(1955), 4.134
Kimble, G.H.T. (1938), 1.2320b
King, L.W., ed (1912), 2.0210c
Kircher, A. (1665), 5.061
Kish, G. (1949), 2.181
(1953), 1.231
(1965), 4.134
(1976), 3.071
(1980), 6.1310e
Klaproth, H.J. von (1823), 2.122
Klemp, E. (1971), 1.231
Klotz, A. (1931), 1.2320a
Koeman, C. (1964), 5.162, 8.0110b
(1967-71, 1985), 8.0110
(1969), 2.081
(1970), 4.133
(1972), 1.0320c
(1975), 7.081, 7.1210d, 7.151, 7.161
(1976), 1.191, 5.1610c
(1980), 1.0320c, 6.1310h
(1983), 5.0910d
Körber, H-G. (1959), 5.0910g
Kraling, C.H., ed (1938), 1.161
Kretschmer, K. (1909), 1.0320d
Krogt, P. van der (1984), 1.0710, 1.0710c, 1.0710d

Krug, M. (1901), 3.151
Krüger, H. (1951), 1.1810d, 2.0210d, 2.122,
5.1610a, 6.032, 6.081, 6.121
(1958), 1.1810d, 6.121
Kuchař, K. (1937), 4.141, 6.121
(1961), 3.131, 6.121
Kume Yasuo (1976), 6.1310f
Kurita Mototugu (1938), 1.161
La Condamine, C.M. de (1751), 4.051
Laertius, D. (1925), 3.011
Laguarda Trias, R.A. (1981), 4.191
Lahontan, L-A. de (1703), 6.1310a
Lalanne, L. (1843), 5.0910f
Lamb, C.M., ed (1968), 7.091
Lamb, R.B. (1961), 2.091
Langlois, V. (1867), 5.032
Larcom, T.A. (1845), 5.121
La Roncière, C. de (1924-25), 1.0320d
Larsgaard, M.L. (1984), 1.203
Lavedan, P. (1954), 1.161
Lawson, J. (1709), 6.1310a
Learmonth, A.T.A. (1969), 2.191
Le Gear, C.E. (1950, 1953), 8.0110b, 8.0110f
Leggett, R. (1975), 1.1810
Legrand, E. (1897), 8.0110c
Lehmann, J.G. (1820), 5.081
(1843), 6.032
Lelewel, J. (1852-57), 2.081
Lemercier, A. (1899), 7.041, 7.1210e
Lemoine-Isabeau, C. (1978), 6.121
(1984), 2.1310
Lende, H. (1953), 1.201
Lestringant, F. (1984), 8.0110c
Levallois, J.J. (1970), 4.071
Levi, A., and Levi, M. (1967), 1.161, 5.033, 6.132
(1974), 1.161, 1.203
Levi, A.C., and Trell, B. (1964), 2.202
Loomis, E. (1846), 3.231, 5.0910c
Lugli, P.M. (1967), 1.161
Lycosthenes, C. (1552), 4.141
Lynam, E. (1953), 5.033, 5.1610a
Lyons, H.G. (1914), 3.081
Macdonald, G. (1933), 2.011
Mackay, A.L. (1975), 2.181
Mackenzie, C. (1954), 1.201
Maffei, P. (1962), 1.0710b
Manguel, A., and Guadalupi, B. (1981), 1.091
Margary, H. (1973), 8.0110b
Mason, K.T. (1976), 8.0110b
Maupertuis, P.L. M. de (1738), 4.051
Maury, M.F. (1848-80), 5.231
(1851-52), 3.262
Mayr, G. (1871, 1874), 1.041
Meek, T.J. (1935), 1.203, 2.0210b, 2.0210c
(1936), 4.151
Merriman, M. (1983), 1.1630a, 2.1310, 2.1310b
McCleary, G.F. (1969), 1.041
McMurtrie, D.C. (1925), 1.204
Meinardus, W. (1899), 5.0910g
Meine, K-H. (1982), 1.0710b

(1967), 1.031, 1.062, 2.203, 5.041, 5.061,
5.1610a, 5.1610b
(1971), 2.161, 3.031, 5.0910, 5.0910d, 5.0910f
(1976), 5.0910d
(1982), 1.031, 1.034, 1.041, 1.044, 1.202, 2.132,
2.161, 2.162, 2.163, 3.022, 3.161, 4.133,
5.041, 5.121, 5.141, 5.1610a, 5.1610b,
5.1610c, 5.191
Robinson, A.H., et al. (1984), 5.123
Robinson, A.H., and Wallis, H. (1967), 5.0910g
Robinson, A.H.W. (1962), 1.0320, 1.0320a
Röhricht, R. (1898), 1.1630
Rosenberg, C. (1846), 5.0910b
Rosińska, G. (1973), 1.0710a
Rowley, G. (1984), 2.091
Rowley, G., and Shepherd, P.McL. (1976), 2.091
Roy, W. (1773), 2.011
Różycki, J. (1973), 4.134
Russell, J. (1797), 1.0710b
Russell, W. (1809), 1.0710b
Rutherfurd (1898), 1.201
Ryan, W.F. (1966), 1.0710b
Salichtchev, K.A. (1960), 8.0110d
ed, (1964, 1976), 8.0110f
Salomon, R. (1936), 2.192
Salviat, F. (1977), 2.0210b
Salviati, G. (1822), 4.151
Santarem, M.F. de Barros (1842-53), 2.021
Sarton, G.A.L. (1927, 1931), 1.164
Saxl, F. (1915), 3.011
Saxl, F., and Wittkower, K. (1948), 3.011
Scaglia, G. (1964), 2.011
Schienert, J.A. (1806), 5.081
Schilder, G. (1979), 2.052, 2.163
(1984), 6.081
Schleussner, K. (1908), 1.201
Schnelbögl, F. (1966), 3.022, 4.151, 6.091
Schofield, J. (1983), 2.0210e
Schoonover, D. (1982), 6.231
Schouw, J.F. (1823, 1833), 3.022
Schramm, P.E. (1958), 1.0710, 1.0710a, 6.1310e
Schrire, D. (1963), 6.1310c
Schulten, A. (1900), 6.132
Schulz, J. (1970), 1.203
(1978), 1.1630, 1.1630b, 1.231
Schütte, G. (1920), 3.011
Schwartz, S.I., and Ehrenberg, R.E. (1980), 2.0210
Scott, H.Y.D. (1863), 5.081
Scrope, G.P. (1833), 1.041
Seaman, V. (1798), 2.041
Seeck, O., ed (1876), 2.0210d
Senefelder, A. (1819), 7.151
Shirley, J. W. (1978), 3.011
Shirley, R.W. (1982), 6.1310b
(1983), 1.192, 1.2320, 1.2320b, 2.192, 4.031
Siborne, W. (1822), 5.081, 6.032
(1844), 7.011
Siegrist, W. (1949), 5.1610a
Simoni, A. E. C. (1985), 8.0110
Skelton, R.A. (1952), 5.1610a, 6.081, 6.091,

8.0110
(1954), 1.0320a, 4.122
(1957), 5.193
(1958), 2.164
(1960), 5.032, 6.081
(1963), 6.0920b, 6.1310h
(1964), 1.0320c
(1965), 1.1630a, 1.1630b, 5.033
(1966), 1.1630b, 4.141, 6.0920b
(1967), 4.132
(1968), 8.0110e
(1969), 4.122
(1970), 1.131, 8.0110d
(1972), 8.0110b
Skelton, R.A., and Harvey, P.D.A. (1986), 1.1630,
1.203, 4.031, 5.033
Skelton, R.A., and Summerson, J. (1971),
1.0320b, 2.0210d, 3.131, 4.133
Slater, G. (1907), 2.051
Smet, A. de (1947), 1.1810a, 5.192, 6.1310c
Smith, B.E. (1899), 4.041
Smith, C.S. (1974), 7.051
Smith, J. (1701), 5.032, 6.081
Smith, J. R. (1986), 4.051
Smith, T.R. (1978), 1.0320c
Snow, J. (1855), 1.044, 2.041
Speck, F.G. (1935), 6.1310d
Spencer, F.J. (1969), 2.191
Stebnowski, J. (1960), 2.1310a
Stegena, L. (1982), 6.121
Steinhauser, A. (1858), 3.081, 5.0910a, 5.121
Stephenson, F.R., and Walker, C.B.F., ed (1985),
4.121
Stevens, H. (1973), 1.0320
Stevenson, A.H. (1967), 6.231
Stevenson, E.L. (1911), 1.0320d
(1921), 1.0710, 1.0710a, 1.0710d, 6.1310,
6.1310e, 6.1310g
Stevenson, L.G. (1965), 1.044, 2.041, 2.191
Steward, H. (1980), 1.203
Stukeley, W. (1743), 2.011
Szaflarski, J. (1959), 3.081, 5.121
Takagi Kikusaburu (1966), 4.122
Tan Chi-hsiang (1975), 6.1310c
Tanner, H.S. (1829), 1.1810c
Tate, W.E. (1967, 1975), 2.051
Taylor, E.G.R. (1941), 1.043
(1951), 4.032
(1956), 1.0320d, 4.031, 4.032, 4.123, 4.181,
5.231
Thiele, G. (1898), 1.0710a
Thompson, H.Y., 6.1310c
Thomson, D.F. (1962), 5.033
Thomson, D.W. (1966), 2.121, 5.031
Thomson, J.D. (1948), 4.121, 4.122, 4.134
Thrower, N.J.W., (1961), 2.0210e, 8.0110a
(1966), 2.0210b, 2.0210e
(1969), 3.031, 3.051, 3.232
(1972), 1.202, 4.231, 8.0110a
(1978), 1.043, 3.201